Principles of Horticulture

Principles of Horticulture

Edited by Mary Moreno

SYRAWOOD
PUBLISHING HOUSE

New York

Published by Syrawood Publishing House,
750 Third Avenue, 9th Floor,
New York, NY 10017, USA
www.syrawoodpublishinghouse.com

Principles of Horticulture
Edited by Mary Moreno

International Standard Book Number: 978-1-68286-739-6 (Hardback)

Cataloging-in-Publication Data

Principles of horticulture / edited by Mary Moreno.
    p. cm.
Includes bibliographical references and index.
ISBN 978-1-68286-739-6
1. Horticulture. I. Moreno, Mary.
SB318 .P75 2019
635--dc23

TABLE OF CONTENTS

PREFACE

This book was inspired by the evolution of our times; to answer the curiosity of inquisitive minds. Many developments have occurred across the globe in the recent past which has transformed the progress in the field.

Horticulture is the science of growing plants such as fruits, vegetables and flowers. Plant conservation, garden design, landscape restoration, soil management, arboriculture, etc. are important aspects of horticulture. It involves plant propagation and cultivation with the objective of enhancing plant growth, quality, nutritional profile, yield, disease resistance and environmental stress tolerance. Horticulture also involves the cultivation of ornamental flowers, plants, fruits and vegetables for aesthetic purposes. Some of the prominent areas of study under this discipline are floriculture, pomology, arboriculture, landscape horticulture and turf management. This book provides significant information of this discipline to help develop a good understanding of horticulture. It explores all the important aspects of horticulture in the present day scenario. The extensive content of this book provides the readers with a thorough understanding of the subject.

This book was developed from a mere concept to drafts to chapters and finally compiled together as a complete text to benefit the readers across all nations. To ensure the quality of the content we instilled two significant steps in our procedure. The first was to appoint an editorial team that would verify the data and statistics provided in the book and also select the most appropriate and valuable contributions from the plentiful contributions we received from authors worldwide. The next step was to appoint an expert of the topic as the Editor-in-Chief, who would head the project and finally make the necessary amendments and modifications to make the text reader-friendly. I was then commissioned to examine all the material to present the topics in the most comprehensible and productive format.

I would like to take this opportunity to thank all the contributing authors who were supportive enough to contribute their time and knowledge to this project. I also wish to convey my regards to my family who have been extremely supportive during the entire project.

Editor

DETERMINATION OF SOME PHYSICAL AND CHEMICAL PROPERTIES OF NATIVE CORNELIAN CHERRY (*CORNUS MAS* L.) DISTRICT OF ALMUS (TOKAT)

Volkan OKATAN[1]

[1]Uşak University, Sivaslı Vocational High School, 64800 Sivaslı, Uşak/Turkey

Corresponding author email: okatan.volkan@gmail.com

Abstract

The present study was realized in 2013-2014 on different type of cornelian cherry grown from seed, in the district of Tokat Almus, Turkey. The determination of some physical and chemical properties was studied. For this purpose, in this region, 40 cornelian cherry trees have been identified and were recorded by GPS. The evaluation was based on the observations and particular type selection criteria. In this regard, nine genotypes of cornelian cherry were subjected to the analysis. The fruit width ranged between values of 8.41 and 10.67 mm, the average fruit height between 13.51 to 18.84 mm, and the average fruit weight between 0.78 to 1.73 g were determined. As results provided by physico-chemical analysis were the pH between 2.60 and 4.02, soluble solids (TSS) 11.4-15.5 % and the titratable acidity were up to 0.37.

Key words: Almus, Cornelian Cherry, pH, pomological, soluble solids.

INTRODUCTION

Thanks to its geographical location, our country is the home of many plant species. Also, the climate is also the homeland of cornelian cherry. Therefore, Anatolia has a very rich cornelian population (Ülkümen, 1973). In 2013, the cornelian cherry production was 11.838 tons, and the production area was 1.675 decares (Anonymous, 2014). The cornelian cherry populate Belgium and Southern Europe, especially Germany. Thereby, the cornelian cherry appear like tall shrubs and small trees and can be found in the form of sorting between 5 to 8 meters (Ercişli, 2004a; Tetera, 2006). Usually, in our country, cornelian is found in the wild form in mountainous areas and riverbeds (Mert and Soylu, 2006). Unlike Tokat Almus, the dogwood populations are scattered in other counties (Gerçekçioğlu, 1998). Quality is a hard core of the fruit species with very different families, the cornelian cherry being a fruit with shape and coloration. Fruit varies from elliptical cylinder and 2-2.5 cm long and is dark red in color. The fruit varies from elliptical cylinder and 2 – 2.5 cm

long and is dark red in color. The fruit contain high amounts of sugar and vitamin C between 7-9 % (Feyzioğlu and Ayan, 2002). Cornelian fruit is very rich in phenolic compounds, anthocyanins and total flavonoids. Although the antioxidant capacity varies according to the maturity period, is very high, as in all red fruits. Therefore the cornelian cherry fruit is a natural source of antioxidants (Gündüz et al., 2013).

In recent years, the cornelian cherry is grown by consumers both for ornamental and its health value, facing a demand from the pharmaceutical industry (Bijelik et al., 2012). Cornelian cherry shows diversity with regard to the fruit shape, massiveness, color, taste and nutritional value due to open pollination. Therefore genotype is a potential genetic resource for breeding programs (Ercişli, 2004a). Antioxidant properties of cornelian cherry fruit with outstanding antimicrobial, antiallergenic and in terms of human health is a very valuable fruit characteristics regarding antihistamines (Çelik et al., 2006). However, herbal preparations are used for patients with diabetes (Jayaprakasa et al., 2005).

Cornelian cherry is used in wood industry because of its powerful and flexible features. It is also an important source of nectar and honey in terms of the flowering period. Cornelian cherry has high levels of air pollution tolerance, protects from erosion and is used as a hedge plant in landscaping in urban areas. In this way it is able to tolerate the high levels of air pollution (Bijelic, 2008). Although it is widely known, cornelian cherry is a fruit species neglected in our country. A large part of the cornelian cherries are grown either in mixed gardens or in breeding places. This constitutes a major problem in reaching the fruit harvest (Ercişli, 2004a). This problem can only be solved successfully with the selection of promising genotypes and their controlled propagation. Because cornelian cherry pollination of the plant, over the centuries many different characteristics occurred. In terms of ensuring the standardization of type, cornelian cherry with desired properties should be selected within populations (Gerçekçioğlu, 1998). Selection is made in our country working in different regions (Access et al., 1992; Turkoglu et al., 1999; Yalcinkaya and Eti, 2000; Karadeniz, 2002; Pırlak et al., 2003). Naturally-grown cornelian cherry promising genotypes in Tokat (Almus) were identified some physical and chemical properties have been investigated in this study.

MATERIALS AND METHODS

This study was conducted to determine the physical and chemical properties of 9 different genotypes of cornelian cherry in Tokat (Almus). Coordinates and altitude of the genotypes are presented in Table 1.

In our research in terms of physical properties the fruit, fruit size and fruit weight was examined. Chemical properties as the juice pH, soluble solids (TSS) have been examined and titratable acidity ratio (Smiley, 1977; Karaçalı, 1990; Cemeroğlu, 2007) also. Neck fruit shape index was determined by the proportion of fruit

transverse (Gundogdu, 2006). Measuring and weighing fruit from trees was made for 30-60 fruits randomized taken.

Table 1. Promising genotypes of Cornelian cherry coordinates and elevations

Genotypes Number	East Coordinates	North coordinates	Altitude
60 ALM 01	37318787	4469704	1069
60 ALM 02	37319061	4469802	1065
60 ALM 03	37319150	4469836	1052
60 ALM 04	37319049	4469806	1041
60 ALM 05	37319060	4469796	1035
60 ALM 06	37318775	4469704	1058
60 ALM 07	37318788	4469737	1060
60 ALM 08	37319009	4469833	1032
60 ALM 09	37319112	4469834	1047

Fruit digital calipers were used in the measurement, with a sensitivity of 0.01, while the precision of the weighing was 0,001 grams. The amount of fruit soluble solids (TSS) was determined using a hand refractometer, and the pH of the juice was identified using a pH-meter. The titratable acidity (TEA) was determined using a titration method (Yarılgaç and Yıldız, 2001).

RESULTS AND DISCUSSIONS

The results of the weighted ratings were determined as genotypes mentioned above. Physical and chemical properties of those genotypes which are presented in tables 2 and 3. The average fruit weight of the genotype is seen as the most promising low 0.78g (60 ALM 04), and the highest 1.73 (60 ALM 01), respectively. The average width of the fruit was between 8,41mm (60 ALM 04) and 10.67 mm (ranged from 60 ALM06). The lowest average fruit size values were 13.51 mm (13 ALM 05) and highest average was 18.84mm (13 ALM 01). Previous research studies on cornelian cherry fruit weight were differing regarding the length and width, the major effect on the pomological characteristics being due by environmental factors as well as genotype (Güleryüz et al., 1998; Demir and Kalyoncu 2003).

We obtained values that are supported by earlier work selection. Another study which was conducted by Erzincan, fruit weight ranged between 1.44 – 4.24 g, most of the fruit size being between 9.6 – 15.8 mm and 14.1 – 22.8 mm (Selçuk and Özrenk, 2011). Oblak (1980), presented that the average fruit weight in the studied population, which grows naturally in Slovenia was 1.78 grams. Bolu, Zonguldak, Karabük and Bartin grown in natural population between 1996-1998 in the provinces and in the manufacturer's garden were conducted to determine the best cornelian cherry. Such studies were found that the average fruit weight was between 1.02 and 4.07 grams (Yalcinkaya and Eti, 1999). Derebucak district of Konya in a study of ten different types of cornelian cherry, naturally grown, showed that the weights were between 3.65 – 4.57 g (Turkoglu et al., 1999). In another study, in Konya the selection of the fruit weight was determined being between 1.496 g and 4.116 g (Demir and Kalyoncu, 2003). Another study, realised by Rural and Koca (2008) in Samsun shows that the weight of the fruit that grows naturally was between 0.39 - 1.03 g, the fruit length between 14.24 mm – 22.20 mm, and most of the fruit varies between 9.59-13.21mm

Table 2. Some physical properties of promising cornelian cherry genotype (2013-2014)

Genotype number	Weight (g)	Width (mm)	Height (mm)	Shape index
60 ALM 01	1.73 ± 0.14	10.64 ± 0.14	18.84 ± 1.36	1.77 ± 0.19
60 ALM 02	1.32 ± 0.10	10.52 ± 1.01	14.94 ± 1.24	1.42 ± 0.06
60 ALM 03	1.02 ± 0.08	09.55 ± 0.96	14.64 ± 1.85	1.53 ± 0.43
60 ALM 04	0.78 ± 0.05	08.41 ± 0.29	13.96 ± 2.06	1.66 ± 0.28
60 ALM 05	1.13 ± 0.07	09.83 ± 0.35	13.51 ± 0.53	1.37 ± 0.17
60 ALM 06	1.46 ± 0.16	10.67 ± 0.44	17.47 ± 1.61	1.64 ± 0.32
60 ALM 07	1.24 ± 0.21	10.24 ± 1.06	15.08 ± 0.80	1.47 ± 0.02
60 ALM 08	1.17 ± 0.11	10.29 ± 0.83	14.44 ± 0.72	1.40 ± 0.41
60 ALM 09	1.01 ± 0.13	09.69 ± 0.18	14.09 ± 1.40	1.45 ± 0.75

Table 3. Promising chemical properties of the cornelian cherry genotype (2013-2014)

Genotype number	pH	TSS (%)	TEA
60 ALM 01	3.12 ± 0.35	11.4 ± 0.24	0.35 ± 0.04
60 ALM 02	3.56 ± 0.40	11.8 ± 0.65	0.37 ± 0.08
60 ALM 03	4.02 ± 0.15	13.8 ± 0.32	0.35 ± 0.03
60 ALM 04	3.88 ± 0.66	14.9 ± 1.28	0.28 ± 0.07
60 ALM 05	3.80 ± 0.10	17.2 ± 0.87	0.30 ± 0.08
60 ALM 06	3.60 ± 0.24	15.2 ± 0.37	0.26 ± 0.15
60 ALM 07	3.92 ± 0.32	15.0 ± 0.20	0.36 ± 0.10
60 ALM 08	2.60 ± 0.18	15.5 ± 1.24	0.33 ± 0.03
60 ALM 09	3.28 ± 0.25	15.1 ± 2.49	0.32 ± 0.06

The results of this study show closeness with our obtained results, the highest values in terms of the weight of the fruit being low in both studies. We consider that the results are dependent on the genotype and the environmental conditions, the shape of the fruit affecting the quality of the fruit. The evaluation showed that the lowest index value was 1.37 for 13 ALM 05 genotype. 60 ALM 01 genotype had the longest length, with the highest value of shape index 1.77. This value is lower than 2.50 found in cornelian cherry genotype from Serbia (Bijelic, 2012).

As well as fresh, the cornelian cherry are consumed and processed in fruit industry, due to their physical and chemical properties. The pH of the promising occurring kızılcık genotypes soluble solids (TSS) and titratable acidity (TA) is given in percentage in Table 3. PH 2.60 in the genotype (60 ALM 08) to 4.02 (60 ALM 03) in between, while the soluble solids ratio of 11.4 (60 ALM 01) to 15.5 (60 ALM 08) ranged. The lowest rate in the amount of titratable acidity 0.28 (60 ALM 04), and the highest ratio 0.37 (60 ALM 02) respectively. Selcuk and Özrenk (2011), the pH 2.9-5.7 in a similar study conducted in cornelian cherry have found in the water soluble dry matter content of 9.0-17.7. In a population study carried out in Slovenia, the average amount of TSS in the cornelian cherry grown naturally was 20.6 %, total sugar was found to be 7.42 % and the pH 3.38 (Oblak, 1980). Another study which was conducted in Trabzon, has been reported to range between 8% and 13.5% of the total dry matter (Karadeniz et al., 2001). In another study conducted in Zonguldak amount of soluble solids it has been reported to be between 12.1 to 16.9%. Tural and Koca (2008), the amount of soluble solids in chemical analysis they have found in cornelian cherry have been obtained through selection between 28.19% and 15.88%. The same researchers have found the total amount of 1.10 to 2.53% for acidity. Total acidity between 4.69% and 1.24% was found in similar studies (Smiley et al., 1998; Strain, et al., 2000; Demir and Kalyoncu 2003). In particular, the same kinds of chemical compounds (or type) of the year even though the ecological differences are greatly influenced by the environment and maintenance requirements (Gerçekçioğlu, 1998).

CONCLUSIONS

Our results have shown that it is consistent with the results provided by other genotypes grown both inside and outside the country. However, it is expected to give better results with a different cultivation technology. Also, characterıstıcs such as phenolic compounds, the antioxidant capacity, the pollen biology and others can provide more information about the genotype's value. Research like this will shed light our study and the genetic material can be considered as the promising genotypes.

REFERENCES

Anonymous, 2014. Türkiye İstatistik Kurumu. www.tuik.gov.tr

Bijelić S., 2008. Cornelian cherry as a food and medicine organic prcoluclion. Agriculture Magazine, 47: 16-21.

Bijelić S., Gološın B., Todorović Nınıć J., Cerović S., 2012. Promising cornelian cherry (Cornus mas L.) genotypes from natural population in Serbia. Agriculturae Consspectus Scientificus, 77:5-10.

Celik S., Bakirci I., Sat I. G., 2006. Physico-chemical and organoleptic properties of yogurt with Cornelian Cherry Paste. Int. J. Food Prop., 9: 401 - 408.

Cemeroğlu B., 2007. Gıda Analizleri. Gıda Teknolojisi Yayınları.No:34 Ankara. S: 168-171.

Demir F., Kalyoncu I. H., 2003. Some nutritional, pomological and physical properties of cornelian cherry (Cornus mas L.). J. Food Eng., vol. 60, 3: 335 - 341.

Didin M., Kızılaslan A., Fenercioğlu H., 2000. Malatya'da Yetiştirilen Bazı Kızılcık Çeşitlerinin Nektara İşlem Uygunluklarının Belirlenmesi Üzerine Bir Araştırma. Gıda, 25; 435-441.

Ercisli S., 2004b. Cornelian cherry germplasm resources of Turkey. J. Fruit Ornam. Plant Res., Spec. vol. 12: 87 - 92.

Eriş A., Soylu A., Barut E., Dalkılıç Z. 1992. Bursa Yöresinde Yetişmekte Olan Kızılcık Çeşitlerinde Seleksiyon Çalışmaları. Türkiye I. Ulusal Bahçe Bitkileri Kongresi (13-16 Ekim 1992) Ege Üniversitesi Ziraat Fakültesi,Bahçe Bitkileri Bölümü, İzmir, 1: 503 –507

Ercişli S., 2004a. A short review of the fruit germplasm resources of Turkey. Genet. Resour. Crop Evol. 51: 419–435.

Feyzioğlu F., ve Ayan S., 2002 Farklı Bitki Büyüme Düzenleyici Maddelerin Kızılcık (Cornus mas L) Planlet Gelişimi Üzerine Etkileri. Gazi Üniversitesi Fen Bilimleri Enstitüsü Dergisi 15 (2):533-556.

Gerçekçioğlu R., 1998. Tokat Merkez İlçede Doğal Olarak Yetişen Kızılcıkların (Cornus mas L.)Seleksiyonu Üzerine Bir Araştırma Gazi Osmanpaşa Üniversitesi Ziraat Fak. Dergisi, 1.

Güleryüz M., 1977. Erzincan'da Yetiştirilen Bazı Önemli Elma ve Armut Çeşitlerinin Pomolojileri ile Döllenme Biyolojileri Üzerinde Araştırmalar. Atatürk Üniv. Yay. Ziraat Fak. Yay. No: 229. Erzurum. s.180.

Gündoğdu M., 2006. Pervari (SİİRT) Yöresi Nar (Punica granatum L.) Populasyonlarında Mahalli Tiplerinin Seleksiyonu Y.Lisans Tezi.

Gündüz K., Saraçoğlu O., Özgen M., Serçe S., 2013. Antioksidant, Physical and chemical characteristics of Cornelian Cherry fruits (Cornus mas L.) at different stages of ripeness. Acta Sci.Pol., Hortorum Cultus 12 (4): 59-66.

Jayaprakasam B., Vareed K. S., Olson L. K., Muraleedharan N. G., (2005). Insulin secretion by bioactive anthocyanins and anthocyanidins. J. Agric. Food Chem. 53: 28 - 31.

Karaçalı İ., 1990. Bahçe Ürünlerinin Muhafazası ve Pazarlanması. Ege Üniv. Zir.Fak.Yay. No: 494, İzmir.

Karadeniz T., 2002. Selection of native cornelian cherries grown in Turkey. J. Amer. Pol. Soc. 56 (3):164-167.

Mert U., ve Soylu A., 2006. Bazı Kızılcık (Cornus mas L.) Çeşitlerinin Döllenme Biyolojileri Üzerine Araştırmalar. Uludağ. Univ. Ziraat Fak. Dergisi 2:21

Oblak M., 1980. Contribution to studying some pomolojical properties of indigenous small fruit species in Slovenja. Productions spontenees. Cooloque. Colmar, Paris-France. 49-57.

Pırlak L., Güleryüz M., Bolat I., 2003. Promising cornelian cherries (*Cornus mas*) from the Northeastern Anatolia region of Turkey. J. Amer. Pol. Soc. 57 (1):14-18.

Selçuk E., ve Özrenk K., 2011. Erzincan Yöresinde Yetiştirilen Kızılcıkların (Cornus mas L.) Fenolojik ve Pomolojik Özelliklerinin Belirlenmesi. Iğdır Üni. Fen Bilimleri Enst. Der./Iğdır Univ. J. Inst. Sci. & Tech. 1(4): 23-30, 2011.

Tetera V., 2006 Ovoce Bilych Karpat. CSOP, Veseli na Moravou, Czech Republic.

Tural S., ve Koca İ., 2008. Physico-chemical and antioksidant properties of cornelian cerry fruits (Cornus mas L.) grown in Turkey.Scientia Horticulturae 116:362-366.

Türkoğlu N., Gazioğlu R. Ş., Kör M., 1999. Konya'nın Derebucak İlçesinde Yetişen kızılcıkların (Cornus mas L.) Seleksiyonu Üzerine Bir Ön Çalışma. Türkiye III. Ulusal Bahçe Bitkileri Kongresi. 14-17 Eylül. 1999. Ankara s: 768-771.

Ülkümen, L. 1973. Bağ-Bahçe Ziraati. Atatürk Üniversitesi Ziraat Fakültesi, Yayın No: 67, Erzurum, 415s.

Yalçınkaya E., Eti S., 1999. Batı Karadeniz Bölgesinin Bazı İllerinde Kızılcık (Cornus mas L.)Seleksiyonu. Türkiye III. Ulusal Bahçe Bitkileri Kongresi. 14-17 Eylül.1999. Ankara s: 781-786.

Yarılgaç ve Yıldız, 2001. Adilcevaz İlçesinde Yetiştirilen Mahalli Armut Çeşitlerinin Bazı Pomolojik Özellikleri. Yüzüncü Yıl Üniversitesi, Ziraat Fakültesi, Tarım Bilimleri Dergisi (J. Agric. Sci.), 2001, 11(2):9-12, www.tuik.gov.tr.

FIRST RESULTS OF TESTING GOJI BERRY (*LYCIUM BARBARUM* L.) IN PLOVDIV REGION, BULGARIA

Hristo DZHUGALOV[1], Valentin LICHEV[1], Anton YORDANOV[1], Pantaley KAYMAKANOV[1], Velmira DIMITROVA[2], Georgi KUTORANOV[2]

[1]Department of Fruit Growing, Agricultural University, Plovdiv, 12 Mendeleev Str., 4000, Bulgaria
[2]Bio Tree Ltd., 7 Bansko road, 1331 Sofia, Bulgaria
Corresponding author email: hristo.djugalov@gmail.com

Abstract

The study was conducted during the period 2013 - 2014 at the experimental field of the Department of Fruit Growing at Agricultural University – Plovdiv, Bulgaria. In vitro propagated plants of two Bulgarian cultivars ('JB 1' and 'JB 2') were planted in 20 June 2013 at distances of 3 x 2 m. The plants were grown under drip irrigation. The plants were formed as trees with 40 cm trunk height. Despite of late planting in the permanent place the plants of both cultivars started to bloom and bear fruits (although single fruits) in the same year. During the second vegetation flowering of tested cultivars started in the beginning of June and finished in the end of November. Better productivity was recorded in cultivar 'JB 1' with 0,56 kg/tree and theoretical yield per decare – 93,52 kg whereas the yield of cultivar 'JB 2' was 0,31 kg/tree and 51,77 kg respectively. The cultivar 'JB 2' was more susceptible to powdery mildew than 'JB 1'.

Key words: cultivar, fruit bearing, growth, Lycium barbarum L.

INTRODUCTION

An increasing interest to growing untraditional crops especially those with high nutritional value has been observed recently. *Lycium barbarum* L. is a frutescent plant which belongs to *Solanaceae*. The fruits of this plant are also known as goji berry and together with other parts of the plant have been used for a long time in traditional chinese medicine (Institute of Chinese Materia Medica, 1997). The place of origin of *L. barbarum* is not definitively determined. Probably it can be found in the Mediterranean Basin (Genaust, 1996). The plant is widely distributed in the Mediterranean area, Southwest and Central Asia. It is also cultivated in North America and Australia (Hänsel et al., 1993). Niculescu M. et al. (2011) reported presence of *Lycium barbarum* L. (*L. halimifolium* Miller.) in the Bistrita - Varatic Valley and Jiu Valley.

Fruits with optimum quality can be obtained in hot summer conditions. Rains during ripening cause cracking of the fruit. There is relationship between the environmental conditions and the yield.

Liu Jing et al. (2004) reported that for good growth and fruit bearing effective temperature of 3450^0 C and light for 1640 h are required.

Other factors such altitude, temperature and sunshine are important for the quality of the fruits. As a result of a study Lin et al. (2013) defined the most suitable regions for growing goji berry in their province.

Nutritionists describe goji berry as "exotic super food" because of its high content of polysaccharides, vitamins and carotenoids. Cultivation of goji berry plants recently increases due to high demand of fresh or dried organic fruits. In the past new plantations have been established using seedlings whereas nowadays vegetative propagated plants of certain cultivars are mostly used.

In Ningxia-China goji berry plants were formed as a shrub or a small tree and were planted at density 1,5 x 1m (Hummer et al., 2012).

Diploid and triploid cultivars were studied in China. The authors reported that triploid cultivars were more vigorous and there were differences in terms of beginning and finish of certain phenophases (Ann et al., 1998).

Wang et al. (2011) had compared the cultivar Ningqi 6 with the standard Ningqi 1. The authors pointed out cultivar Ningqi 6 as more vigorous, better feathered and higher productive. The fruit of this cultivar were bigger, with smaller seeds and higher quality than the standard.

In different study the cultivar Ningqi 1 were compared with two new cultivars of *Lycium barbarum* L. - Ningqi 6 and Ningqi 8. It was found out that the new cultivars have thicker leaves, which explains their vigorous vegetative growth and large size of the fruits (Yan et al., 2014).

Different cultivars were studied at arid and semi arid conditions of Gansu and Ningxia province in China. It was established that cultivars Ningqi 5, Mengqi 1 and Bianguogouqi were the most vigorous during first few years after planning. Therefore, the authors define them as most suitable for growing under such conditions (Zhang et al., 2013).

Five new goji berry strains were studied in three regions of Ningxia province, China. As a result it was established that the environmental conditions influence the growth of different strains (An et al., 2009). Investigating the possibilities for growing goji berry is done in other countries in Asia. The suitability of growing goji berry in typical steppe in Inner Mongolia was investigated and it was reported that *Lycium barbarum* L. can be successfuly grown under such conditions (Liu, 1999).

Two cultivars of *Lycium barbarum* L. were studied near by North Bucharest, Romania (Mencinicopschi et al., 2012). It was established that the cultivars are suitable for growing in this region. The authors reported the differences concerning vegetative and reproductive characteristics of the studied cultivars. Mencinicopschi and Balan (2013) established differences between two cultivars goji berry in terms of beginning of phenophases and pointed out the importance of weather conditions for the beginning of phenophases flowering and fruit bearing. The authors inform that one of the cultivars was more vigorous and less productive but its fruits were of higher quality.

Till now in the available scientific literature we have not found data concerning testing goji berry plants grown on the field in Bulgaria. The aim of the study was to evaluate some of the growth and reproductive characteristics of two bulgarian goji berry cultivars under conditions of Plovdiv region, Southern Bulgaria.

MATERIALS AND METHODS

The study was conducted during the period 2013 - 2014 at the experimental field of the Department of Fruit growing at Agricultural University–Plovdiv, Bulgaria. For that purpose in vitro propagated plants of two Bulgarian cultivars (JB 1 and JB 2) were planted in 20 June 2013 at distance 3 x 2 m. The thickness at the root neck was 2-3 mm and the plant height was 20-30 cm. The plants were formed as trees with 40 cm trunk height, supporting sticks were used and drip irrigation was applied. The experiment was set up in a randomized block design. Six replications and one plant per plot were included in each variant. The data were statistically processed by the method of analysis of variance. The following parameters were evaluated: height of the plants, total growth, trunk thickness, crown diameter and crown volume and yield per tree. Phenological observations concerning the beginning and end of phenophases flowering and fruit bearing also were determinated.

The climate in Plovdiv region is typical of temperate climate zone. The average year active temperature sum is $3900°$ C, mean annual rainfall is 515 mm and pH of the soil is 7,2 – 7,4 (Angelov, 2006).

RESULTS AND DISCUSSIONS

Growth characteristics of tested cultivars are presented in (Table 1). The differences between the two studied cultivars concerning height of the plants in the year of planting (2013) are insignificant. In the end of the second vegetation the plants of cultivar JB 2 are significantly higher than these of JB1. Other authors (Zhang et al., 2013; Mencinicopschi and Balan, 2013; Wang et al., 2011) also reported for differences in terms of vigor of tested cultivars.

For the period 2013 – 2014 the differences between tested cultivars in terms of vegetative growth are significant and the cultivar JB 2 surpasses JB 1.

Table 1. Growth characteristics of two goji berry cultivars.

Cultivar	Height of the plants, cm		Total growth, cm		
	2013	2014	2013	2014	∑ 2013-2014
JB 1	42.0	103.50	86.00	761.50	847.50
JB 2	48.50	238.67	130.00	1061.17	1191.17
LSD 5%	19.92	56.51	69.69	202.63	184.81

Data about other parameters concerning the growth characteristics of the tested goji berry cultivars at the end of the second vegetation (2014) are presented in Table 2. Concerning the trunk thickness there is no significant difference between tested cultivars. In terms of crown diameter and crown volume cultivar JB 2 surpasses JB 1.

Table 2. Other parameters characterizing the growth characteristics of two cultivars goji berry at the end of the second vegetation (2014).

Cultivar	Trunk thickness, mm	Crown diameter, cm	Crown volume, m³
JB 1	8.99	66.25	0.09
JB 2	8.91	110.00	0.63
LSD 5%	2.14	13.04	0.20

Despite of relatively late planting (20 June) the plants started to bloom and bear fruits in the same vegetation - the cultivar JB 1 produced approximately 16 fruits per plant as long as the cultivar JB 2 produced 1-2 fruits per plant.
In the second year after planting the flowering of the cultivars began in the beginning of June. Start of flowering was recorded in JB 1 cultivar as well as JB 2 cultivar started flowering 2-3 days later. End of flowering for both cultivars was recorded in the end of November. Mencinicopschi and Balan (2013) reported for long period of flowering of the tested goji berry cultivars in Romania.
In the second vegetation fruit bearing of JB 1 cultivar was higher in comparison with JB 2 cultivar (Table 3). Other authors (Wang et al., 2011; Mencinicopschi and Balan, 2013) also reported about differences concerning the obtained yield in different tested cultivars.

Table 3. Reproductive characteristics of two goji berry cultivars in second vegetation (2014).

Cultivar	Yield, kg/tree	Theoretical yield, kg/decare
JB 1	0.56	93.52
JB 2	0.31	51.77
LSD 5%	0.28	45.32

In 2014 the first harvest of both cultivars was conducted in late August and the last in mid-November.
In another study in Bulgaria, Stoykov (2012) in laboratory analysis of samples of *Lycium barbarum* L. was discovered the causative agent of powdery mildew on goji berry plants. The author defined it as *Arthrocladiella mougeotii* (Lev.) Vassilkov. Monitoring for attack by pests and diseases was conducted in our experimental plantation. The results confirmed the presence of powdery mildew. Among the two cultivars JB 2 is visibly more susceptible to this disease than JB 1.

CONCLUSIONS

The first results (till the end of the second vegetation) of testing two goji berry cultivars allow the following conclusions:
1. The Bulgarian goji berry cultivars JB 1 and JB 2 can be successfully grown under environmental conditions of Plovdiv region (Southern Bulgaria).
2. Cultivar JB 2 is more vigorous in terms of height of plants, diameter and volume of the crown than JB 1 cultivar.
3. When planting in the permanent place the plants of both cultivars started to bloom and bear fruits (although single fruits) in the same year. During the second vegetation flowering of tested cultivars started in the beginning of June and finished in the end of November. Better productivity was recorded in cultivar JB 1 with 0,56 kg/tree and theoretical yield per decare – 93,52 kg whereas the yield of cultivar JB 2 was 0,31 kg/tree and 51,77 kg respectively.

REFERENCES

Angelov, I., 2006. A study of vegetative and reprodoktive habits of Tempranillo cultivar during the process of formation and initial bearing. Dissertation, Plovdiv.

Ann L., Yoon H., Wei L., Yunxiang J. N., 1998. Breeding new varieties of triploid seedless wolfberry. Journal of Ningxia Agricultural Colege, Vol. 19 (3), 41-43.

An W.,Wang Y. J., Shi Z. G., Zhao J.H., 2009. Growth Characteristics of five new Wolfberry Strains in different Regions. Nortern Horticulture, 1, 5.

Genaust H.,1996.Wörterbuch der botanischen Pflanzennamen,Auflage. Basel: Birkhäuser Verlag.

Hänsel R., Keller K., Rimpler H., Schneider G., 1993. Hagers Handbuch der pharmazeutischen Praxis, Vol. 5, Drogen Berlin, Heidelberg, New York, Springer Verlag.

Hummer K., Pomper K., Postman J., Graham C., Stover E., Mercure E., Aradhya M., Crisosto C., Ferguson L., Thomson M., Byers P., Zee F., 2012. Emerging fruit crops. In fruit breeding. Badens, Springer, New York, USA, Vol. 8, 188-195.

Lin N., Yang Z. X., Lin H. M., Zhang J.G., 2013. Evaluation of the Quality of *Lycium barbarum* from different production areas. Journal of Gansu Agricultural University, 4, 34-39.

Liu J., Zhang X. Y., Yang Y. L., Ma L.W., Zhang X. Y., Ye D., 2004. Research in relationship of yield and it's meteorological conditions of (*Lycium barbarum* L). National climate center of China River, 22–25.

Liu T., 1999. Development of drygrassland area wolfberry resources.InnerPrataculture, 3, 10.

Mencinicopschi I. C., Balan V., 2013. Growth and development characteristics of plant individuals from two *Lycium barbarum* L. Varieties. Scientific Papers, Series A. Agronomy, Vol. LVI, 490-497.

Mencinicopschi I. C., Balan V., Manole C., 2012. *Lycium barbarum* L. - A new spices with adaptability potencial in Bucharest area. Scientific Papers. Series A. Agronomy, Vol. 2, №1, 95-101.

Niculescu M., Luminita B. D., Podeanu L.Nuta I . S., Novu I., 2011. Contributions regarding invasive alien plants in the Valcan mountains. Analele Universitai din Craiova, vol. XLI, 200-204.

Qin K., Dai C. L., Cao Y. L., Tang H. F., Yan Y. M., He J., Li R., 2012. A New Wolfberry Cultivar "Ningqi 7". Acta Horticulturae sinica, vol. 39, 1(11), 2331-2332.

Stoykov D., 2012.Ecological interactions between invasive alien vascular plants, and essential saprophytic and parasitic fungi in Bulgaria. Phytologia balcanica, 18 (2), 113–116.

Yan Y., Wang H., Wang J., 2014. Comperative anatomical studies of 2 new kinds LB. Xian Nong Ye Keji, № 5, 29.

Wang J. X., Wang Y. L., Chang H.Yu., Xiong X., Tian Y., 2011. Report on Superior Characteristics of Ningxia Woolfberry New Variety Ningqi № 6. Modern Agricultural Science and Technology, 23, 150–155.

Zhang B., Cai G., Wang S., Zhang G., Zhong L., Wu L., Hu B., 2013. Vegetative Growth Evaluation and Selection of Different Varieties of Lycium for Dry Sand Land. Chinese Agricultural Science Bulletin, 29 (13), 40-43.

***The Institute of Chinese Materia Medica, China Academy of Traditional Chinese Medicine, (1997).Medicinal Plants in China – A selection of 150 commonly used species, WHO Regional Publications,Western Pacific Series 2, 169.

3

EVALUATION OF SOME AUTOCHTHONOUS PEACH AND NECTARINE CULTIVARS AT RESEARCH STATION FOR FRUIT GROWING CONSTANȚA

Corina GAVĂȚ, Liana Melania DUMITRU, Cristina MOALE, Alexandru OPRIȚĂ

Research Station for Fruit Growing Constanta, No.1 Pepinierei Street, 907300,
Valu lui Traian, Constanța, Romania

Corresponding author email: corina_gavat@yahoo.com

Abstract

The peach is one of the species for which runs a very active worldwide genetic improvement programme, both in public and in private. At Research Station for Fruit Growing Constanta (RSFG Constanta) an efforts have been made to identify, collect and preserve a rich germplasm found of this species. Currently, the national collection of peach, rejuvenated in the period 2011-2014 has a total of 505 genotypes of peaches and nectarines, of which 14 are old, native varieties. This paper describes these varieties in terms of morphological, biochemical, pathological by grouping according to their biological characteristics, in order to choose the initial material for hybridization schemes, the knowledge of the existing gene sources and promotes valuable varieties.

Key words: Prunus persica, varieties, blooming, yields.

INTRODUCTION

Collection, conservation and rational use of germplasm fund (Cociu V. and Oprea S., 1989) is an essential condition for the creation of new peach varieties with better qualities (Cepoiu N. and Manolache C., 2006) regarding productivity, diseases resistances, superior quality of the fruit and destination (Monet R., 1992). At Research Station for Fruit Growing (RSFG) Constanta peach germoplasm preservation was started in 1977 (Cociu V., 1993, 1999).

At RSFG Constanta the main objectives in peach breeding are the following:
- Creation of peach and nectarine varieties with superior quality, with early ripening (18-25 of June) and late ripening (25 September-5 October), for a longer fresh fruit assortment consumption;
- Creation of peach and nectarine varieties with pest and diseases resistance;
- Obtaining of varieties with longer winter latency period and resistance to the return frosts and late hoar-frosts;
- Diversification of fruits in size, shape, color, firmness;

- Obtaining of peach and nectarine with reduced habit (full dwarf and semidwarf) (Dumitru, 2003);
- Obtaining of peach varieties for canning (clingstone);
- Creation of some oranamental varieties (red leaf and abundant flowers)

Currently, the national collection of peach, rejuvenated in the period 2011-2014 has a total of 505 genotypes of peaches and nectarines, of which 14 are old, native varieties that could be used as genitors for creation of new peach varieties.

MATERIALS AND METHODS

RSFG Constanta is located in the south-eastern part of Romania, in the area between the Danube River and the Black Sea, and has specific steppe climatic conditions, with a semi-arid character. Frosts return is a quite often phenomena in spring and affect fruit trees with early blooming as nectarine. Absolute temperature beyond the limits of resistance of peach and nectarine species, e.g. -25°C or above +40°C is rare (1/20 or 1/30 years). Rainfall is deficient to the requirements of the trees; the average amount of rainfall is around 400 mm, with unequal time distribution in the active growing season (April 1 to September 30). Chernozem soil type is deep, well supplied with humus, showing proper conditions for water circulation.

The trees were observed from the phenological point of view.

There were made biometrical measurements on fruits and trees and physico-chemical analyses on fruit. The crown form was the improved vase. The orchard density was 833 trees/ha (4/3 m).

Phenological observations and measurements, and physical and chemical analyses on plants were done.

The trees and fruit characteristics were evaluated according to the Methodology for trying new varieties of fruit trees, shrubs and rootstock in order to approve the homologation and International Union for the Protection of New Varieties of Plants (UPOV) guidelines.

The peach yield was appreciated by weighing the tree crop (kg/tree) and reporting per hectare the average yield recorded in the years of study.

RESULTS AND DISCUSSIONS

All studied varieties have proved a superior quality and are suitable both for fresh consumption and for processing. Most of the peach varieties have medium vigor, except De Cândeşti variety that has high vigor (table 1).

The ripening time started with Nectarine superintensiv variety (5-20.07) and finished with Excelsior (20.09-5.10). Regarding the fruit ripening time one of the variety is early, two are medium, five are late and one extralate. The fruit shape is spherical, simetrical or slightly asymmetric; Băneasa 1 turtite genotype has flat fruit.

The fruit are attractive, colored, with white flesh (De Voineşti, Superbă de toamnă, De Cândeşti, Băneasa 1 turtite).

The average weight of the fruit (table 2) is between 70 g (Băneasa 1 turtite) and 180 g (Flacăra clon 1). Dry matter ranged between 9% (De Cândeşti) and 14% (Nectarin superintensiv). Acidity was between 0.41 mg % (Miorița) and 0.83 mg % (Nectarin superintensiv).

Table 2. Quality test of some genotypes of peach and nectarine (multiannual data) at the Research Station of Fruit Growing Constanta, Romania

No	Variety	Average weight of a fruit (g)	% of kernel	Dry matter (%)	Acidity (mg%)	Yield	
						kg/tree	t/ha**
1.	De Voineşti	105	9,2	11	0.6	37	23
2.	Superbă de toamnă	175	8,7	10	0.53	22	14
3.	Flacăra clon 1	180	7,2	11.5	0.53	27	17
4.	Nectarin superintensiv	80	7,7	14	0.83	25	15,7
5.	Miorița	130	8,5	11	0.41	30	18,8
6.	Cluj 1112	145	8,3	10	0.56		
7.	Băneasa 1 turtite	80	7,8	10	0.51	23,5	15,0
8.	De Cândeşti	100	9,1	9	0.54	30	18,75
9.	Superbă de vară	120	9,3	10	0.47	25	15,6

*Acidity: mg malic acid / 100 g flesh fruit

Table 1. Charactersitics of some authtenous peach varieties from National Peach Collection of RSFG Constanta

N.	Variety/group	Vigour	Ripening time	Pollination	Fruit shape	Fruit skin and flesh colour	Use
1	De Voineşti / peach	medium	late (10.08-23.08)	autofertile	Spherical, asymmetrical,	White greenish, white-cream flesh	Fresh consumption and proccesing
2	Superbă de toamnă / peach	medium	late (25.08-15.09)	autofertile	Spherical, asymmetrical, slightly elongated	White greenish, white flesh	Fresh consumption and proccesing
3	Flacăra clon 1 / peach	medium	late (01.09-20.09)	autofertile	Spherical, slightly asymmetric	Orange, red streaked on 10% of fruit; yellow	Fresh consumption and proccesing
4.	Nectarin superintensiv / nectarine	medium	early (5.07-20.07)	autofertile	Spherical, slightly asymmetric	Orange, yellow-orange flesh	Fresh consumption and proccesing
5.	Miorita / peach	medium	medium (12.07-27.07)	autofertile	Spherical, slightly asymmetric	Yellow -green, red streaked on 10% of fruit; white-cream flesh	Fresh consumption and proccesing
6.	Cluj 1112 / peach	medium	late (20.08-20.09)	autofertile	Spherical, slightly asymmetric	Yellow, red 30 % of the surface; yellow flesh	Fresh consumption and proccesing
7.	Băneasa 1 turtite / flat peach	medium	late (10.08-30.08)	autofertile	Flattened	Greenish white, pink streaked; greenish-white flesh	Fresh consumption and proccesing
8.	De Cândeşti / peach	high	late (01.08-20.08)	autofertile	Spherical, slightly asymmetric	Greenish white , pink streaked 5% of the fruit; greenish-white flesh	Fresh consumption and proccesing
9.	Superba de vară / peach	medium	medium (25.07-10.08)	autofertile	Spherical, slightly elongated	Yellow, red streaked 30% of the fruit; yellow	Fresh consumption and proccesing

CONCLUSIONS

The studied varieties, can be used in breeding programmes due to their qualities such as: rusticity, precocity of fruiting, annual high and constant productivity, superior fruit taste and flavour.

REFERENCES

International Union for the Protection of New Varieties of Plants, 1995. Guidelines for the conduct of tests for distinctness, uniformity and stability. Peach, Nectarine [*Prunus persica* (L.) Batsch]. Geneva, Switzerland.

Cepoiu N., Manolache C., 2006. Piersicul - sortimente şi tehnologii moderne. Ed. Ceres, Bucureşti, 296.

Cociu, V., Oprea, S., 1989. Research methods of fruit breeding. Ed. Dacia, Cluj-Napoca.

Cociu, V. 1993. Peach culture. Ed. Ceres, Bucuresti.

Cociu, V. 1999. Advances in breeding of horticulture fruit in Romania. Fruit trees. The Peach: 113-128. Ed. Ceres, Bucuresti

Dumitru, 2003. Studii şi cercetări privind crearea şi cultivarea piersicului şi nectarinului dwarf. Teză de doctorat, USAMV Bucureşti.

Monet R., 1992. Le pecher. Objectifs et criteres de selection. Ed. A. Gallais, H. Bannevot-INRA Paris-France (595-604).

4

RESEARCH AND STUDIES REGARDING THE BEHAVIOR OF CERTAIN RASPBERRY VARIETIES WITHIN BUCHAREST REGION

Dorel HOZA, Adrian ASĂNICĂ, Ligia ION

University of Agronomic Sciences and Veterinary Medicine of Bucharest, 59 Mărăşti Blvd,
District 1, 011464, Bucharest, Romania
Corresponding author email: dorel.hoza@gmail.com

Abstract

The experiment was conducted within the teaching field of the Horticulture Faculty during 2012-2014, with 5 genotypes of raspberry: 'Heritage', 'Opal', 'Gustar', 'Elite 89' and 'Malling Promise', out of which the first four with biannual fructification, planted at a distance of 2.2 m between rows, with a plant management system in the shape of lane, with espalier with two rows of double wire. The culture technology applied was specific for the raspberry plantations. The analyzed varieties had a different behaviour from the point of view of the vegetative growth, production capacity and fruit quality. The 'Gustar' variety had the highest capacity to produce root shoots; 'Elite 89' had the highest number of inflorescences per strain, while 'Malling promise' had the highest number of fruit per inflorescences. From the quality point of view, the 'Heritage' and 'Elite 89' varieties recorded higher content in dry substance; 'Opal' recorded the highest content in vitamin C, while 'Gustar' and 'Malling Promise' recorded higher content in carbohydrates.

Key words: raspberry, production, quality.

INTRODUCTION

The raspberry (*Rubus idaeus* L.) is a rustic species that grows spontaneously in the mountain region and in clearings in forests, but also one that is cultivated both within individual gardens and within commercial plantations (Hoza, 2000). The fruit obtained from an organic culture have a higher antioxidant capacity than the ones from a conventional culture (Jin et al., 2012). It represents a species of a major food and sanogeneous importance, the fruit have a complex biochemical composition, and its antioxidant components contribute to the protection of cells against serious diseases. The biochemical composition of fruit (soluble dry substance, acidity, phenol and antioxidant content) is influences by the cultivar variety (Zhang et al. 2010, Mazur et al., 2014). The raspberry based extracts have a neuroprotective role by inhibiting the peroxynitrite that determines the effects on the DNA (peroxynitrite-induced DNA damage) (Chen W. and al., 2012). Raspberry is a species that bearly deals with the soluble salts from the soil and especially with chlorine (Neocleous D., and Vasilakakis M., 2007); the soils with secondary salinization must be avoided. It is one of the species with high capacity to produce root shoots and with an anti-erosion protection role (Chira, 2000).

MATERIALS AND METHODS

The objective of this study was to evaluate the behavior of several Romanian varieties ('Opal', 'Gustar' and 'Elite 89-15-3') compared to the varieties 'Heritage' and 'Malling Promise', in the pedo-climatic conditions from the Bucharest area, a region with temperate-continental climate, with temperatures low during winter and high during spring and an amplitude of more than 50°C. The soil from the plantation is reddish brown, with a moderate degree of mineral supply and a weakly acid pH. The experiment was conducted within the teaching field of the Faculty of Horticulture Bucharest, during 2013-2014, within a four-year-old raspberry culture, with a distance between rows of 2.5 m and a width of 40 cm for the band along the row. Five raspberry varieties were analyzed: 'Heritage', 'Opal', 'Gustar', 'Malling Promise' and 'Elite 89-15-3', out of which only the variety Promise. The length of the fruit growing strains was limited to 120 cm through pruning.

The study method used was in field stationary and several biometric parameters were determined, while for the laboratory measurements related to quality the HPLC system was used HPLC (Agilent Technologies). The measurements evaluated the growth and fructification of the five genotypes in terms of: capacity to produce root shoots, capacity to form fertile shoots, number of flowers per inflorescence, fruit size, production and quality.

RESULTS AND DISCUSSIONS

The capacity to produce root shoots was influenced by the variety, number of root shoots per linear meter of band was between 14.3 shoots/linear meter for the variety 'Malling Promise' and 45.1 shoots/linear meter for 'Gustar', compared to an average of the varieties of 32.12 shoots/linear meter (Figure 1). Three varieties, 'Heritage', 'Opal' and 'Gustar', had a higher capacity to produce root shoots, above average (32.12 shoots/linear meter), while two varieties, 'Malling Promise' and 'Elite 89-15-3', had a weak capacity, less than 26 shoots/linear meter.

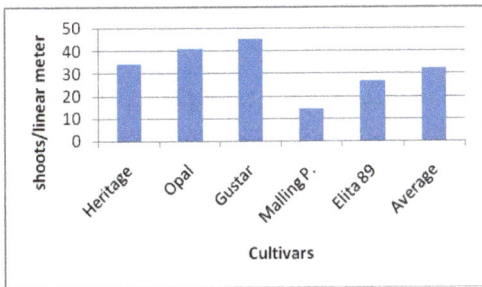

Figure 1. The capacity to produce root shoots for several raspberry varieties

The number of fertile shoots per strain was influenced by the variety, being higher for the varieties 'Gustar', 14.7 fertile shoots/strain, and 'Malling Promise', 11.3 fertile shoots/strain. Comparing the capacity of the varieties to produce fertile shoots to the average, three varieties, 'Gustar', 'Heritage' and 'Elite 89-15-3', had higher values, while 'Opal' and 'Malling Promise' had lower values (Figure 2). The number of flowers per inflorescence varied among the studied varieties, with values from 7.15 flowers for 'Opal' and 10.56 flowers for 'Elite 89-15-3', compared to the average of 8,68 flowers (Figure 3).

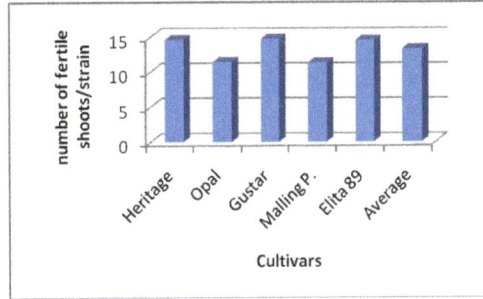

Figure 2. Capacity to form fertile shoots for several raspberry varieties

It was interesting that for these varieties, a strong positive correlation was highlighted between the number of fertile shoots and the number of flowers per inflorescence, expressed through $r^2 = 0.683$ (Figure 4), which showed that these two characteristics are dependent on the variety and biologically controlled.

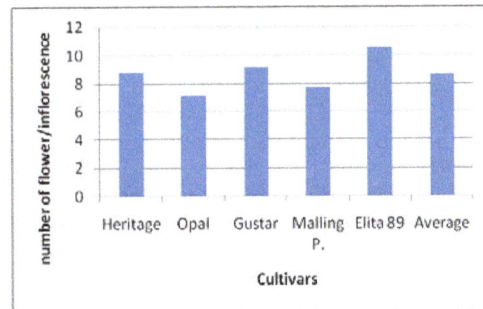

Figure 3. Number of flower per inflorescence for several raspberry varieties

It was interesting that for these varieties, a strong positive correlation was highlighted between the number of fertile shoots and the number of flowers per inflorescence, expressed through $r^2 = 0.683$ (Figure 4), which showed that these two characteristics are dependent on the variety and biologically controlled.

The production capacity was variable and recorded values between 1.45 kg/l.m. (linear meter of fruit growing band) for the variety Opal, a variety with smaller fruits, and 3.15 kg/l.m. for 'Gustar', as maximum value. The average of the experiment was 2.3 kg/l.m.; the varieties 'Gustar', 'Elite 89-15-3' and 'Heritage' had values higher than the control, while the varieties 'Opal' and 'Malling Promise' recorded lower values (Figure 5).

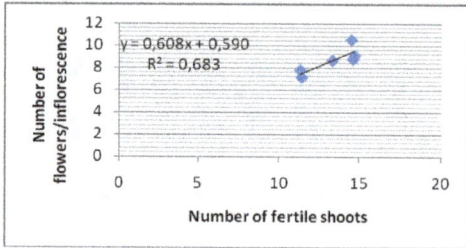

Figure 4. Correlation between the number of fertile shoots and the number of flowers per inflorescence

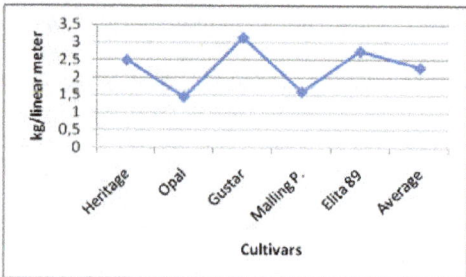

Figure 5. Production capacity for several raspberry varieties (kg/l.m.)

Fruit quality is important for capitalization, but also for the consumer. Differences were recorded among the varieties related to the content in soluble dry substance (SDS), content in vitamin C and carbohydrates (table 1). From the point of view of the content in dry substance, a higher content was recorded for the varieties 'Heritage' and 'Elite 89-15-3', 14.2 respectively 13.7%, the differences being statistically ensured by the variance analysis as very significant, while a lower content was recorded for the varieties 'Gustar' and 'Malling Promise', with very significant negative differences compared to the average. The content in vitamin C was higher for the varieties 'Gustar' (28.4 mg/100 g p.p.), the difference being very significant positive, and for 'Opal' (24.1 mg/100 g p.p.) with a significant positive difference from the control, compare to the other varieties (Table 1). The carbohydrate content varied according to the variety.

Table 1. Biochemical characteristics of the fruit for several raspberry varieties

Variety	S.U.S. (%)	Vitamin C (mg/100 g p.p.)	Fructose (mg/100 g p.p.)	Glucose (mg/100 g p.p.)	Sucrose (mg/100 g p.p.)
'Heritage'	14.2***	17.7°°	3.7	3.0	1.2
'Opal'	12.1 N	24.1**	4.0	2.7	1.2
'Gustar'	10.7°°°	28.4***	4.0	3.0	1.6
'Malling promise'	11.2°°°	18.5°°	3.7	2.9	1.8
'Elite 89-15-3'	13.6***	19.9°	3.63	3.12	1.3
Average	12.36 Control	21.72	3.80	2.94	1.42
DL 5%	0.32%	1.31 mg/100 g p.p.			
DL 1%	0.54%	2.17 mg/100 g p.p.			
DL0.1%	1.01%	4.07 mg/100 g p.p.			

CONCLUSIONS

From the present study, the following conclusions can be drawn:
The analyzed varieties had a good behaviour in the specific conditions of Bucharest area; they can be recommended for industrial cultures or cultivating them as a hobby
The capacity to produce root shoots was different and dependent on the variety; the varieties 'Opal' and 'Gustar' had a high capacity, over 40 root shoots per linear meter of band, while the rest of the varieties recorded lower values.

The number of fertile shoots per strain was higher for the varieties 'Heritage', 'Gustar' and Elite 89-15-3', over 14 shoots per strain, while lower numbers were recorded for 'Opal' and 'Malling Promise'.
The production capacity was higher for the varieties 'Gustar', 'Elite 89-15-3' and 'Heritage', with more than 2,5 kg per linear meter of band, while the varieties 'Opal' and 'Malling Promise' recorded smaller productions.
Fruit quality depended on the variety. The varieties 'Opal' and 'Gustar' had higher values for the content in vitamin C.

REFERENCES

Chen W., Su H., Huang Z., Feng L., Nie H., 2012. Neuroprotectiv effect of raspberry extract by inhibiting peroxinitrate-induced DNA damage and hydroxyl radical formation. Food Rechearch International, Volume 49, Issue 1, p. 22-26.

Chira L., 2000. Cultura arbuştilor fructiferi. Editura MAST, p.80-89.

Hoza D. 2000. Pomologie. Editura Prahova Ploieşti, p.240.

Jin P., Wnag S.Y., Gao H., Chen H., Zheng Y., Wang C., 2011. Effect of cultural system and essential oil treatment on antioxidant capacity in raspberryes. Food Chemistry, 2012, 132, p.399-404.

Kruger E., Districh H., Schopplein E., Rasim S., Kurbel P., 2011. Cultivar, storage conditions and ripining effects on phisical and chemical qualities of red raspberry fruit. Postharvest Biology and Technology 60, p. 31-37.

Mazur S.P., Nes A., Wold A.B., Remberg S.F., Aaby K., 2014. Quality and chemical compozition of ten red raspberry (*Rubus idaeus* L.) genotypes during three harvest seasons. Food Chemistry, 160, p.233-240.

Neoccleos D., Vasilakakis M. 2007. Effects of NaCl stress on red raspberri (*Rubus idaeus* L. Autumn Bliss). Scientia Horticulturae, 112, Issue 3, p. 282-289.

Neocleous D., Vasilakakis M., 2007. Effects of NaCl stress on red raspberry (Rubus idaeus L. 'Autumn Bliss'). Scientia Horticulturae, Volume 112, Issue 3, 23, Pages 282-289.

5

SOIL CHARACTERISTICS INFLUENCE THE FATTY ACID PROFILE OF OLIVE OILS

Hakan CETINKAYA[1], Muhittin KULAK[2]

[1]Kilis 7 Aralık University, Faculty of Agriculture, Department of Horticulture, Kilis, Turkey
[2]Kilis 7 Aralık University, Faculty of Arts and Sciences, Department of Biology, Kilis, Turkey
Corresponding author email: hcetinkaya67@gmail.com.com

Abstract

Olive tree is a typical Mediterranean plant grown in Marmara, Aegean, Mediterranean and South East Anatolian regions of Turkey and important oil sources for Mediterranean countries, fulfilling 90 % of the world olive oil production. Turkey is one of the important producer and stakeholders of olive oils after Spain, Italy, Greece and Tunisia. Cultivated olive cultivars in Turkey represent high genetic diversity, which may result in a standardization problem in terms of olive production and their fatty acid composition because the constituents of the fruit and the composition of its oil mainly depend on several factors such as climate, maturity, index variety, etc. Hence, the present study was designed to investigate the optimal soil characteristics for the favourable oil quality. In the study, the olive oils from South-Eastern region in Turkey according to their fatty acid profiles using gas chromatography of their fatty acid methyl esters were characterized and compared using chemo-metrics techniques. In this context, fatty acid profiles characterization was determined on ten olive samples collected from ten different locations. For the statistical evaluation, principal component analysis, variance analysis and correlation analysis were used. Accordingly, physical properties of soils influence the chemical composition and subsequently the quality of olive oils.

Key words: Soil, olive oil, fatty acid, chemo metrics

INTRODUCTION

Traditionally, plants have been extensively used for medicinal, nutritional, flavouring, cosmetically and industrial purposes. Of those plants, *Olea europaea* L. (olive) belonging to the Oleaceae family is one of the most important crops especially in Mediterranean countries on which they cover around 8 million hectares on the worldwide (Guinda et al., 2004) and its fruit and oil have a major agricultural importance in Turkey. Besides its fruits as table olive, its fatty oil is characterized with distinguished fatty acid composition, of which sanitary importance has been proven by a number of studies (Leon et al. 2004; Matson and Grundy, 1985). The important property of olive oil, the odour, as well as flavours association with oil quality have been found to be correlated with fatty acid composition (Maestro and Borja, 1990; Leon et al., 2004). Moreover, the oil obtained from olive fruits have essential key roles of reactive oxygen species (ROS) which are associated with pathology of some diseases including cancer, diabetes, cardiovascular, age related, and neurological disorders has been well documented (Chacraborty et al., 2009; Ishii, 2007; Burhans and Weinberger, 2007; Polidori et al., 2007; Halliwell and Guteridge, 1999; Soholm, 1998).

The chemical and physical properties of the soil influence plant growth, development and subsequently main primary and secondary metabolite production, secretion and accumulation. It is worth to note that the produced metabolites transport among the organs of the plants is also significantly affected by soil properties.

Uptake of an element from soil to plant depends on not only on the structure of the element, but also on different physicochemical factors of soil. Herein, transfer factor presents important information with respect to the certain amount of element transport from soil to the plant. Physicochemical parameters such as pH, $CaCO_3$ content, conductivity, organic matter content and soil texture are important factors affecting the transport of elements from the soil to plant species and consequently influence the plant growth, development and subsequently biochemical and physiological aspects of plant (Adriano, 2001; Lindsay and

Norvell, 1978; Kabata-Pendias and Mukherjee, 2007).

Application of chemo-metric approach in characterization of experimental samples has been extensively applied to quantitative evaluation of discrimination of variable results. In the current study, olive oil samples collected from different ten locations were compared for their fatty acid profile using analysis of variance (ANOVA) followed by the multiple comparison test of Duncan using SPSS. Furthermore, some characteristics of the sampling soils including pH, (CaCO₃), total salt, P (P₂O₅), K (K₂O) and organic matter were also determined and subsequently correlated with the fatty acid components by Pearson correlation matrix in Excel. Due to the existence of different experimental factors, chemo- metric techniques including Principal Component Analysis (PCA) were applied for analytical evaluation of fatty acid components between locations.

MATERIALS AND METHODS

Experimental Material:
The olive fruits were sampled from the Kilis Yaglik cv. (approximately the same aged trees) from Kilis district of south-eastern part of Turkey. Fruits were also harvested in the same ripening period (mid-December 2015) from the same position on the sampled trees.

Analysis of soil characteristics:
Organic matter by a modification of the Walkley and Black, calcium carbonate (CaCO₃) contents by Scheibler Calcimeter, total dissolved salts by Saturation Extract Method, soil phosphorus content by Olsen method, soil potassium content by flame photometer were determined for each samplings locations (Ure, 1990; Kaçar, 1995; Falciani et. al., 2000; Kaçar and İnal, 2008; Marin et al., 2008) The measurements were done in three replicates.

Oil extraction and fatty acid composition analysis:
The oils were extracted from olive fruits (each 10 g sample) with n-hexane for four hours using a Soxhlet Extraction Apparatus (Thermal). Then the solvent was evaporated under reduced pressure and temperature using a Rotary Evaporator (Heidolph). 0.5 g of olive oil

was added 10 ml n-heptanes into a screw-capped tube for esterification. The fatty acid analyses were conducted according to the official method COI/T.20/Doc.no.24 2001. 0.1 g of olive oil was taken into screw-capped tube. 2 ml n-heptanes were added to it and shaken. After 0.2 ml methanolic potassium hydroxide was added for esterification, tubes were vigorously shaken for 30 sec. after the vials were closed. The supernatant of the solution was taken followed after one hour of incubation at room temperature. Then, the supernatant was put in 2 ml vials for injection. Gas chromatography with flame ionization detector (GC-FID) analyses of fatty acids methyl esters was carried out on a Shimadzu gas chromatography (GC-2010 series) equipped with an Supelco SP 2380 fused silica capillary column (100 m, 0.25 mm i.d., 0.2 μm film thickness). Helium was used as carrier gas, at a flow rate of 3 mL/min. The injection and detector temperature were 140 °C and 240 °C, respectively. The oven temperature was held isothermal at 140 °C for 5 min, then raised to 240 °C

Statistical analysis
SPSS statistical program was used to determine statistical significance levels by employing the independent one-way ANOVA followed by Duncan multiple range test and the differences between individual averages were considered to be statistically important at $p < 0.05$. The results were expressed as mean.

RESULTS AND DISCUSSIONS

The fatty acid composition is a quality indicator parameter of olive oils and hence the component profile of fatty acids should be monitored. For all the samples, 13 fatty acids were identified and quantified but the major fatty acid components including arachidic acid, behenic acid, linoleic acid, linolenic acid, oleic acid, palmitic acid, palmitoleic acid and stearic acid were compared using variance analysis and correlated with soil characteristics (Table 1-2). Accordingly, oleic and palmitic acid were the major fatty acids and ranged between 68.77–73.32% and 12.74–14.64%, respectively. No statistical differences were found in sampling locations for oleic acid and palmitic acid different for each location ($p < 0.05$).

Table 1. Fatty acid profile of the samples olive fruits and soil characteristics of the sampling locations

	(C20:0)	(C22:0)	(C18:2)	(C18:3)	(C18:1)	(C16:0)	(C16:1)	(C18:0)	pH	CaCO₃ (%)	Total salt (%)	P (P₂O5, %)	K (K20, %)	Organic matter (%)
1	0.57 a	0.15 bc	8.94 b	0.86	69.71	14.02 ab	0.85 bcd	3.65 cd	7.85	19.10 e	0.017 bc	2.21 g	87.23 d	2.75 d
2	0.53 c	0.14 d	8.64 c	0.81	70.27	13.98 ab	0.92 abc	3.54 d	7.83 bcd	28.43 b	0.033 ab	2.29 g	75 f	3.50 b
3	0.50 c	0.13 d	8.60 c	0.89	69.67	14.81 a	1.08 a	3.20 e	7.87 bc	46.175 a	0.011 a	0.46 h	52.42 h	2.30 f
4	0.60 a	0.17 a	6.41 f	0.78	73.32	12.74 b	0.64 ef	3.88 ab	7.67 d	11.44 h	0.033 ab	3.30 e	92.5 b	0.99 h
5	0.59 a	0.16 bc	7.29 e	0.81	72.24	13.06 ab	0.69 def	3.86 ab	7.76 cd	13.35 h	0.034 ab	2.79 f	102 a	3.06 c
6	0.58 a	0.15 ab	9.30 a	0.90	68.77	14.64 ab	0.80 cde	3.63 cd	7.86 bcd	24.98 c	0.031 ah	5.51 c	63.15 g	2.56 e
7	0.60 a	0.16 ab	8.70 c	0.86	69.39	14.40 ab	0.84 cd	3.73 bc	7.86 bcd	23.11 d	0.037 a	4.52 d	21.5 j	2.57 e
8	0.60 a	0.17 a	7.43 e	0.94	71.19	13.54 ab	0.61 f	4.11 a	8.10 a	4.725 i	0.035 a	5.82 b	90.42 c	1.51 g
9	0.60 a	0.16 ab	7.41 e	0.81	72.31	12.78 b	0.66 ef	3.91 ab	7.99 ab	18.66 f	0.025 bc	6.88 a	78.32 e	4.00 a
10	0.43 c	0.11 e	7.88 d	0.78	72.06	13.40 ab	1.02 ab	3.18 e	7.76 cd	25.145 c	0.027	2.195 g	43.5 i	2.16 f

C20: 0; Arachidic acid, **C22: 0;** Behenic acid, **C18:2;** Linoleic acid, **C18:3;** linolenic acid, **C18:1;** Oleic acid, **C16:0;** Palmitic acid, **C16:1;** Palmitoleic acid, **C18:0;** Stearic acid

Table 2: Correlation matrix (Pearson (n)) for the fatty acid components and soil properties

Variables	C20:0	C22:0	C18:2	C18:3	C18:1	C16:0	C16:1	C18:0	pH	CaCO₃ (%)	Total salt (%)	P (P₂O5, %)	K (K₂0, %)	Organic matter (%)
C20:0	1	**0.972**	-0.229	0.265	0.041	-0.226	**-0.838**	**0.898**	0.274	-0.602	0.437	**0.635**	0.421	0.023
C22:0	**0.972**	1	-0.405	0.227	0.196	-0.346	**-0.896**	**0.944**	0.258	**-0.699**	0.495	0.598	0.499	-0.139
C18:2	-0.229	-0.405	1	0.455	**-0.952**	**0.889**	0.606	-0.483	0.144	0.594	-0.325	-0.173	-0.511	0.356
C18:3	0.265	0.227	0.455	1	**-0.655**	0.623	-0.042	0.181	**0.700**	0.046	-0.115	0.242	-0.082	-0.187
C18:1	0.041	0.196	**-0.952**	**-0.655**	1	**-0.950**	-0.477	0.322	-0.251	-0.523	0.246	0.097	0.487	-0.196
C16:0	-0.226	-0.346	**0.889**	0.623	**-0.950**	1	**0.640**	-0.506	0.136	**0.682**	-0.334	-0.294	-0.576	0.034
C16:1	**-0.838**	**-0.896**	0.606	-0.042	-0.477	**0.640**	1	**-0.967**	-0.257	**0.883**	-0.576	**-0.721**	**-0.635**	0.121
C18:0	**0.898**	**0.944**	-0.483	0.181	0.322	-0.506	**-0.967**	1	0.384	**-0.850**	0.581	**0.713**	0.576	-0.065
pH	0.274	0.258	0.144	**0.700**	-0.251	0.136	-0.257	0.384	1	-0.165	-0.039	0.579	0.025	0.196
CaCO₃ (%)	-0.602	**-0.699**	0.594	0.046	-0.523	**0.682**	**0.883**	**-0.850**	-0.165	1	**-0.647**	-0.573	-0.581	0.260
Total salt (%)	0.437	0.495	-0.325	-0.115	0.246	-0.334	-0.576	0.581	-0.039	**-0.647**	1	0.495	0.084	-0.149
P (P₂O5, %)	**0.635**	0.598	-0.173	0.242	0.097	-0.294	**-0.721**	**0.713**	0.579	-0.573	0.495	1	0.096	0.148
K (K₂0, %)	0.421	0.499	-0.511	-0.082	0.487	-0.576	**-0.635**	0.576	0.025	-0.581	0.084	0.096	1	-0.030
Organic matter(%)	0.023	-0.139	0.356	-0.187	-0.196	0.034	0.121	-0.065	0.196	0.260	-0.149	0.148	-0.030	1

Values in bold are different from 0 with a significance level α= 0,05

C20: 0; Arachidic acid, **C22: 0;** Behenic acid, **C18:2;** Linoleic acid, **C18:3;** linolenic acid, **C18:1;** Oleic acid, **C16:0;** Palmitic acid, **C16:1;** Palmitoleic acid, **C18:0;** Stearic acid

The average linolenic acid level of olive oil samples ranged between 0.78-0.94% in south-eastern region of Turkey, below the maximum value fixed by the IOOC (1.0%) (International Olive Oil Council, 2003); but within the ranges proposed by the Turkish Codex (0.9%).

However, it is worthy to note that no statistical differences were determined under different growing conditions.

Pearson correlation coefficients among fatty acid profiles are presented in Table 2. The maximum positive correlations were found between C22:0 and C20:0 (r=.972), C18:0 and C22:0 (r=.944), C18:0 and C20:0 (r=.898), C16:0 and C18:2 (r=0.889) whereas strong and negative ones were observed between C18:1 and C18:2 (r=−.952), C18:1 and C16:0 (r=−.819). The major component, oleic acid (C18:1) was negatively correlated with pH (r=-.251), CaCO₃ (r=−.523) and organic matter (r=-.0196) but positively moderate associated with K content (r=.487).

On the other hand, salt content also positively-but weak-correlated with oleic acid content (r=.246). Of those major components, palmitic acid (C16:0) composition significantly varied with CaCO₃ content (r=.682) but negatively affected with salt content (r=.-334), P content (r=−.294) and K content (r=−.576). Linoleic acid (C18:2) displayed similar reaction with palmitic acid against CaCO₃ content (r=.594) and positive moderate relation with organic matter content (r=.356) but negative correlation with salt content (r=−.325), P content (r=−.173), K content (r=−.511) were found with linoleic acid.

Data of the fatty acids compositions corresponding to all olive oil samples were submitted to Principal Component Analysis (PCA) to transform a number of possibly correlated variables into a smaller number of uncorrelated variables called principal components (PC). Only eigenvalues of greater than 1.0 are considered significant descriptors of data variance, according to Kaiser's rule (Kammoun and Zarrouk, 2012). Eigen analysis of the correlation matrix loadings of the significant principal components were summarized in Table 3.

Table 3. Eigen analysis of the correlation matrix loadings of the significant principal components

	PC1	PC2	PC3	PC4
Eigenvalue	6.701	3.294	1.247	1.073
Variability (%)	47.867	23.531	8.909	7.662
Cumulative (%)	47.867	71.398	80.308	87.970

The first four components (PC1, PC2, PC3, and PC4) had eigenvalues of 6.701, 3.294, 1.247 and 1.073, and accounted for 47.867 %, 23.531 %, 8,909 % and 7,662 % of the variance in the data, respectively.

Table 4. Correlations between variables and factors

	PC1	PC2	PC3	PC4
Arachidic	-0.771	0.463	0.007	0.043
Behenic	-0.858	0.348	-0.118	-0.005
Linoleic	0.707	0.595	0.120	0.138
Linolenic	0.068	0.873	-0.327	-0.283
Oleic	-0.591	-0.756	0.064	-0.111
Palmitic	0.733	0.613	-0.226	0.073
Palmitoleic	0.985	-0.130	-0.010	0.019
Stearic	-0.947	0.297	0.014	-0.017
pH	-0.201	0.710	0.229	-0.318
CaCO₃	0.910	0.024	0.116	-0.062
Total Salt	-0.612	0.039	-0.131	0.685
P	-0.657	0.473	0.283	0.241
K	-0.641	-0.186	0.047	-0.565
Organic matter	0.153	0.157	0.944	0.045

The first PC accounted for more 47.867 % of total explained variance. Linoleic acid. palmitic acid. CaCO₃ content were the most important factors in PC1 whereas linolenic acid. pH were the most important factor in PC2 (Table 3).

The ten sampling locations are successfully discriminated by their fatty acid compositions and physicochemical factors of soil. Oleic acid, palmitic acid and linoleic acid-major fatty acid components- were discriminated with K content, CaCO₃ content, organic matter content, respectively (Figure 1-3).

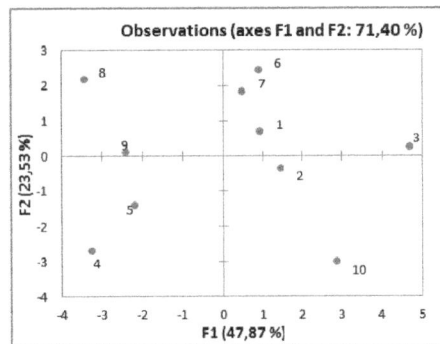

Figure 1. Observations (axes F1 and F2: 71.40 %)

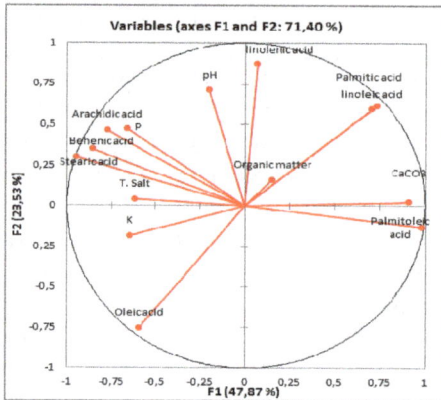

Figure 2. Variables (axes F1 and F2: 71.40 %)

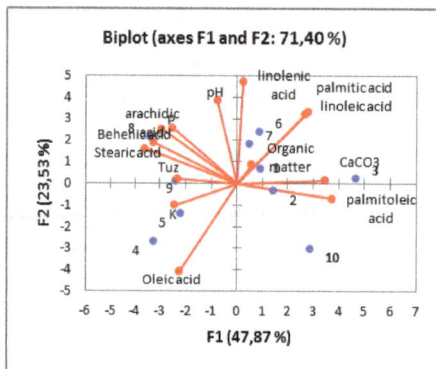

Figure 3. Biplot (axes F1 and F2: 71.40 %)

CONCLUSIONS

Of major fatty acids, oleic acid content was not affected in relation to the sampling locations. The possible effects of physicochemical characteristics of the sampling soils on the fatty acid profiles of olive oil were investigated by correlation test and then principal component analysis was performed to reduce the dimension of the experimental samples.

Oleic acid, palmitic acid and linoleic acid were more pronounced under K. $CaCO_3$ and organic matter content rich soils, respectively.

REFERENCES

Adriano D.C., 2001. Trace elements in the terrestrial environments: Biogeochemistry bioavailability and risks of heavy metals. Springer-Verlag. New York. NY.

Burhans W.C., Weinberger M., 2007. DNA replication stress. genome in stability and aging. Nucleic Acids Res;35:7545-7556.

Chakraborty A., Ferk F., Simic T., Brantner A., Dusinska M., Kundi M., Hoelzi C., Nersesyan A., Knasmüller S., 2009. DNA-protective of sumach (*Rhus coriaria*

l.). a common spice. results of human and animal studies. Mutat Res; 661:10–17.

Falciani R., Novaro E., Marchesini M., Gucciardi M., 2000. Multi-element analysis of soil and sediment by ICP-MS after a microwave assisted digestion method. Journal of analytical atomic spectrometry. 15(5). 561-565.

Guinda A., Albi T., Camino M.C.P., Lanzo´n A., 2004. Supplementation of oils with oleanolic acid from the olive leaf (*Olea europaea*). European Journal of Lipid Science and Technology.106. 22–26.

Halliwell B., Gutteridge J.M.C., 1999. Free radicals in biology and medicine (3rd ed.). Oxford: Oxford University Press.

International Olive Oil Council, 2003. Trade standard applying to olive oils and olive-pomace oils. COI/T.15.NC No3/Rev. 1.

Ishii N., 2007. Role of oxidative stress from mitochondria on aging and cancer. Cornea 200; 26:S3–S9.

Kabata-Pendias A., Mukherjee A. B., 2007. Trace elements from soil to human. Springer Science & Business Media.

Kacar B., 1995. Soil analysis. Ankara University. Faculty of Agriculture. Ankara. Turkey.

Kaçar B., Inal A., 2008. Plant analysis. Nobel Pres. (1241). 891.

Kammoun N.G., Zarrouk. W., 2012. Exploratory chemometric analysis for the characterisation of Tunisian olive cultivars according to their lipid and sterolic profiles. International Journal of Food Science and Technology. 47. 1496–1504. doi:10.1111/j.1365-2621.2012.02997.

Leon L., Uceda M., Jimenez A., Martin L.M., Rallo L., 2004. Variability of fatty acid composition in olive (*Olea europea* L.) progenies. Spanish Journal of Agricultural Research.2(3). 353-359.

Lindsay W. L., Norvell W. A., 1978. Development of a DTPA soil test for zinc. iron. manganese. and copper. Soil science society of America journal.42(3). 421-428.

Maestro D., Borja P., 1990. La calidad del aceite de oliva la composicion y maduracion de la aceituna. Grasas y Aceites; 41:171–178

Marín-Spiotta E., Swanston C.W., Torn M. S., Silver W. L., Burton S.D., 2008. Chemical and mineral control of soil carbon turnover in abandoned tropical pastures. Geoderma. 143(1). 49-62.

Matson F.M., Grundy S.M., 1985. Comparison of effects of dietary saturated. mono unsaturated and poly unsaturated fatty acids on plasma lipids and lipoproteins in man. Journal of Lipid Research. 26. 194-202.

Polidori M.C., Griffiths H.R., Mariani E., Mecocci P., 2007. Hall marks of protein oxidative damage in neurodegenerative diseases: Focus on Alzheimer's disease. Amino Acids; 32:553–559.

Soholm B., 1998. Clinical imrovement of memory and other cognitive functions by gingko biloba: Review of relevant literature. Adv Ther;15:54–65.

Ure A.M., 1990. Trace elements in soils. Their determination and speciation. Fresenius Journal of Analytical Chemistry. 337(5). 577-581.

6

A BRIEF OVERVIEW OF HAND AND CHEMICAL THINNING OF APPLE FRUIT

Alina Viorica ILIE[1], Dorel HOZA[1], Viorel Cătălin OLTENACU[2]

[1]University of Agronomic Sciences and Veterinary Medicine of Bucharest,
59 Mărăşti Blvd., District 1, Bucharest, Romania
[2]Research Station for Fruit Trees Growing Băneasa, Bd. Ion Ionescu de la Brad, No. 4,
District 1 Bucharest, Romania
Corresponding author email: alisa_ilie@yahoo.com

Abstract

A brief overview of the fruit thinning effect on apple fruit quality is presented in the current paper. Handy and chemical fruit thinning has been previously studied, existing many published reports in literature that examine various aspects of this technological procedure. The state of the art in the field of apple thinning reveal that this operation can improve fruit size, increase return bloom and reduce alternate bearing habit of apple trees.

Key words: apple, quality, thinning, fruit, blossom.

INTRODUCTION

Apple thinning can be done in different ways: handy, chemicaly or mechanicaly.
Apple is characterized by heavy bloom and heavy set of fruits throughout the growing season with several negative consequence associated with small, poorly coloured, low quality fruits. Furthermore, flower bud formation for the following year is significantly reduced, resulting in low cropping and inferior quality fruit that has a reduced postharvest storage life. Thinning of the fruitlets is the removal of a portion of the crop before it matures on the tree to increase the marketability of the remaining fruit and to break the biennial bearing tendency of the tree (Greene, 2002).
Fruit thinning can improve fruit size, increase return bloom and reduce alternate bearing habit of apple trees.
Chemical and hand contribution on maintain the physiological balance between growth and fruiting and increasing the quantity and quality of fruit (Vămăşescu and Bălan, 2014).
Frequently, apple trees bloom abundantly and set too many fruits to optimize fruit size and return bloom. Therefore, most producers attempt to increase fruit weight by reducing the number of fruits on a tree (Treder, 2008).

Chemical thinners include different chemicals, but plant growth regulators and some insecticides are used for thinning in most cases. Result from the recently published literature point to advantages of combining certain growth regulators and insecticides for thinning. Combination of carbaryl and 1-naphtaleneacetic acid has given good results in some years but caused excessive thinning in others (Rogers, 1977).
Chemical thinning is important measure for the profitable agricultural production of fruits. Chemical thinning provides a good yield potential for the following vegetation. Thinning the apple crop during the post bloom period is absolutely essential to ensure large fruit size, superior fruit quality, and reliable annual cropping (Peşteanu, 2015).
Most of the thinning literature is focused on the effect of thinners on yield or other production aspects, while the fruit quality becomes a secondary issue (Jemrić et al., 2005).
Jemrić et al. (2003) cited on Link (2000) have reviewed the Germany experience with chemical thinning accumulated over three decades and concluded that there are two groups of quality components. First group includes size, colour, skin performance, firmness, sugar and acid content. The second group includes calcium and potassium levels which are important for storability and

occurrence of physiological disorders. Thinning intensity differently affects these two groups, therefore it is important to select an optimal thinning strategy by growing and local market conditions.

Hand thinning of the apple trees can be very accurate, but it is extremely expensive and requires skilled labor inputs (Costa et. al, 2001).

MATERIALS AND METHODS

A literature search strategy was used, starting on the older to the recent scientific papers on the hand and chemical thinning of apple fruit.

RESULTS AND DISCUSSIONS

Modern fruit thinning studies focus on the relationships between the thinning agent or combination of agents, thinning parameters such as the rate and timing of the application, and the effects on crop load (a measure of fruiting density typically expressed as number of fruit per trunk or branch cross-section area, or per tree canopy volume), fruit size (expressed as weight or diameter), fruit quality, and return bloom (the crop for the following year) (Davis, 2004).

Hand thinning of fruits is a very important process in fruit culture because it is a method of obtaining the optimum quantity of fruits, which have high physical and chemical qualities (Iordanescu et al., 2009).

Hand thinning is a common and high cost practice, not only due to the labor involved, but also because seedling and viguros clonal rootstock result in large trees (Reyes et al., 2008).

Iwanami et al. (2015) examined the efficiency of hand thinning on some apple cultivars. They reported that time required for hand thinning were very similar among clusters with four, five, or six flowers/fruitlets, which was twice as long as that required for clusters with two or three flowers/fruilets. They also concluded that time required for hand thinning cluster of axillary buds became significantly longer from bloom to 7 days after bloom and then decreased gradually from 7 to 25 days after bloom.

Chemical fruit thinning methods were tested by different researchers in many countries.

Generally, plant growth regulators are used such as NAA, NAD, BA and ethephon for fruit tinning (Peşteanu, 2015).

In the literature there is information on the effect of different plant regulators in the chemical thinning of apple.

Naphthaleneacetic acid (NAA) is an auxin type thinner and was the first hormone type thinner used commercially (Stopar et al., 2007). The most effective time to apply NAA as a chemical thinner is when fruit diameter is 10-12 mm.

Synthetic cytokinin 6-benzyladenine (BA) has been found to be a good thinning agent. BA it is most effective if applied when fruit diameter averages about 10 mm and has a positive influence on return bloom, reduced the crop load and increased the fruit size (Greene, 2002; Robinson et al., 1998).

Naphthaleneacetamide (NAD) has long been in use, it acts as a synthetic auxin. NAD is considered to be a weak thinning agent who does not give satisfactory results if used alone (Stopar, 2006).

NAD induces early mild thinning, which starts the differentiation process among the flower cluster (Peşteanu, 2015).

Ethephon has long been used as a thinning agent. The main advantages of ethephon are that it greatly improves return bloom and that it can be applied over a longer period (Stopar, 2006).

The thinning response to chemical agents can be highly variable among cultivars and is strongly influenced by environmental conditions.

Weather, timing, choice of chemicals and concentration affect chemical thinning (Autio et al., 2005).

Šebek (2014) tested several thinning procedures comparing applications of NAA and BA in different concentration on three cultivars of apples. All treatments of NAA and NAA + BA had a positive effect on thinning of fruits, in terms of fruit weight and number of fruits per unit cross-sectional area of the trunk.

The highest reduction in 'Golden Delicious' fruit set was found on the NAA – sprayed trees, when evaluated at 45 days after full bloom (Reyes et al., 2008).

Peşteanu (2015) tested NAD (Geramid-New) and ethephon agents on 'Idared' apple variety.

The effect of the treatments with Geramid-New in dose 1.5 l/ha and Cerone 480 SL in dose of 0.3 l/ha had a positive effect on the number of inflorescences ant the placement of fruits in the trees crown. Also, these agents had an essential influence on quality of production.

NAD did not reduce fruit set. NAD also had no effect on fruit growth, and did not cause an increase in the proportion of pygmy fruit. NAD applied alone can not be recommended for thinning 'Fuji' apple trees (Stopar, 2006).

Wertheim (2000) mentions that NAD cannot be used on 'Delicious' trees because it induces the formation of a high number of pygmy fruit which stay on the tree until harvest time.

Radivojević et al. (2011) reported several experiments thinning two apple cultivars 'Gala' and 'Granny Smith', by hand and NAA and carbaryl. Average fruit size was consistently increased, especially in cv. 'Granny Smith'. A high return bloom was recorded in cv. 'Granny Smith' than cv. 'Gala'.

Šebek (2015) cited on Marini (2002) states that it is very difficult to adequately apply the procces of thinning in spur varieties from the Red Delicious cultivar, and that the use of NAA obtained satisfactory result in terms of fruit size and share of small fruits.

Milić et al. (2011) reported that the number of fruits per native branchet, fruit weight, fruit diameter and height yields, the best results in the chemical thinning of apple cultivar 'Golden Delicios' were obtained using BA in the concentration of 200 mg/L.

Turk and Stopar in 2010, studied the effect of 6-benzyladenine on apple thinning of cultivars 'Golden Delicious' and 'Idared'. The authors reported that BA sprayed at 10 mm did not support the findings contained in many reports which indicated that the best application time coincides with about 10 mm stage of fruit development (Greene, 2002). They concluded that BA can be active as a thinner in a wider period of phonological stages, from the end of bloom up to 20 mm of fruit diameter.

Single application of thinning agents ethephon 200 ppm at ballon stage and NAA 10 ppm or BA 100 ppm at 10 mm fruitlet diameter did not cause thinning response of 'Golden Delicious' (Stopar et al., 2007).

Recently, Peşteanu (2015) tested combination 6-BA and NAA. Therefore, it was established that the combined application of BA + NAA influenced on the nymber of inflorescences formed in the tree crown. Simultaneously the weight of single fruits increased from 16.8 to 28.2%, registering more favorable values where it was tested the combination of BA 100 ppm + NAA 10 ppm.

Significantly increased fruit weight resulted in the combined use of BA + NAA (Šebek, 2014). Šebek (2015) reported the effect of single or combined use of products based on NAA and BA in thinning of the cultivars 'McIntosh', 'Jonathan' and 'Prima'. The author reported a significant reduction in the yield of 'Prima' apple fruits had the NAA 17.82 ppm treatment. Significantly higher yield compared to the control was determined by combined application of BA 100 ppm + NAA 4.29 ppm. Cultivar 'Prima' show a positive relation between the crop load and yield efficiency. The application of all treatments (individual application of NAA and combination of NAA + BA) in thinning cultivar 'McIntosh' has led to an increase in fruit weight. The application of NAA and NAA + BA combination of all treatments in thinning of fruits of 'Jonathan' cultivar has led to a statistically significant increase in fruit weight and statistically significant decrease in yield and total number of fruits per unit cross-sectional area of the trunk in relation to the control.

Yildirim-Akinci (2014) noticed that the combination of BA+GA_{4+7} is highly effective on fruit size when applied before cell division completed. It also increased fruit diameter, fruit lengh and fruit shape. The application of 100 ppm BA + GA_{4+7} was significantly increased fruit weight by 37% and improved the fruit shape.

Greene et al. (2006) applied a new BA thinning product, respectively MaxCel. This contains no gibberellins, contains more BA, the label was changed to apply more active ingredient, and the formulation was changed to improve foliar penetration and increase formulation stability. MaxCel significantly increased fruit weight at harvest.

Thinning showed no effect on fruit shape of apple cultivars 'Empire', 'Jon-A-Red', 'Braeburn" (Ouma and Matta, 2003).

Meland (2009) mentioned that it was more difficult to adjust the crop load at first bloom than at the 20-mm fruitlet stage due to higher levels of fruit drop at first bloom. This was confirmed, where thinning at first bloom, rather than at the 20 mm fruitlet stage, improved the mean fruit weight and fruit quality, when comparing similar crop levels. Thinning at first bloom gave an annual crop of high quality.

The author Basak (2006) reported that on 'Gala' trees need to be intensively thinned. Because they blossom over a long period, the fruitlets are at different stages of development at thinning time. Thinning is most effective when too or more preparations are used in sequence. Thus, fruit quality was particularly good when BA was mixed with carbaryl or applied after NAA. In case of mixture of BA and carbaryl, yield and fruit size were better, but color was worse. When BA was used with NAA, fruit color was the same but refractive index was lower.

Peșteanu (2015) reports that in case of using BA 100 ppm + NAA 10 ppm application it was registered a decrease of fruit production, but and increase of their quality. Increasing the dose of BA 140 ppm + NAA 14 ppm does not permit fruit development in comparison with the previous variant. This is explained by the fact of pygmy fruit type appearance, diminishing the fruit number per a tree, average weight and diameter decreases, finally influencing the fruit production quantity and quality.

NAA has shown a strong thinning effect in spur-type 'Red Delicious' trees, but it may also induce an excessive development of pygmy fruits (Marini, 1996).

Milić et al. (2012) reported several experiments thinning three apple cultivars 'Braeburn' and 'Camspur' with NAA and BA. Thinning with NAA and BA has a potential risk of oversized fruits in 'Braeburn' and abnormally small (pygmy) fruit occurrence in 'Camspur'. The average fruit weight was increased, while effects of thinning on fruit parameters in 'Braeburn' were not consistent.

When ethephon was sprayed at a dose of 200 ppm did not reduce fruit set. When used alone, ethephon had no significant effect on mean fruit weight and fruit size distribution (Stopar, 2006).

Stopar et al. (2007) reported that spraying of ethephon at 200 ppm at the balloon stage slightly reduced the fruit number per tree at harvest time.

Al-Absi (2009) related that the ethephon at 500 ppm in absence or presence of BA at 200 ppm significantly increased the number of fruits per cm^2 trunk cross sectional area. A positive correlation existed between intensity of fruitlet thinning and fruitlet retention. These results are in agreement with that of Stopar and Lokar (2003), who reported that ethephon, BA and their combination significantly increased flower bud retention of 'Summerred' apples.

When ethephon treatment was followed up by BA treatment after blossoming, the thinning rate did not increase. Results were best when ethephon treatment was followed up by NAA treatment (Basak, 2006).

Marini (2004) applied combination of ethephon and Accel for thinning 'Delicious' apple trees. BA applied to 'Delicious' and 'Golden Delicious' at 11 to 12 mm fruit diameter reduced fruit set, and fruit weight and length/diameter ratio increased with concentration. The direct effect of Accel and ethephon on fruit weight appears to be inconsistent. Accel did not improve return bloom, even when trees were adequately thinned.

Iordănescu et al. (2009) cited on Cepoiu (1978) which reports that when hand thinning is done sometime in June, after the physiological fall of apples when they have 3-4 cm diameter, it has a good influence upon fruits physical and chemical qualities, but a very small impact upon buds differentiation.

Stopar et al. (2007) has noted that hand thinning performed at the end of June drop resulted in a reduced final number of fruit per tree and an increased mean fruit weight, but they were not significant. Hand thinning should be done more rigorously to reach the commercial fruit size about 150 g per fruit.

CONCLUSIONS

Fruit thinning is intended to address both horticultural and economic concerns as the grower simultaneously seeks to protect the tree from damage due to excess cropping, ensure adequate return bloom, and increase the number of larger, more valuable, fruit both in the current year and subsequent years (Davis, 2004).

The few studies summarized in this review illustrate that the effects of fruit thinning differ widely and successful thinning, resulting in increase fruit size, increased yield and improved certain parameters of the apple quality.

REFERENCES

Al-Absi K.M., 2009. Thinning intensity of 'Ace Spur Delicious' and 'Idared' apples with ethephon, benzyladenine and thei combination. Jordan Journal of Agricultural Sciences, 5(3):237-250.

Autio W.R., Krupa J., Greene D.W., 2005. Late-season chemical thinning of apples. Fruit Notes, 70:1-5.

Basak A., 2006. The effect of fruitlet thinning on fruit quality parameters in the apple cultivar 'Gala'. Journal of Fruit and ornamental Plant Research. 14(2):143-150.

Costa G., Corelli-Grappadelli L., Bucchi F., 2001. Studies on apple fruit abscission and growth as affected by cytokinins. Acta Horticulturae, 557:213-249.

Davis K., Stover E., Wirth F., 2004. Economics of fruit thinning: a review focusing on apple and citrus. Hortechnology, 14(2):282-289.

Greene D.W., 2002. Chemicals, timing, and environmental factors involved in thinner efficacy on apple. HortScienec, 37:477-481.

Greene D.W., Krupa J., Vezina M., 2006. Effect of MaxCel on fruit set, fruit size, and fruit characteristics of Summerland McIntosh apples, 2005 results. Fruit Notes, 72:12-15.

Iordănescu O., Szonyi I., Micu R., Mihuţ C., 2009. Researches concerning the influence of manual thinning of Romus 2 apples in conditions of Timisoara. Journal of Horticulture, Forestry and Biotechnology, 13:295-298.

Iwanami H., Moriya-Tanaka Y., Honda C., Wada M., 2015. Efficiency of hand-thinning in apple cultivars with varying degrees of fruit abscission. The Horticulture Journal, 84(2):99-105.

Jemrić T., Pavičić N., Blašković D., Krapac M., Pavičić D., 2003. The effect of hand and chemical fruit thinning on 'Golden Delicious Cl. B' apple fruit quality. Current Studies of Biotechnology, 3:193-198.

Jemrić T., Pavičić N., Skendrović M., 2004. Influence of thinning method on postharvest quality of ‚Golden Delicious Cl. B' apple (Malus domestica Borkh.). Agriculturae Conspectus Scientificus, 70(1):11-15.

Link H., 2000. Significance of flower and fruit thinning on fruit quality. Plant Growth Regul, 31:17-26.

Marini R.P., 1996. Chemically thinning spur ‚Delicious' apples with carbaryl, NAA, and ethephon at various stages of fruit development. HortTechnology, 6:241-246.

Marini R.P., 2002. Thinning 'Golden Delicious' and spur 'Delicious' with combinations of carbamates and NAA. HortScience, 37(3):534-538.

Marini R.P., 2004. Combinations of ethephon and accel for thinning ‚Delicious' apple trees. J. Amer. Soc. Hort. Sci. 129(2):175-181.

Meland M., 2009. Effects of different crop loads and thinning times on yield, fruit quality, and return bloom in Malus domestica Borkh. ‚Elstar'. Journal of Horticultural Science &Biotechnology, 171-121.

Milić B., Keserović Z., Magazin N., 2011. Production effects of chemical fruit thinning in young apple orchards. Agricultural Economics, 58:133-146.

Milić B., Keserović Z., Magazin N., Dorić M., 2012. Fruit quality and bearing potential of chemically thinned 'Braeburn' and ‚Camspur' apples. Žemdirbysté Agriculture, 99(3):287-292.

Ouma G., Matta F., 2003. Response of several apple tree cultivars to chemical thinner sprays. Fruits, 58(5):275-281.

Peșteanu A., 2011. Effect of thinning 'Idared' apple variety using NAD and ethephon. Lucrări Ştiinţifice Seria Horticultură, 58(1):129-134.

Peșteanu A., 2015. The influence of thinning agent on base of 6-BA and NAA on productivity and fruit quality of ‚Gala Must' variety. Bulletin UASVM Horticulture, 72(1):151-153.

Radivojević D., Zabrkić G., Milivojević M.J., Veličković M., Oparnica Č., 2011. Effect of chemical and hand thinning young apple tree on yield and fruit quality. Proceedings.46th Croatian and 6th International Symposium on Agriculture. Opatija, Croatia, 1044-1047.

Radivojević D., Milivojević M.J., Oparnica Dj. Ć., Vulić B. T., Djordjević S.B., Ercişli S., 2014. Impact of early cropping on vegetative development, productivity, and fruit quality of Gala and Braeburn apple trees. Turkish Journal of Agriculture and Forestry, 38:773-780.

Reyes D.I.B., Chaćon A.R., Campos Á.R.M., Prieto V.M.G., 2008. Apple fruit chemical thinning in Chihuahua, Mexico. Rev.Fitotec.Mex., 31(3):243-250.

Robinson T., Lakso A., Stover E., Hoying S., 1998. Practical apple thinning programs for New York. New York fruit quarterly, 6:14-18.

Rogers BL, Williams GR., 1977. Chemical thinning of spur-type Delicious apple fruit. Virginia Fruit, 65:23-28

Stopar M., Lokar V., 2003. The effect of ethephon, NAA, BA and their combinations on thinning intensity of ‚Summered' apples. Journal of Central Agriculture, 4:399-403.

Stopar M, 2006. Thinning of ‚Fuji' apple trees with ethephon, NAD and BA, alone and in combination. Journal of Fruit and Ornamental Plant Research, 14:39-45.

Stopar M., Schlauer B., Turk Ambrožič B., 2007. Thinning ‚Dolden Delicious' apples using single or combining application of ethephon, NAA or BA. Journal Central European Agriculture, 8(2):141-146.

Šebek G., 2014. Application of NAA and BA in chemical fruit thinning of autochthonous cultivars of apple. Journal of Agricultural Science and Technology, 4:21-28.

Šebek G., 2015. Application of NAA and BA in chemical thinning of some commercial cultivars of apple. Acta Agriculturae Serbica, XX, 39:3-16.

Treder W., 2008. Relationship between yield, crop density coefficient and average fruit weight of 'Gala' apple. Journal of Fruit and Ornamental Plant Research, 16:53-63.

Treder W., Mika A., Krzewińska D., 2010. Relations between tree age, fruit load and mean fruit weight. Journal of Fruit and Ornamental Plant Research, 18(2):139-148.

Turk B.A., Stopar M., 2010. Effect of 6-benzyladenine application time on apple thinning of cv. 'Golden Delicious' and cv. 'Idared'. Acta agricuturae Slovenica, 95(1):69-73.

Vămăşescu S., Bălan V., 2014. Thinning and foliar fertiliyation influence on the zield of Idared apple cultivar. Scientific Papers. Seria B, Horticulture, LVIII:107-110.

Wertheim S.J., 2000. Developments in the chemical thinning of apple and pear. Plant Growth Regulation, 31:85-100.

Yildirim-Akinci F., Kepenek G., Şan B., Zildirim A.N., Kacal E., 2014. Effects of BA+GA $_{4+7}$ treatments on fruit qualitz in 'Fuji' apple variety. Turkish Journal of Agricultural and Natural Sciences, 2:1387-1390.

7

EFFECTS OF CLIMATE CHANGE ON OLIVE CULTIVATION AND TABLE OLIVE AND OLIVE OIL QUALITY

Yasin OZDEMIR[1]

[1]Ataturk Central Horticultural Research Institute, Department of Food Technology, Yalova,Turkey
Corresponding author email: yasin.ozdemir@gthb.gov.tr

Abstract

Climate change is undoubtedly the most imminent environmental issue the world is facing today. Climate changes could heavily affect olive oil producing areas, especially in the Mediterranean basin. Olive trees are tougher than vines and can thrive on many different terrains and under various climate conditions. Researchers reported that a key area of Spanish olive oil production in Catalonia may become unviable within 20 years due to these increasing temperatures and water shortages. Extreme temperatures pose risks for olive production. The Mediterranean basin region, where more than 90% of the world's olive oil is produced is expected to be exposed to higher temperatures in the future due to global climate changes, which caused unfavorable growing conditions and showed negative effects on oil production and quality. According to result of studies there is an urgent need for approaches to estimate the consequence of climate change and its effects on olive quantity and quality of olive oil and table olive. This research was aimed to present result of studies which focused on effects of climate change on olive growing and production and also quality of table olive and olive oil products.

Key words: global warming, olive industry, table olive, olive oil.

INTRODUCTION

Climate change is undoubtedly the most imminent environmental issue the world is facing today. The rise in climate temperature will have certain major effects on ecosystems, wildlife, food chains and eventually human life (Appels et al., 2011). According to result of studies, in less than forty years, three-quarters of the wine producing areas on earth will not be suitable for vine farming due to the effects of climate change. Soon enough those vineyards will move to other territories that will have the conditions to grow the grapes, like northern Europe, northwestern America and areas of central China (Hannah et al., 2013; Vasilopoulos, 2013). Similarly, the climate changes could heavily affect olive oil producing areas, especially in the Mediterranean basin (Vasilopoulos, 2013). Climate change alters both average and extreme temperatures and precipitation patterns which in turn influence crop yields, pest and weed ranges and introduction and the length of the growing season (Anon., 2008). Olive trees are tougher than vines and can thrive on many different terrains and under various climate conditions. They give olive oil with little effort

and care throughout the year, often without much watering. This is why countries like India, Libya and Australia are planting more olive trees; it is relatively easy to grow them and they can yield a profit (Vasilopoulos, 2013). According to result of studies climate chage have caused effects on olive cultivation and alteration on quality of table olive and olive oil (Ponti et al., 2014; Dag et al., 2014; Tupper 2012). This review aimed to present these effects and alterations on olive cultivation and olive product quality.

EFFECTS OF GLOBAL WARMING ON OLIVE CULTIVATION

Climate change threatens agro-ecosystems of olive which is an ancient drought-tolerant crop of considerable ecological and socioeconomic importance in the Mediterranean Basin (Ponti et al., 2014). Researchers reported that a key area of Spanish olive oil production in Catalonia may become unviable within 20 years due to these increasing temperatures and water shortages. Spain is thought to be highly susceptible to climate change. Studies have shown that the flowering period of olive trees is highly dependent on the yearly spring temperatures, which are rising steadily over time (Tupper, 2012).

The most immediate issue for the olive cultivation is rainfall which is highly affected by climate change. 2013 has been reported as the driest year on record since dating back over 150 years for California (Moran, 2014). Less rainfall means low olive oil production, with few options for farmers when water prices remain high (Moran, 2014; Tupper, 2012). This pertains to the bulk of American olive oil production, considering that 90 percent of olives grown domestically come from California (Moran, 2014). If Spain is to continue its supremacy as an olive oil producing nation, new and innovative irrigation alternatives will have to be created to combat the constantly changing climate. This is no easy task however, as increasing irrigation can have negative effects on water supplies for the area, leading to desert like areas and water shortages for other purposes, as has previously been seen in Greece, Italy and Portugal when irrigation demands increased (Tupper, 2012).

When the weather becoming warmer, olive groves on high hills or slopes will probably suffer less, but groves located on low altitude areas or plains could become totally unproductive. There are already signs of the oncoming change, with this year's harvest in Spain crippled by the drought and the phenomenal weather variations (Vasilopoulos, 2013). Water scarcity affects every continent and countries such as Greece and Italy have already suffered the devastating effects of drought, with olives dying at high temperatures and from lack of water. In addition to the direct effects of a changing climate on the olive population, variations in weather can also cause changes in other environmental factors such as insects and disease. These may then influence the olive tree population, as an indirect effect of changing climates (Tupper, 2012).

Extreme temperatures, pose risks for olive production. In 1998, severe cold temperatures caused significant losses for olives in California. Olive trees can normally handle brief cold snaps, but sub- freezing temperatures that last longer than a few hours will damage new and small branches and may prevent fruit production (Moran, 2014).

Olive phenology has been reported as a good indicator of future climatic change (Osborne et al., 2000). This could be explained by the fact that photoperiod would also affect the start of flowering in the late spring flower species. Moreover, one of the most expected consequences of climate change will be the increase in minimum temperatures, especially in winter and early spring (Ahmad, 2001).

The variability in chilling hours, which garner less attention than frost, are equally important to overall olive vitality. Accumulated winter chill hours are declining in the growing regions of California, which affects a range of crops from olives to almonds. A substantial amount of chilling hours (32-45°F) are necessary for olive flower bud development, which facilitate the plant's movement out of its vegetative state so fruit can be produced (Moran, 2014).

Inability to determine reliably the direction and magnitude of change in natural and agro-ecosystems due to climate change poses considerable challenge to their management. This level of climate warming will have varying impact on olive yield and fly infestation levels across the Mediterranean Basin, and result in economic winners and losers (Ponti et al., 2014).

Mild temperatures in summer are reported as a reason for increased olive fly infestations in the region without stretches of summer heat to reduce the fly population. Together, these effects create a complex web of changing climate and olive oil production, whose future will require further scientific research, careful monitoring (Moran, 2014; Marshall et al., 2011). Recycling to soil olive mill waste has the potential to improve soil fertility, thus reducing CO_2 emission associated to global warming (Altieri and Esposito, 2010).

Climate change will impact the interactions of olive and the obligate olive fruit fly (*Bactrocera oleae*) and alter the economics of olive culture across the Mediterranean Basin. The effects of climate change on the dynamics and interaction of olive and the fly using physiologically were estimated based on demographic models in a geographic information system context as driven by daily climate change scenario weather (Ponti et al 2014).

Climate does not only affect olive trees directly, but changing temperatures also influences insect diversity and frequency for a given area. Rising carbon dioxide levels will exacerbate most insect and pest problems. This is particularly relevant to the olive fly, olive's most notorious and costly pest, but studies show that this effect may actually operate in a counter intuitive way (Moran 2014).

Decreased level of production may become common place if continued scarcity of water and increased temperatures start to effect groves in Spain. While high temperatures are optimal for growth and development of olives, heavy rain is also necessary to complete the ripening process (Tupper 2012). Irrigation of olives with saline water will inevitably increase in the future in the Mediterranean due to negative effects of population growth and climate change on the availability and quality of existing fresh water supplies. As a consequence, the risk land salinisation will exacerbate threading the agricultural production particularly in countries with a semi-arid or arid climate (Chartzoulakis 2005).

Spanish olive oil production has doubled in the last ten years, but ongoing drought and climate change may mean a setback for the global leader in olive oil production. Spain may fall to the same fate as fellow olive oil producing power houses Greece and Italy due to the effects of climate change. Italy has seen a drop of 50 % in production since 2001 and Greece has also seen its annual production levels decline by half with climate change thought to be an important factor (Tupper 2012).

EFFECTS OF GLOBAL WARMING ON TABLE OLIVE AND OLIVE OIL PRODUCTION AND QUALITY

Climate warming will affect olive yield and oil quality across the Mediterranean Basin, resulting in economic winners and losers at the local and regional scales. At the local scale, profitability of small olive farms in many marginal areas of Europe and elsewhere in the Mediterranean Basin will decrease, leading to increased abandonment (Ponti et al., 2014).

Emerging players of the olive industry like China and India with vast lands for cultivating olive trees could challenge European produ-cers. Because of rapid changing weather, reduction of olive oil production is on the way for traditional olive oil powerhouses such as Spain, Italy and Greece (Vasilopoulos, 2013).

Decline of olive oil production in Italy and Greece has had a temporarily positive effect on Spain, which is now producing twice the joint production of Greece and Italy. Current harvest in Spain will be a poor one, with a 40 % drop in production due to drought, leading to a huge leap in market prices for olive oil (Tupper, 2012).

More than 90% of the world's olive oil is produced in the Mediterranean basin where is expected to be exposed to higher temperatures in the future due to global climate changes which caused unfavorable growing conditions and showed negative effects on oil production and quality (Dag et al., 2014). Early harvest (relatively low ripening index) is reported as one of the major findings to prevent from those climate changes (Dag et al., 2014; Vasilopoulos, 2013,).

Determination of the optimal fruit ripening stage for the production of olive oil represents a critical choice based on the best combination of oil quantity and quality (Dag et al., 2014). As olive ripening proceeds, fruit characters such as weight, pulp to stone ratio, color, oil content, enzymatic activities and profiles of various phytochemicals, including fatty acids and total polar phenol content, are constantly changing. These changes in fluence fruit firmness, olive oil chemical composition and sensory qualities (Beltrán et al., 2004). Climate has a major in fluence on the ripening process and hence on oil accumulation and its chemical composition (Aparicio and Luna, 2002).

As well as the quantity of fruit, the qualitative components of olive oil produced can be influenced by the environmental conditions of the growing year (Lombardo et al., 2008). This relates to the absolute variations in fatty acids and the relationships between these individual components such as the oleic acid/linoleic fatty-acid ratio, and the ratio between oleic acid and the sum of palmitic and linoleic acids (D'Imperio et al., 2007). Getting a high quality olive oil requires several different factors which are healtly trees, appropriate climate and proper farming. Also the ground morphology and the moisture levels of the area play an important role in shaping the oil characteristics. European olive oils in fifty years from now could be very different in terms of their qualities and organoleptic characteristics (Vasilopoulos, 2013).

One of the most important aspects is the use of irrigation while monitoring for potential water stress caused by water deficit in the summer months. Thus, greater attention needs to be paid on the part of the olive growers for evaluation of the vegetative–productive state of the trees and their reproductive cycle, and of the hydro-pedological conditions of the terrain that vary with the seasonal meteorological trends (Orlandi et al., 2012).

A water deficit during initial development of the fruit (in June in the northern hemisphere) can result in a decrease in the size of the cells of the mesocarp that cannot be recovered, except, at least in part, if the plants are regularly irrigated in the following stages (Servili et al., 2004). Water deficit affects fruit maturation, which occurs earlier and more rapidly, and can result in more intense pre-harvest fruit fall (Inglese et al., 1996). However, a number of studies have shown that the water state of the plant has marginal, if any, effects on free acidity and peroxide value of the olive oil produced (Servili et al., 2007). A direct relationship connects the water stress to the levels of linoleic and linolenic acids, where higher stress corresponds to high levels of these fatty acids. This therefore provides further support for the concept of qualitative irrigation (Servili et al., 2004).

The phase of maturation of the olive fruit influences not only its acid composition, but also the composition of its minor constituents, and particularly its phenolic and volatile compounds. Thus, factors that affect the evolution of maturation of the drupe can also affect the qualitative characteristics of the resulting olive oil (Fiorino and Nizzi Griffi, 1991).

Variability in acid composition has been correlated to the temperature sum of the period from fruit setting to fruit maturation. The high temperatures during this phase that arise in hot seasons and environments can result in decreased oleic acid content, which is accompanied by increased palmitic and/or linoleic acids (Lombardo et al., 2008). A very high temperature sum also tends to reduce total polyphenol content (Ripa et al., 2008). Similarly, in cooler areas, a positive correlation has been shown between the temperature sum from August to October and the total

polyphenol content of olive oil (Tura et al., 2008).

The water state of the plant also has marked effects on concentrations of volatile compounds in the oil. Thus, oil from plants grown without irrigation, as opposed to those with, can be more bitter and biting to the taste (Servili et al., 2007). Plants grown under conditions of water stress therefore tend to produce oils that are more full bodied and strong in their taste, with strong bitter and biting notes, but that are relatively less aromatic (Orlandi et al., 2012).

CONCLUSIONS

According to result of these mentioned studies there is a urgent need for approaches to estimate the consequence of climate change and its effects on olive quantity and quality of olive oil and table olive. Water uses for irrigation should be good planned for olive orchard to reach maximum olive yield and olive oil or table olive quality with minimum water consumption. Global warming and rising carbon dioxide levels will increase the most of the insect and pest populations and change their life cycle. So that olive orchard should be kept under constant observation to determine the time and use the type and amount of pesticide and insecticide uses. According to climatic changes, harvest time of olives should be redefined which should be provide a balance between for high yield and final product quality. New breeding studies should be focused on the behavior of olive genotypes with respect to climate change. Water stres and diseases resistance of olive trees and olive oil quaity or table olive quality under increased temperatures and water stres should be use as a dominant advanced selection criteria for new olive cultivar candidates.

ACKNOWLEDGMENT

This work was supported by Ataturk Central Horticultural Research Institute (Yalova, Turkey).

REFERENCES

Altierl R., Esposito A., 2010. Evaluation of the fertilizing effect of olive mill waste compost in short-term crops. International Biodeterioration & Biodegradation, 64:124–128.

Anonimus, 2008. 2009 California Climate Adaptation Strategy - Final Report A report to the Governor of the State of California in Response to Executive Order S-13-2008. California Natural Resources Agency. California, USA.

Aparicio R., Luna G., 2002. Characterization of monovarietal virginoliveoils. European Journal of Lipid Science and Technology, 104:614–627.

Appels L., Lauwers J., Degrève J., Helsen L., Lievens B., Willems K., Impe J.V., Dewil R., 2011. Renewable and Sustainable Energy Reviews, 15:4295–4301.

Dag A., Harlev G., Lavee S., Zipori I., Kerem Z., 2014. Optimizing olive harvest time under hot climatic conditions of Jordan Valley, Israel. Eur. J. Lipid Sci. Technol, 116:169–176.

Beltran G., Del Rio C., Sanchez S., Martinez L., 2004. Seasonal changes in olive fruit characteristics and oil accumulation during ripening process. Journal of the Science of Food and Agriculture, 84:1783–1790.

Chartzoulakis K.S., 2005. Salinity and olive: Growth, salt tolerance, photosynthesis and yield. Agricultural Water Management, 78:108–121.

D'Imperio M., Dugo G., Alfa M., Mannina L., Segre A., 2007. Statistical analysis on Sicilian olive oils. Food Chemistry, 102:956–965.

Fiorino P., Griffi N., 1991. Olive maturation and variations in certain oil constituents. Olivae, 35:25–33.

Gómez-Rico A., Salvador M.D., Moriana A., Pérez D., Olemdilla N., Ribas F., Fregapane G., 2007. Influence of different irrigation strategies in a traditional Cornicabra cv. olive orchard on virgin olive oil composition and quality. Food Chemistry, 100:568–578.

Hannah L., Roehrdanz R., Ikegami M., Shepard A.V., Shaw M.R., Tabor G., Zhi L., Marquet P.A., Hijmans R.J., 2013. Climate change, wine, and conservation. Proceedings of the National Academy of Sciences of the United States of America, USA.

Inglese P., Barone E., Gullo G., 1996. The effect of complementary irrigation on fruit growth and ripening pattern and oil characteristics of olive (*Olea europea* L.) Cv. Carolea. Journal of Horticultural Science and Biotechnology, 71:257–263.

Johnson M.W., Wang X.G., Nadel H., Opp S.B., Lynn-Patterson K., Stewart-Leslie J., Daane K.M., 2011. High temperature affects olive fruit fly populations in California's Central Valley. California Agriculture, 65(1):29-33. 2011.

Lombardo N., Marone E., Alessandrino M., Godino G., Madeo A., Fiorino P., 2008. Influence of growing season temperatures in the fatty acids (FAs) of triacilglycerols (TAGs) composition in Italian cultivars of Olea europaea. Advances in Horticultural Science, (1), 49-53.

Moran M.E., 2014. The toll of climate change on california olive oil. Olive oil times, January 14, 2014.

Orlandi F., Bonofiglio T., Romano B., Fornaciari M., 2012. Qualitative and quantitative aspects of olive production in relation to climate in southern Italy. Scientia Horticulturae, 138:151–158.

Ponti L., Gutierrez A.P., Ruti P.M., Dell'Aquila A., 2014. Fine-scale ecological and economic assessment of climate change on olive in the Mediterranean Basin reveals winners and losers. Proceedings of the National Academy of Sciences, 111(15):5598-5603.

Ripa V., De Rose F., Caravita M.L., Parise M.R., Perri E., Rosati A., Pandolfi S., Paoletti A., Pannelli G., Padula G., Giordani E., Bellini E., Buccoliero A., Mennone C., 2008. Qualitative evaluation of olive oils from new olive selections and environment on oil quality. Advances in Horticultural Science, 22: 95–103.

Servili M., Selvaggini R., Esposto S., Taticchi A., Montedoro G.F., Morozzi G., 2004. Health and sensory properties of virgin olive oil hydrophilic phenols: agronomic and technological aspects of production that affect their occurrence in the oil. Journal of Chromatography A, 1054:113–127.

Servili M., Esposto S., Lodolini E., Selvaggini R., Taticchi A., Urbani S., Montedoro G.F., Serravalle M., Gucci R., 2007. Irrigation effects on quality, phenolic composition and selected volatiles of vergin olive cv Leccino. Journal of Agricultural and Food Chemistry, 51:6609–6618.

Stefanoudaki E., Chartzoulakis K., Koutsaftakis A., Kotsifaki F., 2001. Effect of drought stress on qualitative characteristics of olive oil of cv Koroneiki. Grasas Aceites, 52:202–206.

Tupper N.,2012. Spanish olive oil under constant threat from climate change. Olive Oil Times, October 26, 2012.

Tura D., Failla O., Pedò S., Gigliotti C., Bassi D., Serraiocco A., 2008. Effects of seasonal weather variability on olive oil composition in northern Italy. Acta Horticulture, 791:769–776.

Vasilopoulos C., 2013.Climate change effects on vines should alarm olive oil producers. Olive Oil Times, April 22 2013.

8

A NEW APPROACH OF SWEET CHERRY *(PRUNUS AVIUM* L.) POLLINATION: CORIANDER *(CORIANDRUM SATIVUM* L.) ESSENTIAL OIL

Sultan Filiz GUCLU[1], Ayşe Betül AVCI[2]
[1]Süleyman Demirel University Atabey Vocational School, Isparta, Turkey
[2]Ege University Ödemiş Vocational School, İzmir Turkey
Corresponding author email: sultanguclu@sdu.edu.tr

Abstract

The objective of this study was to examine the effect of coriander essential oil applied as an alternative method to increase fruit set of sweet cherry which is an important fruit for the Isparta region. In the trial, coriander essential oil was applied to the branches in blooming period and comparisons were made with branches on which essential oil was not applied (control). When average values are considered, it has been determined that the application increased fruit set from 27.86 to 53.28 in comparison with the control group.

Key words: 0900 Ziraat, pollination, linalool, aspect, fertilization, bee

INTRODUCTION

The main factors determining fertilization and yield of the sweet cherry include temperature and rainfall during blooming period, the period of overlapping blooming of the main and pollinizer cultivars and their self and cross incompatibility status. Other environmental factors such as bees are referred to as pollen vectors (Sutyemez and Eti, 1999; Choi and Andersen 2001; Pırlak, 2002). 0900 Ziraat is an export cultivar grown in Turkey which has dark colour, sweet taste and is crack resistant. Fertilization is essential for fruit set and yield in sweet cherry growing (Tosun and Koyuncu, 2007; Beyhan and Karakas, 2009). Successful fertilization of sweet cherries depends on the transfer of compatible pollen by honeybees, as most commercial cultivars are self-incompatible (Janıck and Moore, 1996; Thomson, 2004). Flowers are usually sufficient to obtain a product in cherries 25-50% of the flowers must become fruits. 0900 Ziraat sweet cherry cultivar has some problems in terms of efficiency. The biggest problem that causes decrease in yielding can be caused by self-sterility and late bloom, pollination and outcrossing (Bekefi, 2004). Sweet cherry pollen must come from another – and compatible - cultivar; therefore, a high degree of bee activity on the tree and between trees is required

(Tonitti et al., 1991). Entomophily is a form of plant pollination whereby pollen is distributed by insects, particularly bees, Lepidoptera (e.g. butterflies and moths), flies and beetles. Honey bees will pollinate many plant species that are not native to their natural habitat but are often inefficient pollinators of such plants. Pollinators of sweet cherry are honey bees, bumblebees, solitary bees and flies (Bosch and Kept, 2002). Hussein and Abdel-Al (1982) also reported that honey bee consisted more than 67% of the total bees are visiting coriander and others. Diederichsen (1996) attributed that coriander is also a good melliferous plant; one hectare of coriander allows honey bees to collect about 500 kg of honey.

Coriander (*Coriandrum sativum*) is a valuable weed and spice plant from the Umbelliferae family (Baydar, 2009). The essential oil ratio in its fruits varies between 0.03-2.60% (Kaya et al., 2000). Coriander oil is among the 15 most produced essential oils. The previous researches indicate that linalool (1,6-octadien-3-ol, C10H18O, Mr=154.25, d20= 0.862 and boiling point 196– 198°C) is the main component in coriander essential oil, and it has potential usage as antispasmadic immunostimulatory antinociceptive (Peana et al., 2003; Telci et al., 2006). α-pinene, γ-terpinene, geranyl acetate, p-cymene and hexadeconoic acid in the coriander essential oil are also other important components (Anitescu et al., 1997; Baydar, 2009). In our study, the effect of using the indirect bee activity of essential coriander oil has been examined as an alternative method to increase fruit set in sweet cherry.

In this study, coriander essential oil which is used as an activator to increase bee activity has been applied to cherry branches during blooming period and the number of fruits on the coriander applied branches along with that on the branches with no coriander application have been compared.

MATERIALS AND METHODS

Plant material

In this study, *Coriandrum sativum* essential oil has been applied to Gisela 5 rootstocks of the 0900 agricultural cherry type 10 years old which belong to an individual at the Atabey district of the city of Isparta in order to increase fruit set ratio. Starks Gold species at a ratio of 1/8 is used as a pollinator in the garden where the application was done.

Statistical analysis

The study was planned as a randomized block design with 3 repetitions. Three trees were randomly selected during each repetition and coriander essential oil was applied in 4 directions (North, South, East and West) to the fruit buds on the branches of each tree selected at the same level. Whereas the buds that were left as control were isolated during the application. Fruit set ratio in the study was calculated by counting the number of fruits on the essential oil applied branch along with the number of fruits on the control branch.

Data analyses

Data were subjected to analysis of variance for the essential oil application. All data were analyzed by computer software (Standard ANOVA analysis). The means were compared by using the LSD test described by Steel and Torrie 1980. Mean percentage and standard deviation of essential oil values of the collected samples were calculated by MS Excel program.

Isolation of essential oil

The essential oils were extracted by hydrodistillation for 3 h using Clevenger type apparatus using 10 g of the air-dried aerial parts of the plant samples. The volatile oils were stored in dark glass bottles at $4^{\circ}C$ until analysis (British Pharmacopoeia 1980).

GS-MS analyses of essential oil

Essential oil constituents were analyzed by (%) gas chromatography method, and GC-MS analysis was carried out by utilizing Shimadzu GC/MS-QP 5050 A in Suleyman Demirel University Experimental and Observational Student Practice and Research Center. CP Wax 52 CB (50 m x 0.25 mm i.d., film thickness 1.2 µm) capillary column and Helium as a carrier gas were used. The temperature program reached from $60^{\circ}C$ to $22^{\circ}C$ with $2^{\circ}C$ increases in temperature in a minute, and was applied by maintaining $22^{\circ}C$ for 20 minutes. Temperature of the injector was of $24^{\circ}C$. Mass spectra were used at 70 eV. After the compounds were ionized in gas chromatography column and separated, mass spectrum of each of them were obtained. Evaluation procedures were conducted using "Wiley, Nist and Tutor" libraries.

RESULTS AND DISCUSSIONS

The effect of coriander essential oil application to increase fruit set in cherry can be seen in Table 1.

Accordingly, it is observed that the application has increased the number of fruits and that this increase is statistically significant.

When the average values are examined, it is determined that the application has doubled the number of fruits.

Table1. Effect of coriander essential oil on cherry fruit set

| | Number of Fruits | | | | |
	North	South	East	West	Average
Control	26.22	27.11	29.22	28.89	27.86 a
Application	48.06	58.99	51.11	54.44	53.28 b
Average	37.44	43.00	40.17	41.67	40.57
LSD$_{app (\%1)}$:	4.244				
LSD$_{asp}$: ns					

Even though it was not determined to be statistically significant, it has been observed when the number of fruits in the directions is evaluated that the highest number is 43 on average for the south direction which is an expected result since blooming is expected early and fruit set is expected to be high on trees that face south (Janick and Moore, 1996). This is followed respectively by west, east and south directions. When directions and application are evaluated together, it is observed that the highest number of fruits is found on essential oil applied on trees and that the increase was almost double in comparison with the control.

It is thought that this increase is due to linalool which is the main component in coriander essential oil (Table 2).

Table 2. Essential oil composition of *Coriandrum sativum.*

Element	Rt	Area (%)
2-dodecenal	59.9	1.80
Geraniol	58.5	2.91
Geranyl acetate	53.3	14.39
Linalool	40.0	80.9

Coriander plant is named in some resources as the honey plant and its attractive effect can be thought of as the reason for this. Raguso and Pichersky (1999) stated that the monoterpene alcohol, linalool, is present in the ßoral fragrance of diverse plant families and is attractive to a broad spectrum of pollinators, herbivores and parasitoids.

Priority in such applications is not only to increase fruit set; the effect on the quality parameters of the fruit should also be examined. Even though there are some studies on the effect of essential oils on seed germination, the number of studies on pollen germination and tube growth is scarce. The effect of coriander oil on pollen germination and pollen tube growth will be examined in future studies and we believe that this will put forth a new perspective for the solution of fertilization problems in cherry.

CONCLUSIONS

Coriander essential oil application increase fruit set in cherries express in the number of fruits is examined.

It has been determined that the application has doubled the number of fruits and linalool which is the main component of the coriander essential oil has been presumed as the reason for this.

In the light of these results, it is thought to carry out the application using different doses of coriander essential oil and that the increase

in the number of fruits is due Linalool which is the primary component of the coriander essential oil; carrying out the application with all the other components can be thought of as well.

In addition, carrying out the application of different essential oils on different trees instead of only coriander essential oil on cherry trees will enrich the further studies.

REFERENCES

Anıtescu G., Doneanu C., Radulescu V., 1997. Isolation of Coriander Oil: Comparision between steam distillation and supercritical CO2 Extraction. Flavour and Fragrance Journal, 12:173–176.

Baydar H., 2009. Tıbbi ve Aromatik Bitkiler Bilimi ve Teknolojisi (Genişletilmiş 3. Baskı). SDÜ Yayınları No: 51, Isparta.

Bekefi Z., 2004. Self-fertility studies of some sweet cherry (*Prunus avium* L.) cultivars and selections. International Journal Horticulture Science, 10 (4):21–26.

Beyhan N., Karakaş B., 2009. Investigations of the fertilization biology of some sweet cherry cultivars grown in central Northern Anatolian Region of Turkey. Scientia Horticulturae, 121; 320–326.

Bosch J., Kept W.P., 2002. Developing and establishing bee species as crop pollinators: the example of Osmia spp. (Hymenoptera: Megachilidae) and fruit trees. Bulletin of Entomological Research, 92:3–16.

British Pharmacopoeia. 1980. Pharmacopoeia, Vol II.H. M. Stationary Office, London. P A 109.

Choı C., Andersen R.L., 2001. Variable fruit set in self-fertile sweet cherry. Canadian Journal of Plant Science 81: 753–760.

Choı C., Tao R., Andersen R.L., 2002. Identification of self-incompatibility allelesand pollen incompatibility groups in sweet cherry by PCR based s-allele typing and controlled pollination. Euphytica 123: 9–20.

Dıederıchsen A., 1996. Promoting the Conservation and Use of Underutilized and Neglected Crops.3. Coriander (Coriandrumsativum L.). Institute of Plant Genetics and Crop Plant Research Gatersleben/International Plant Genetic Resources Institute, Rome. ISBN: 92-9043-284-5.

Hussein M.H., Abdel-Aal S.A., 1982. Wild and honey bees as pollinators of 10 plant species in Assiut area, Egypt. Zeitschrift für Angewandte Entomologie. 93(1-5): 342-346.

Janick J., Moore N.J., 1996. Fruit Breeding, Tree and Tropical Fruits. Vol. 1. New York, John Wiley and Sons, Inc.

Kaya N., Yılmaz G., Telci I., 2000. Agronomic and technological properties of coriander (Coriandrum sativum L.) population planted on different Dates. Tr. J. of Agr. Forestry 24, 355–364.

Peana A.T., D'aquila P.S., Chessa M.L., Morettil M.D., Serra G., Pıppıa P., 2003. (-)-Linalool produces antinociception in two experimental models of pain. European Journal Pharmacology, 460:37–41.

Pırlak L., 2002. The effects of temperature on pollen germination and pollen tube growth of apricot and sweet cherry. Gartenbauwissenchaft, 67 (2): 61–64.

Raguso R.A., Pıchersky E., 1999. New Perspectives in Pollination Biology: Floral Fragrances a day in the life of a linalool molecule: Chemical communication in a plant-pollinator system. Part 1: Linalool biosynthesis in flowering plants. Plant Species Biology, 14: 95-120.

Stell R.G.D., Torrie J.H., 1980. Principles and procedures of statistics: A Biometric Approach, 2nd Ed., Mc Graw-Hill, NY, USA.

Sütyemez M., Eti S., 1999. Investigations on the fertilization biology of some sweet cherry varieties grown in Pozantı Ecological Conditions. Turkish Journal AgricultureFor. 23 (3): 265–272.

Telcı I., Bayram E., Avcı B., 2006. Changes in Yields, Essential Oil and Linalool Contents of Coriandrum sativum Varieties (var. vulgare Alef. and var.

microcarpum DC.) Harvested at Different Development Stages. European Jornal of Horticultural Science 71 (6).

Thompson M., 2004. Flowering, pollination and fruit set. In: Webster a.d.,Looney n.e. (eds.), Cherries, Crop Physiology, Production and Uses. Wallingford, CABI Publishing: 223–243.

Thomson J.D., 1989. Germination schedules of pollen grains: implications for polen selection. Evolution 43 (1), 220–223.

Tonutti P., Ramına A., Cossıo F., Bargıonı G., 1991. Affective pollination period and ovule longevity in Prunus avium L. Adv. Hortic. Sci. 5, 157–162.

Tosun (Guçlu) F., Koyuncu F., 2007. Investigations of suitable pollinator for 0900 Ziraat sweet cherry cv.: pollen performance tests, germination tests, germination procedures, in vitro and in vivo pollinations. Horticulturae Science (Prague) 34: 47–53.

9

EFFECT OF DIFFERENT DOSES OF POTASSIUM ON THE YIELD AND FRUIT QUALITY OF 'ALBION' STRAWBERRY CULTIVAR

Mehmet POLAT[1], Volkan OKATAN[2], Burak DURNA[3]
[1]Süleyman Demirel University, Faculty of Agriculture, Isparta/turkey
[2]Usak University, Vocational High School, Sivaslı/Uşak/Turkey
[3]Provincial Directorate of Agriculture, Çorum/Turkey
Corresponding author e-mail: mehmetpolat@sdu.edu.tr

Abstract

This study was carried out to determine effect of the potassium fertilizer rate on yield and fruit quality of day-neutral strawberry cultivar 'Albion' during 2013-2015 years. Frigo seedlings of Albion cultivar were planted in June 2013. In the experiment, 0-5-10-15 kg/da K_2O fertilizer was applied and identified as K-0, K-5, K-10 and K-15. Fruit weight, fruit length, fruit width, yield per plant, total soluble solid (TSS), titratable acidity (TA), pH, total phenolic and anthocyanin were determined. Applications exhibited a range of 10.95-27.94g for fruit weight, 24.95-37.91mm for fruit width, 34.44-52.18mm fruit length, 3.87-4.26 for pH, 7.00-8.10% for total soluble solids and 0.97-1.20% for titratable acid. Total phenolic content was observed in 650.28-900.02mg (Gallic acid equivalents (GAE)/100g) and anthocyanin content was observed 155.06-204.18mg/100g. We can recommend the K-10 and K-15 applications according to obtained data in our research for producers.

Key words: Strawberry, albion, K_2O, phenolic, anthocyanin

INTRODUCTION

Although commercial strawberry (*Fragaria x ananassa* Duch.) cultivation started towards the end of 1970 in Turkey, it is currently one of the biggest strawberry producers in the world with 372.498 tons production annually (FAO, 2015). Strawberry cultivation is generally carried out in open field or walk-in plastic tunnels to provide earliness in Turkey. There is a big demand for greenhouse strawberry production, during the recent years to extend the harvest period. The cultivation of strawberries in greenhouses in Turkey was also increasingly expanded in the Mediterranean Region during the last few decades. (Ercişli et al., 2005). The quality characteristics of strawberry fruit is concerned with: sugar rate, acidity, durability, brittleness, verjuice, aroma, allure and nutritive value (Kader, 1991; Salame-Donoso et. al., 2010).
Albion is a day neutral strawberry cultivar. Day neutral strawberry varieties show better performance in areas with highland climate. Because the day-neutral strawberry varieties are not affected by the length of day and consistently produces fruit. Therefore, they are recommended for areas with highland climate.

A study published in 2001 showed that strawberries actually contain in three basic flowering structures: short-day, long-day, and day-neutral. These refer to the day-length sensitivity of the plant and the type of photoperiod that cause flower formation. Day-neutral cultivars produce flowers regardless of the photoperiod (Hokanson and Maas., 2001).
The objective of this study was to determine the effect of different doses of potassium on the yield and fruit quality in 'Albion' strawberry variety.

MATERIALS AND METHODS

This study was carried out in the Corum region during 2013-2015. Albion strawberry seedlings were planted on black mulch for the commercial plantation, in June 2013. The soil properties of experimental area are presented in Table 1. Potassium applications were made as 0-5-10-15 kg on decares (K-0, K-5, K-10 and K-15). Pomological and chemical characteristics that is fruit weight, fruit length, fruit width, yield per plant, total soluble solid content (TSS), titrable acidity (TA) and pH, total phenolic (mg/100g) and anthocyanin (mg/100g) were investigated.

Table 1. Physical and chemical characteristics of the soil before cropping of strawberry.

Soil Properties	Analysis Results
Sand (%)	67
Silt (%)	19
Clay (%)	14
Salinity (mmhos/cm)	140.80
pH (1:2,5)	7.95
Calx (%)	34.23
Organic Matter (%)	2.28
Azote (ppm)	1141.96
Phosphor (ppm)	8.97
Potassium (ppm)	128.40
Calcium (ppm)	4078.60
Magnesium (ppm)	60.94
Iron (ppm)	2.74
Cupper (ppm)	0.45
Manganese (ppm)	4.75
Zinc (ppm)	0.47

The fruit characteristics of the strawberries were determined and cluster samples were randomly selected in 40 units of fruits. The weight and yield of per plant were determined using a 0.01g-sensitive weighing. The measurements of both the length and width (diameter) of fruits were made using a 0.01 mm-sensitive digital compass.

Total soluble solids contents (TSS): Samples of the examined cultivars were pooled to obtain a composite sample and analyzed for SSC using a digital refractometer. (Atago Model PR-1, Tokyo).

pH measurements were performed using a Hanna HI 98103 pH meter at 20 °C.

Titrable acidity (TA) was determined with potentiometrically using 0.1M NaOH to the end point of pH 8.1 and expressed as grams of citric acid per litre. (AOAC, 1984).

Determination of total phenolic content: Total phenolics content of strawberry were determined by using the Folin-Ciocalteu phenol reagent method. (Singleton and Rossi, 1965). Absorbance was measured on a spectrophotometer (MRX Dynex Technologies, USA) at 765 nm. The total phenolic contents were expressed as mg of Gallic acid equivalents (GAE)/l of extract.

Anthocyanins: Total anthocyanin content was measured with the pH differential absorbance method, as described by Cheng and Breen (1991). Briefly, absorbance of the extract was measured at 510 and 700 nm in buffers at pH 1.0 (hydrochloric acid–potassium chloride, 0.2 M) and 4.5 (acetate acid–sodium acetate, 1 M).

Anthocyanin contents were calculated using a molar extinction coefficient of 29,600 (cyanidin-3-glucoside).

$A = (A \lambda510 - A \lambda700) \, pH \, 1.0 - (A \lambda510 - A \lambda700) \, pH \, 4.5$

Results were expressed as mg cyanidin-3-glucoside equivalents $100g^{-1}$ fw.

RESULTS AND DISCUSSION

In this research, fruit weight, fruit length, fruit width, yield per plant, total soluble solid content (TSS), titratable acidity (TA) and pH were determined in 2014 to 2015 growing seasons. Some pomological and chemical properties of Albion cultivar are presented in Table 2. The Fruit weight was found to vary from 10.95 to 27.94g according to the applications. K-10 application was the highest at 27.94g. K-15 application was followed by 26.71 g. Both of these applications were included in the same group. There was not significantly different between K-10 and K-15 applications. But there were observed significant differences between this group and other applications (K-5 and K-0) (P<0.05). The highest fruit length was obtained from K-15 application (52.18 mm). Other applications were followed by 40.48 mm, 38.11 mm and 34.44 mm (K-0, K-10 and K-5 respectively). Both of these three applications were included in the same group and there were significant differences (P<0.05) between this group and K-15 application (Table 2). Applications K-10 and K-15 were obtained (37.91 and 34.05 mm respectively) higher value of fruit width from others. There were not significant differences between these applications. The lowest value of fruit width was obtained from K-5 application (21.83 mm), but difference was not statistically significant between K-5 and K-0 application (Table 2).

The weight of albion fruits were reported by Hughes et. al. (2010) as 15.7 to 16.4 g. Our results were much higher than those values. These are considered to be caused by potassium treatments. The same researchers (Huges et. al., 2010) and Ballington et al., (2008) were declare the yield value as 2.1 to 2.4 kg/m and 309-775 g/plant respectively. Our findings were lower than those reported in the literature.

In our research, the highest value in terms of yield per plant was obtained from K-15 application (285.20g) taken from the lowest yield K-0 applications (146.09g). However, differences between treatments were not statistically significant (Table 2). The highest SSC value was obtained from K-5 application (8.10%). The others applications followed by 7.90, 7.60 and 7.0 (K-0, K-10 and K-15 application respectively). Differences between K-15 (lowest SSC value) and other treatments were statistically significant (Table 2) (P<0.05).

SSC values were reported by Ruan et al., (2013) and Ornelas-Paz et al., (2013) as 8.35% and 6.6 to 9.0% respectively. Along with partially coincide, the values obtained in this research were lower than reported values in the literature generally. The reason for this the fruits used in this study were very large and they contain large amounts of water.

The highest value in terms of titratable acid was obtained from K-5 application (1.20%). This treatment was followed by the K-15 and K-0 applications. The differences between these three treatments were not statistically significant. The differences between K-10 (having the lowest value) applications and other treatments were statistically significant (P<0.05). Titratable acidity values were reported by Akhato and Recameles, (2013) and Ornelas-Paz et al., (2013) as 0.50 to 0.84% and 0.70 to 1.2% respectively in strawberries. Our findings were upper than these values.

When our findings were observed with respect to pH, the highest value was determined in K-15 treatment. This was followed by application of K-5 and K-10 respectively.

Differences between other applications and K-0 treatment (the last rank) were statistically significant (P<0.05). The determined pH values were reported by Akhatou and Recamales, (2013) and Ornelas-Paz, et al., (2013) as 3.51 to 3.82 and 3.39-3.80 respectively. The data determined in our study were higher than those reported in the literature.

Table 2. Effect of application on fruit weight, fruit length, fruit width, yield per plant, soluble solid content (TSS), titratable acidity (TA) and pH.

	Fruit Weight (g)	Fruit Length (mm)	Fruit Width (mm)	Yield Per Plant (g)	TSS (%)	Titratable Acidity (%)	pH
K-0	10,95±0,64* c	40,48±2,25* b	24,95±0,91* b	146,09±7,32	7,90±0,10* a	1,08±0,02** ab	3,87±0,13** b
K-5	17,26±1,79 b	34,44±0,57 b	21,83±0,34 b	181,20±11,90	8,10±0,15 a	1,20±0,07 a	4,06±0,04 ab
K-10	27,94±1,58 a	38,11±2,29 b	37,91±2,51 a	215,70±54,10	7,60±0,12 a	0,97±0,04 b	4,02±0,04 ab
K-15	26,71±1,05 a	52,18±0,65 a	34,05±0,21 a	285,20±46,00	7,00±0,06 b	1,16±0,01 ab	4,26±0,03 a

In this research, total amount of phenolic were determined as 650.28-900.02 mg/100g in Albion fruits (Table 3 and Figure 1).

The differences between applications were statistically insignificant in terms of total phenolic and anthocyanin amount (Table 3, Figure 1 and Figure 2) (P<0.05).

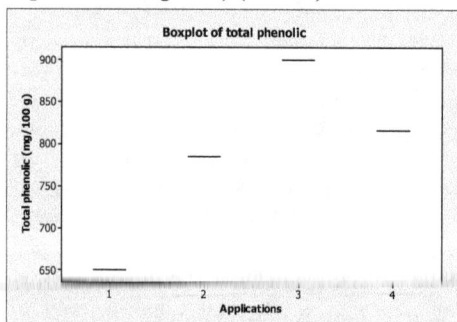

Figure 1. Boxplot of total phenolic

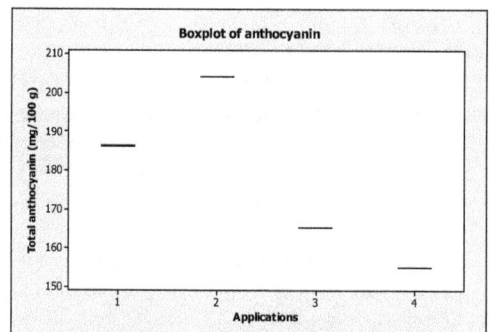

Figure 2. Boxplot of total anthocyanin

The highest phenolic contents were obtained from K-5 application. Application K-15 took place at the last in ranking

The total amount of phenolic and anthocyanin were reported by Diamante. et al. (2012) as

10

THE EFFECT OF ROOTING MEDIUM TEMPERATURE AND MOISTURE ON ROOTING OF BLACK MULBERRY HARDWOOD CUTTING

Mehmet KOÇ[1], Kenan YILDIZ[2], Saadettin YILDIRIM[3]

[1]Kilis 7 Aralık University, Faculty of Agriculture, Department of Horticulture, Kilis
[2]Gaziosmanpaşa University, Faculty of Agriculture, Department of Horticulture, Tokat
[3]Adnan Menderes University, Faculty of Agriculture, Department of Biosystems Engineering, Aydın
Corresponding author email: mehmetkoc@kilis.edu.tr

Abstract

In this study, the effect of different temperature and humidity content of rooting medium on rooting percentage of black mulberry hard wood cutting was investigated. For this purpose, different temperature and humidity contents were applied. In the result of the study, the effect of interaction of temperature x humidity and temperature on rooting ratio was not significant. On the other hand, humidity content caused significant changes on rooting percentage. The highest rooting ratio (63,11%) was observed in cutting planted in medium with 40% humidity. The lowest rooting percentages were obtained from two other media contained higher humidity compared to medium with 40% humidity. Both temperature and humidity treatment caused significant changes in root length. Among humidity treatment, 80% humidity produced the longest roots. The longest roots were obtained from 22°C of basal heating treatment. In term of the number of root per cutting, significant difference was not found among humidity treatments while temperature treatment caused significant changes. The highest root number per cutting was observed in cutting planted at 22°C.

Key words: mulberry, Morus nigra, hardwood cuttings.

INTRODUCTION

Mulberries (*Morus* spp.) belonging to the *Moraceae* family are widely distributed in many parts of the world, mainly in the northwest of South America and some parts of Africa (Datta, 2002) and their fruits are used for fresh consumption and syrup production (Davis, 1972). Their cultivation in Anatolia has been known since ancient times (Özbek, 1977). Hence, Turkey has a significant genetic potential for mulberries but the number and production of mulberry fruit decreased in the last decades due to the shift away from farming, increase of cutting and lack of sufficient maintenance. The anthocyanin found in berries and red fruits have been recently reported to possess preventive and curative properties against mouth, larynx, oesophagus, stomach and colon cancer (Prior, 2003; Zarfa et al., 2007). An excess accumulation of cyanidin-3-glucoside and cyanidin-3-rutinozit are of anthocyanin found in mulberries (Chen et al., 2005) and the contents are higher than other red fruits and some berries (Özgen et al., 2009). Expanding in application area and growing interest in the nutritional value of mulberry are increasing day by day and subsequently this interest leads to an increase in demand for mulberry seedlings and resulting higher costs. Also, the demands are not fully met due to some practical difficulties experienced in the black mulberry seedlings. Therefore, in the study, the effect of different temperature and humidity content of rooting medium on rooting percentage of black mulberry hard wood cutting was investigated. For this purpose, three temperature and humidity contents were applied. This study was performed under greenhouse conditions.

MATERIALS AND METHODS

Hardwood cuttings taken from an old branch of the rest period were prepared as 15 cm and then firstly soaked 0.3% fungicide for precaution against fungal infection. Then cuttings were submerged in indole butyric acid (IBA, 5000 ppm prepared in ethanol) for 5 seconds about 1 cm below the basal part of the cuttings. After holding the cuttings for 1-2 minutes in order to evaporate the alcohol, cuttings were planted in

1576-2466 mg gallic acid /kg FW and 385-470 mg Pel-3-GL/kg FW respectively.
The total amount of phenolic was reported by Capocasa et al., (2008) as 1.8-3.2mg GAE/g in strawberry. Medina, 2011 is reported the total phenolic content as 282 mg GAE/100g in organically grown strawberries.

The total amount of phenolic and anthocyanin were reported by Ornelas-Sun 2013 as 195.6-325.0 mg GAE/100g and 0.9-56.4 mg/100 g respectively.
Our findings were higher than the value reported in the literature in terms of both the total amount of phenolic and anthocyanins.

Table 3. Total phenolic and anthocyanin contents

Application	Total Phenolic (mg/100 g)		Anthocyanin (mg/100 g)	
	Mean	Ave Rank	Mean	Ave Rank
K-0	650.28±0.01 d	2.0	186.27±0.01 b	8.0
K-5	785.37±0.01 c	5.0	204.18±0.00 a	11.0
K-10	900.02±0.01 a	11.0	165.34±0.01 c	5.0
K-15	816.87±0.01 b	8.0	155.06±0.00 d	2.0

CONCLUSIONS

We can recommend the K-10 and K-15 applications according to the obtained data in our research for producers. Because, these applications showed higher values than control applications and K-5 application in terms of both yield and some quality characteristics and the total phenolic.

REFERENCES

AOAC, 1984. Official Methods of Analysis. 14th edn, Virginia, USA: Association of Official Analytical Chemists.

Akhatou I., Recamales A. F., 2013. Influence of cultivar and culture system on nutritional and organoleptic quality of strawberry. J Sci Food Agric 2014; 94: 866–875

Ballington J.R., Poling B., Olive K., 2008. Day-neutral Strawberry Production for Season Extension in the Midsouth. Hortscience 43(7):1982–1986. 2008.

Capocasa F., Scalzo J., Mezzetti B., Battino M., 2008. Combining quality and antioxidant attributes in the strawberry: The role of genotype. Food Chemistry 111 (2008) 872–878

Cheng G.W., Breen P.J., 1991. Activity of phenylalanine ammonialyase (PAL) and concentrations of anthocyanins and phenolics in developing strawberry fruit. Journal of the American Society for Horticultural Science, 116, 865–869.

Diamanti J., Capocasa F., Denoyes B., Petit A., Chartier P., Faedi W., Maltoni M.L., Battino M., Mezzetti B., 2012. Standardized method for evaluation of strawberry (Fragaria ananassa Duch.) germplasm collections as a genetic resource for fruit nutritional compounds. Journal of Food Composition and Analysis 28 (2012) 170–178

Ercisli S., Sahin U., Esitken A., Anapali O., 2005. Effects of some growing media on the growth of strawberry cvs. 'Camarosa' and 'Fern'. Acta Agrobotanica, 58(1), 185-191.

FAO, 2015. Statistical Database. , www.fao.org.

Hokanson S.C., Maas J.L., 2001. Strawberry biotechnology. Plant Breeding Reviews: 139–179. ISBN 978-0-471-41847-4.

Hughes B., Zandstra J., Dale A., 2010. Effects of Length of Blossom Removal on Production of Albion and Seascape Dayneutral Strawberries. Berry Notes, Vol. 22, No. 10, Machachutes

Kader A.A., 1991. Quality and its maintenance in relation to the postharvest physiology of strawberry. Timber press, Portland, 152p.

Medina M.B., 2011. Determination of the total phenolics in juices and super fruits by a novel chemical method. Journal of Functional Foods 3 (2011) 79 –87

Ornelas-Paz J.J., Yahia E.M., Bustamante N.R., Perez-Martinez J.D., Escalante-Minakata M.P., Ibarra-Junquera V., Acosta-Muniz C., Guerrero-Prieto V., Ochoa-Reyes E., 2013. Physical attributes and chemical composition of organic strawberry fruit (Fragaria x ananassa Duch, Cv. Albion) at six stages of ripening. Food Chemistry 138 (2013) 372–381

Ruan J., Lee Y. H., Hong S.J., Yeoung Y.R., 2013. Sugar and Organic Acid Contents of Day-neutral and Ever-bearing Strawberry Cultivars in High-elevation for summer and Autumn Fruit Production in Korea. Hort. Environ. Biotechnol. 54(3):214-222

Salame-Donoso T.P., Santos B.M., Chandler C.K., Sargent S.A., 2010. Effect of high tunnels on the growth, yields, and soluble solids of strawberry cultivars in Florida. Int. J. Fruit Sci., 10(3):249-263.

Singleton V.L., Rossi J.A., 1965. Colorimetry of total phenolics with hosphomolybdic-phosphotungstic acid reagents. American Journal of Enology and Viticulture, 16, 144–158.

Table 1. Physical and chemical characteristics of the soil before cropping of strawberry.

Soil Properties	Analysis Results
Sand (%)	67
Silt (%)	19
Clay (%)	14
Salinity (mmhos/cm)	140.80
pH (1:2,5)	7.95
Calx (%)	34.23
Organic Matter (%)	2.28
Azote (ppm)	1141.96
Phosphor (ppm)	8.97
Potassium (ppm)	128.40
Calcium (ppm)	4078.60
Magnesium (ppm)	60.94
Iron (ppm)	3.71
Cupper (ppm)	0.45
Manganese (ppm)	4.75
Zinc (ppm)	0.47

The fruit characteristics of the strawberries were determined and cluster samples were randomly selected in 40 units of fruits. The weight and yield of per plant were determined using a 0.01g-sensitive weighing. The measurements of both the length and width (diameter) of fruits were made using a 0.01 mm-sensitive digital compass.

Total soluble solids contents (TSS): Samples of the examined cultivars were pooled to obtain a composite sample and analyzed for SSC using a digital refractometer. (Atago Model PR-1, Tokyo).

pH measurements were performed using a Hanna HI 98103 pH meter at 20 °C.

Titrable acidity (TA) was determined with potentiometrically using 0.1M NaOH to the end point of pH 8.1 and expressed as grams of citric acid per litre. (AOAC, 1984).

Determination of total phenolic content: Total phenolics content of strawberry were determined by using the Folin-Ciocalteu phenol reagent method. (Singleton and Rossi, 1965). Absorbance was measured on a spectrophotometer (MRX Dynex Technologies, USA) at 765 nm. The total phenolic contents were expressed as mg of Gallic acid equivalents (GAE)/l of extract.

Anthocyanins: Total anthocyanin content was measured with the pH differential absorbance method, as described by Cheng and Breen (1991). Briefly, absorbance of the extract was measured at 510 and 700 nm in buffers at pH 1.0 (hydrochloric acid–potassium chloride, 0.2 M) and 4.5 (acetate acid–sodium acetate, 1 M).

Anthocyanin contents were calculated using a molar extinction coefficient of 29,600 (cyanidin-3-glucoside).

$$A = (A \lambda 510 - A \lambda 700) \text{ pH } 1.0 - (A \lambda 510 - A \lambda 700) \text{ pH } 4.5$$

Results were expressed as mg cyanidin-3-glucoside equivalents $100g^{-1}$ fw.

RESULTS AND DISCUSSION

In this research, fruit weight, fruit length, fruit width, yield per plant, total soluble solid content (TSS), titratable acidity (TA) and pH were determined in 2014 to 2015 growing seasons. Some pomological and chemical properties of Albion cultivar are presented in Table 2. The Fruit weight was found to vary from 10.95 to 27.94g according to the applications. K-10 application was the highest at 27.94g. K-15 application was followed by 26.71 g. Both of these applications were included in the same group. There was not significantly different between K-10 and K-15 applications. But there were observed significant differences between this group and other applications (K-5 and K-0) ($P<0.05$). The highest fruit length was obtained from K-15 application (52.18 mm). Other applications were followed by 40.48 mm, 38.11 mm and 34.44 mm (K-0, K-10 and K-5 respectively). Both of these three applications were included in the same group and there were significant differences ($P<0.05$) between this group and K-15 application (Table 2). Applications K-10 and K-15 were obtained (37.91 and 34.05 mm respectively) higher value of fruit width from others. There were not significant differences between these applications. The lowest value of fruit width was obtained from K-5 application (21.83 mm), but difference was not statistically significant between K-5 and K-0 application (Table 2).

The weight of albion fruits were reported by Hughes et. al. (2010) as 15.7 to 16.4 g. Our results were much higher than those values. These are considered to be caused by potassium treatments. The same researchers (Huges et. al., 2010) and Ballington et al., (2008) were declare the yield value as 2.1 to 2.4 kg/m and 309-775 g/plant respectively. Our findings were lower than those reported in the literature.

In our research, the highest value in terms of yield per plant was obtained from K-15 application (285.20g) taken from the lowest yield K-0 applications (146.09g). However, differences between treatments were not statistically significant (Table 2). The highest SSC value was obtained from K-5 application (8.10%). The others applications followed by 7.90, 7.60 and 7.0 (K-0, K-10 and K-15 application respectively). Differences between K-15 (lowest SSC value) and other treatments were statistically significant (Table 2) (P<0.05).

SSC values were reported by Ruan et al., (2013) and Ornelas-Paz et al., (2013) as 8.35% and 6.6 to 9.0% respectively. Along with partially coincide, the values obtained in this research were lower than reported values in the literature generally. The reason for this the fruits used in this study were very large and they contain large amounts of water.

The highest value in terms of titratable acid was obtained from K-5 application (1.20%). This treatment was followed by the K-15 and K-0 applications. The differences between

these three treatments were not statistically significant. The differences between K-10 (having the lowest value) applications and other treatments were statistically significant (P<0.05). Titratable acidity values were reported by Akhato and Recameles, (2013) and Ornelas-Paz et al., (2013) as 0.50 to 0.84% and 0.70 to 1.2% respectively in strawberries. Our findings were upper than these values.

When our findings were observed with respect to pH, the highest value was determined in K-15 treatment. This was followed by application of K-5 and K-10 respectively.

Differences between other applications and K-0 treatment (the last rank) were statistically significant (P<0.05). The determined pH values were reported by Akhatou and Recamales, (2013) and Ornelas-Paz, et al., (2013) as 3.51 to 3.82 and 3.39-3.80 respectively. The data determined in our study were higher than those reported in the literature.

Table 2. Effect of application on fruit weight, fruit length, fruit width, yield per plant, soluble solid content (TSS), titratable acidity (TA) and pH.

	Fruit Weight (g)	Fruit Length (mm)	Fruit Width (mm)	Yield Per Plant (g)	TSS (%)	Titratable Acidity (%)	pH
K-0	10,95±0,64* c	40,48±2,25* b	24,95±0,91* b	146,09±7,32	7,90±0,10* a	1,08±0,02** ab	3,87±0,13** b
K-5	17,26±1,79 b	34,44±0,57 b	21,83±0,34 b	181,20±11,90	8,10±0,15 a	1,20±0,07 a	4,06±0,04 ab
K-10	27,94±1,58 a	38,11±2,29 b	37,91±2,51 a	215,70±54,10	7,60±0,12 a	0,97±0,04 b	4,02±0,04 ab
K-15	26,71±1,05 a	52,18±0,65 a	34,05±0,21 a	285,20±46,00	7,00±0,06 b	1,16±0,01 ab	4,26±0,03 a

In this research, total amount of phenolic were determined as 650.28-900.02 mg/100g in Albion fruits (Table 3 and Figure 1).

The differences between applications were statistically insignificant in terms of total phenolic and anthocyanin amount (Table 3, Figure 1 and Figure 2) (P<0.05).

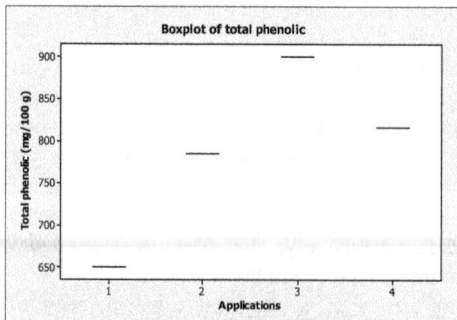

Figure 1. Boxplot of total phenolic

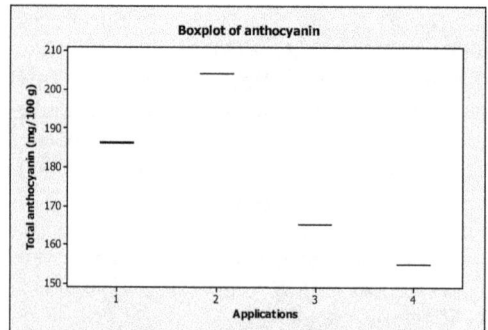

Figure 2. Boxplot of total anthocyanin

The highest phenolic contents were obtained from K-5 application. Application K-15 took place at the last in ranking.

The total amount of phenolic and anthocyanin were reported by Diamante. et al. (2012) as

rooting medium. The experimental design was composed of nine rooting chamber, of which temperature and humidity were separately controlled (Figure 1).

Figure1. Rooting chambers under control conditions

Temperature and humidity level for the study were determined as follows. The data, collected from Ministry of Food, Agriculture and Livestock, have been statistically processed and interpreted, building the trend line and setting up the forecast based on simulation models for the period 2012-2015. Subsequently, the rooting media were applied three different temperature degrees and humidity levels. Root zone humidity value was based on the field capacity. The desired temperature and humidity values of the table based on real-time measurements were performed with a computer-controlled system (Figure 2).

Figure 2. Computer-controlled system for temperature and humidity

The study was performed according to the complete randomized experimental design with three replicates and each replicate corresponded to 20 cuttings. The cuttings were kept in rooting media for 90 days. At the end of this period, the following measurements were made.

Callus formation: Callus was detected from cuttings and the results are expressed as a percentage of total cuttings.

Rooting rate: Adventitious root formation was detected from cuttings and the results are expressed as a percentage of total cuttings.

Root length and diameter: Adventitious root length and diameter were measured with a calliper.

Root number: Root number was expressed as the total number of adventitious formed per number of cuttings forming root.

Statistical Analysis: SPSS 17.00 statistical programme was used to determine statistical significance levels. The independent one-way ANOVA followed by Duncan multiple range test and the differences between individual averages were considered as statistically important at p≤0.05.

RESULTS AND DISCUSSIONS

Callus formation rates were collectively represented in Table 1. Whereas the highest callus formation rate (86.1 %) was obtained under 40 % field capacity and 18°C, the lowest rate was determined under 60 % field capacity and 26°C.

Concerning callus formation, no statistical significant difference was found among temperature and moisture content interaction but the effects of temperature and moisture content were significant (p≤0.05). Compared the means of temperature applications but not considering the moisture content, the highest callus formation was (71.72 %) under 22°C and the average results were statistically different and significant for each temperature application.

Table 1. Callus formation under different growth conditions

Temp.(°C)	Humidity (%)			Mean
	40	60	80	
18	86.10	60.53	55.13	67.26 AB
22	70.66	67.60	76.90	71.72 A
26	79.43	47.80	51.80	59.68 B
Mean	78.73 A	58.64 B	61.28 B	

Mean in the same column by the same letter are not significantly different to the test of Duncan (p≤0. 05).

For the comparison of moisture content without evaluation of temperature, the highest value (78.73 %) was ascertained under 40 % moisture content but no statistical differences were found for 60 % (58.64 %) and 80 % (61.28 %) moisture contents (Table 1).

The highest root formation rate (69.06 %) was obtained under 40 % field capacity and 18°C, the lowest rate was determined under 80 % field capacity and 18°C. With respect to the root formation, no statistical significant difference was found among temperature and moisture content interaction but comparison of moisture content without evaluation of temperature, the highest value (63.11 %) was ascertained under 40 % moisture content but no statistical differences were found for 60 % (43.47 %) and 80 % (49.57 %) moisture contents (Table 2).

Table 2. Root formation under different growth conditions

Temp. (°C)	Humidity (%)			Mean
	40	60	80	
18	69.06	38,60	36.30	47.99
22	64.16	48,46	66,96	59.87
26	56.10	43,33	45.43	48.29
Mean	63.11 A	43.47 B	49.57 B	

Mean in the same column by the same letter are not significantly different to the test of Duncan (p≤0. 05).

Concerning with root length, no significant difference was found among temperature and moisture content interaction but the effects of temperature and moisture content were significant (p≤0.05).

Compared the means of temperature applications but not considering the moisture content, the highest root length (43.21 cm) was under 22°C but the average results were not statistically different at 18°C and 26°C.

Also, the highest root length value (42.64 cm) was obtained from 80 % but moisture content (40 % and 60 %) did not elicit statistically significant differences on root length (Table 3).

Table 3: Root length (cm) under different growth conditions

Temp.(°C)	Humidity (%)			Mean
	40	60	80	
18	29.47	30.07	26.71	28.75 B
22	33.90	37.72	58.02	43.21 A
26	28.23	28.47	43.21	33.30 B
Mean	30.53 B	32.08 B	42.64 A	

Mean in the same column by the same letter are not significantly different to the test of Duncan (p≤0. 05).

Regardless moisture content, average results for temperature were significant for root

diameter but no difference between 18 °C and 22 °C. In agreement with root length, the widest diameter was determined under 80 % moisture content (Table 4).

Table 4: Root diameter (mm) under different growth conditions

Temp. (°C)	Humidity (%)			Mean
	40	60	80	
18	1.76	2.12	2.11	2.00 A
22	1.85	2.13	2.26	2.08 A
26	1.30	0.95	1.60	1.29 B
Mean	1.65 B	1.73B	1.99 A	

Mean in the same column by the same letter are not significantly different to the test of Duncan (p≤0. 05).

Table 5: Root number per cutting

Temp. (°C)	Humidity (%)			Mean
	40	60	80	
18	4.99	4.40	4.80	4.72 B
22	6.37	5.57	7.70	6.54 A
26	4.87	7.27	4.33	5.49 AB
Mean	5.41	5.74	5.61	

Mean in the same column by the same letter are not significantly different to the test of Duncan (p≤0. 05).

Average results related with temperature were significant for root number per cutting for each applications but soil moisture content did not elicit any significant changes.

In cutting propagation, no direct relationship with the rooting and callus formation, which occurs as a response to injury but increase the survival time of cuttings by preventing the decay in rooting medium has been reported (Kaşka and Yılmaz, 1990; Koyuncu et al., 2003a). However, in the present study, there was a parallel variation between callus formation and rooting percentage. These results are agreement with the report by Yıldız and Koyuncu (2000) but disagree with the studies (Sezgin 2009, Koyuncu et al., 2003a). In addition to delaying the decay of cuttings through formation of a protective layer, callus tissue was reported to help the water intake in some cases (Hartman and Kester, 1974). Herein, the highest callus formation percentage was found less than 22°C but no difference was determined between 18°C and 22°C. An increase up to 26°C slowed down callus formation. In general, keeping temperature in rooting media around 24°C promote cell division and callus formation (Ağaoğlu et al., 1995) but it is worthy to note that those temperature values might vary according to the different plant species.

Moisture content in rooting media effected callus formation and 40 % of field capacity was the most favourable for callus formation. Up to our best knowledge, there is no study on the relation between moisture content and callus formation but in general, moisture content at a level is required in order to prevent ventilation (Hartman and Kester, 1974). Basal heating of rooting media can enhance root formation percentage (Alexandrow 1988); Yıldız and Koyuncu (2000) recorded an increase from 60 % to 89 %.

Even the highest average of root length was obtained from the rooting media with highest moisture content; there was decay in many cutting samples. This condition is probably caused by obstruction of the ventilation due to the high moisture content. Indeed, a good aeration of the rooting medium, in addition to good moisture retention, is stated to be the best (Hartmann and Kester, 1974).

CONCLUSIONS

In the study, the best root width was obtained under 18°C and 22°C but it decreased by 26°C. An increase in temperature of the rooting media decreased the width of root. Those coincided with the studies by Yıldız et al. (2009) and Sezgin (2009). In this study, moisture content in rooting media did not cause any change in number of root formation. The present results concerned with root formation are in good agreement with the previous studies (Koyuncu et al., 2003b, Yıldız et al., 2009; Sezgin, 2009, Erdoğan and Aygün, 2006).

REFERENCES

Ağaoğlu Y.S., Çelik H., Çelik M., FidanY., GülşenY., Günay A., Halloran N., Köksal A.İ., Yanmaz R., 1995. Genel Bahçe Bitkileri. Ankara Üniversitesi Ziraat Fakültesi Eğitim,Araştırma ve Geliştirme Vakfı Yayınları, No:4.

Alexandrov A., 1988. İnvestigations of the Rooting Process in Ripe Mulberry Cutting Taken From Various Parts of Overwintered Shoots Plant Science 35(9), 86-93.

Chen P.N, Chu H.L, Kuo W.H, Chiong C.L, Hsief Y.S., 2005. Mulberry Anthocyanins, Cyanidin-3-rutinoside and Cyanidin-3- glucoside, Exhibited and Inhibitory Effect on The Migration and of a Human Long Cancer Cell Line. Cancer Letters 1–12.

Datta R.K., 2002. Mulberry Cultivation and Utilization in India. Mulberry for Animal Production, FAO Animal Production and Healt Paper 147: 45-62.

Davis P.H., 1972. Flora of Turkey IV. Edinburg Uni. Press. Edinburg, 657 p. De Candolle, A., 1967. Origin of Cultivated Plants. New York and London. P. 149- 153.

Erdoğan V., Aygün, A., 2006. Karadutun (Morus nigra L.) yeşil çelikle çoğaltılması üzerine bir araştırma. II. Ulusal Üzümsü Meyveler Sempozyumu 172-175 (14 16 Eylül 2006).

Hartman H.T, Kester, D.E., 1974 Bahçe Bitkileri Yetiştirme Tekniği. (Çevirenler: N. Kaşka ve M. Yılmaz) Ç.Ü.Z.F. Ders kitabı no: 79 Adana.

Kaşka N. ve Yılmaz M., 1990 Bahçe Bitkileri Yetiştirme Tekniği. Çuk. Üniv. Zir. Fak. No: 52 Adana.

Koyuncu F, Vural E. Çelik M., 2003b. Karadut (Morus nigra L.) Çeliklerinin Köklendirilmesi Üzerine Araştırmalar. Ulusal Kivi ve Üzümsü Meyveler Sempozyumu Kitabı, s: 424-427, Ordu.

Koyuncu F. Şenel E., 2003a. Rooting of Black Mulberry (Morus nigra L.) Hardwood Cuttings. J. Fruit Ornam. Plant Res., 11: 53-57.

Özbek S., 1977. Genel Meyvecilik. Çukurova Üniv. Ziraat Fak. Yay.111, Ders Kitapları:6, Adana. 386.

Özgen M., Serçe S. Kaya C., 2009. Phytochemical and Antioxidant Properties of Anthocyanin-rich Morus nigra and Morus rubra fruits. Scientia Horticulturae 119: 275–279.

Prior R.L., 2003. Absorption and Metabolism of Anthocyanins: Potential Health Effects. In: Phytochemicals: Mechanisms of Action. CRC Press Inc., Boca Raton, FL.

Sezgin O., 2009. Genotipik farklılığın karadut odun çeliklerinin köklenmesi üzerine etkisi. Yüksek Lisans Tezi, GOÜ, Fen Bilimleri Enstitüsü.

Yıldız K., Koyuncu F, 2000. Karadutun (M. nigra L.) Odun Çelikleri ile Çoğaltılması Üzerine Bir Araştırma. Derim, 17(3): 130-135.

Yıldız K., Çetin Ç., Güneş M., Özgen M., Özkan Y., Akça Y., Gerçekçioğlu R, 2009. Farklı Dönemlerde Alınan Kara Dut (Morus nigra L.) Çelik Tiplerinde Köklenme Başarısının Belirlenmesi. GOÜ Ziraat Fakültesi Dergisi, 26(1), 1-5.

Zarfa -Stone S., Yasmin T., Bagchi M., Chatterjee A., Vinson J.A., Bagchi D., 2007. Berry Anthocyanins as Novel Antioxidants in Human Health and Disease Prevention. Mol. Nutr. Food Res. 51, 675–683.

11

CHANGES IN THE PHYTOCHEMICAL COMPONENTS IN WINE GRAPE VARIETIES DURING THE RIPENING PERIOD

Gultekin OZDEMIR[1], **Akile Beren SOGUT**[1], **Mihdiye PIRINCCIOGLU**[2],
Göksel KIZIL[2], **Murat KIZIL**[2]

[1] Dicle University, Faculty of Agriculture, Department of Horticulture, Diyarbakir, Turkey.
[2] Dicle University, Faculty of Science, Department of Chemistry, Diyarbakir, Turkey

Corresponding author email: gozdemir@dicle.edu.tr

Abstract

The aim of this study was to determine phytochemical components of Tannat, Cabernet Sauvignon, Malbec, Merlot and Shiraz wine grape varieties during the ripening period. As amounts of total phenolic compounds in different parts of the grape varieties, the highest total phenolic values for berry peel were found to be 300.58 µg GAE/mg in Cabernet Sauvignon, 974.23 µg GAE/mg in Malbec for pulp, 447.01 µg GAE/mg in Merlot for seed. The total flavonoid content in peel, pulp and seeds of varieties were found to be varied between 46.95 µg QUE/mg and 148.01 µg QUE/mg. In conclusion, total bioactive compounds of the grape differed significantly based on variety and grape part. Since higher bioactive compounds were found in pulps for all grape varieties, grapes should be consumed as a whole grape. This study also showed that these grapes are a potential source of natural bioactive compounds. It can be concluded that selected grape varieties and their parts can be considered a good source of phenolic and antioxidants.

Key words: Grape, Diyarbakir, phenolic, flavonoid, cluster, berry, seed

INTRODUCTION

Grapes are considered as a significant source of antioxidants in fruit species in the world (Pirinccioglu et al., 2012). Due to this property, it is a kind of fruit with the importance increasing day by day (Macheix et al., 1990). Among the types of grapes, especially red grapes, grape juice and major phenolic compounds found in red wine are called as flavonoids anthocyanins and flavonols (Rice-Evans et al., 1996; Singleton 1982; Palomino et al., 2000). It is reported that (Morris and Cawthon 1982; Bravdo et al., 1985; Matthews and Anderson 1988; Iland, 1989; Nadal and Arola 1995; De La Hera Orts et al., 2005) these substances that are important in terms of human health and found in grape vary according to the varieties of the grapes (Landrault et al., 2001), climate and soil conditions of the place where it grows (Spayd et al., 2002 ; Mateus et al., 2001), the maturity levels (Cangi et al., 2011), cultural practices (Babalik et al., 2009) and post-harvest transactions (Revilla et al., 2001).

Among the climate features of the vineyard areas, the place and vector issues especially the temperature, humidity and insolation are encountered as the important factors affecting the synthesizing the of phenolic compounds and antioxidant substances and the other phytochemicals. Considering these factors having very important impact on the ripeness and all compounds of ripeness, it is seen that some studies have been carried out to identify the phytochemical features of grape varieties grown in different ecologies in our country recently. (Deryaoglu and Canbas 2003; Karadeniz et al., 2005; Aras, 2006; Orak, 2007; Kelebek, 2009; Ozden and Vardin 2009; Uluocak, 2010; Bayir, 2011; Toprak, 2011; Kaplama, 2012).

Increased competition in the wine sector has led to the increasing demand for quality wine grapes and thus the emergence of the concept of quality in the grapes for wine to the forefront. In growing quality grapes, as well as the effects of the ecological factors in growing grapes especially the climate factors, the cultural applications performed in the viticulture have very significant effects on the the phytochemicals properties. By identifying the effects of the range of cultural applications like pruning (Pehlivan and Uzun, 2015), cultivation (Babalik et al., 2009) and irrigation (Bravdo et al., 1985; Matthews and Anderson

1988; Nadal and Arola, 1995) performed to receive production with the highest efficiency and quality on the phytochemical properties such as phenolic compounds, tannin and antioxidants and performing due to the information obtained will be useful.

Except for the effects of cultural applications performed in the vineyards on the physical changes in the bunch and berry features of the grape varieties and the properties such as brix, pH and total acidity in unfermented grape-juice (Tangolar et al., 2002; 2005 and 2010), it is very important to identify its effects on the amount of phenolic substance that is one of the most important quality indicators, and phytochemical properties including anthocyanin and antioxidant capacities. In the literature, it is seen that the studies to identify the effects of the cultural applications such as pruning, irrigation and fertilization in the varieties grown in the ecological conditions of our country are not too much (Pehlivan and Uzun, 2015).

This study was conducted to determine phytochemical components of Tannat, Cabernet Sauvignon, Malbec, Merlot and Shiraz wine grape varieties during the ripening period in the ecological conditions of Diyarbakır/Turkey.

MATERIALS AND METHODS

The research was carried out in the Dicle University Faculty of Agriculture Department of Horticulture in 2011 and 2012 years. In the research, Cabernet Sauvignon, Merlot, Shiraz, Malbec and Tannat red wine grape varieties are used as materials.

Grape varieties are grown as grafted into 110 R rootstocks. Vines are 7 years old. Planting distances rows are 1 m and intra-row is 2.5 m. Mid wire cordon training system was applied to the vines from 60 cm height. While making yield pruning to the vines, it was loaded as 18 buds/vine stock. Vineyard area has clay loam soil type. The applications such as irrigation, fertilizers and disease and pest control are done regularly in the vineyard.

Within the scope of the research, the chemical change occurred in Total Soluble Solids (TSS), total acidity and pH values during veraison and maturity periods of grape varieties has been determined (Ozdemir et al., 2006; Tangolar et al. 2010).

Besides, in order to determine the phytochemical change in peel, pulp and seeds of the grape samples, total phenolic compound amount and total flavonoid amount were detected (Chandler, 1983; Slinkard et al., 1997; Baydar et al., 2007).

RESULTS AND DISCUSSIONS

As a result of the study, total soluble solids (TSS) values of grape varieties showed significant differences according to the types and years. When TSS values of the grape values during veraison period of 2011 were examined (Table 1), it was determined that the highest value was in the Merlot variety (14.00%) and the lowest value was in Malbec varieties (11.33%). The difference between TSS values of the varieties was not found to be significant in 2012. TSS values were determined between 12.00% and 13.66%. When TSS values of the maturity times of the grape variety between the years of 2011 and 2012 were examined, the highest value was determined in Merlot grape varieties (25.33%) and the lowest value was determined to be in the Malbec grape variety (20.66%) in 2011 among the varieties included in the same letter group.

In all varieties, TSS amount has rapidly increased since the period of veraison and the yields reached to the period of harvest time after 3-6 week maturation process according to the varieties. In both years, TSS amount that was low in the period of veraison was determined to reach 20-25% of values that is desired during the harvest time.

In the trial made in Adana by Tangolar et al., (2005) in the years of 2002-2003, the TSS amounts in Chardonnay, Cabernet Sauvignon, Narince and Okuzgozu were determined to be 24.1-23.2-20.7 and 19.9% respectively in the second year. The researchers reported that there may be significant changes in the rate of TSS according to the varieties and the years. Indeed, as a result of this study, while the difference between TSS values in the veraison time of the grape varieties was important in the year of 2011, it was insignificant in 2012.

Ozdemir and Tangolar (2005) examined phenological stages (EST), temperature total values and some quality features in some table grape varieties in Diyarbakır and Adana conditions, for two years. TSS was determined to be 12.6% and 12.7% respectively for Diyarbakır and Adana provinces in 1997 and 12.5% and 12.5% in 1998.

In both provinces, it has been reported that EST values were at the level that will not create problems for viticulture, the physiological activity in the vicinity of Diyarbakır started earlier but fruit ripening occurred in Adana earlier. Cluster, berry and grape-juice characteristics vary according to the varieties but the values among the provinces are close to each other.

As a result of a study carried out in the region of Kazova during two years, it has been reported that TSS increased in Bogazkere, Cabernet Sauvignon, Chardonnay, Emir, Merlot, Narince, Okuzgozu and Riesling varieties from the period of veraison until the

harvest and the varieties except for Bogazkere and Okuzgozu reached to the sufficient level according to the desired TSS criteria in the harvest period (Sen, 2008).

As a result of the study carried out by Cangi et al., (2011) on wine grape varieties grown in Kazova (Tokat) region (Gewurtztraminer, Pinot Noir, Narince and Shiraz) TSS was determined to change 20.2% (Narince) and 22.3% (Shiraz) in the harvest period.

In maturation period, the findings related to the amount of TSS indicated that the differences cmcrged according to the years and varieties were related to the general characteristics of climate conditions and varieties according to the years and varieties in different years. In addition, in the cultural applications such as summer pruning performed in the vineyards in that year, the varieties can be seen to have significant impacts on TSS accumulation.

Table 1. Total soluble solids (%) during the veraison at harvest stages 2011 and 2012 year for varieties evaluated

Varieties	Veraison			Harvest		
	2011	2012	Average	2011	2012	Average
CabernetSauvignon	12.33bc	12.00	12.16c	23.33a	23.33ab	23.33b
Tannat	13.00ab	12.33	12.66bc	24.00a	24.00a	24.00b
Merlot	14.00a	13.66	13.83a	25.33a	25.33a	25.33a
Malbec	11.33c	12.66	12.00c	20.66b	21.33b	21.00c
Shiraz	13.33ab	13.33	13.33ab	24.66a	23.33ab	24.00b
LSD %5	1.01	N.S.	0.7	1.98	2.09	1.26

The difference between the means with different letters in same column was significant (P<0.05)

When the values obtained as a result of total acidity analysis of grape samples received in the process of veraison and maturity process of the varieties were examined, the acidity values were determined to be from 14.58 g/l

(Merlot) to 19.02 g/l (Tannat) in the veraison process in the trial years and they ranged from 5.86 g/l (Cabernet Sauvignon) to 8.25 g/l (Tannat) in the maturity process (Table 2).

Table 2.Total acidity (g/L) during the blooming at veraison stages 2011 and 2012 year for varieties evaluated

Varieties	Veraison			Harvest		
	2011	2012	Average	2011	2012	Average
CabernetSauvignon	15.83b	15.74bc	15.83b	5.86c	5.89b	5.86c
Tannat	19.02a	18.84a	19.02a	7.92a	8.25a	7.92a
Merlot	14.58c	14.57d	14.58c	6.23bc	6.37b	6.23bc
Malbec	15.41bc	14.79cd	15.41bc	6.40b	6.43b	6.40b
Shiraz	16.18b	16.28b	16.18b	6.30bc	6.28b	6.30bc
LSD %5	1.17	1.08	1.17	0.43	0.6	0.4

The difference between the means with different letters in same column was significant (P<0.05)

Looking to the acidity values in the maturity values in 2011 and 2012, it was determined to be an increase generally and the highest acidity in 2011 and 2012 was seen in Tannat grape variety (respectively 7.92 and 8.25 g/l), the

lowest value was seen to be in Cabernet Sauvignon grape variety (respectively; 5.86 and 5.89 g/l). In all varieties from the period of veraison, a great amount of decrease was observed in the total acidity amount with the maturation in all varieties and the total acidity value has varied according to the varieties and years in the period of harvest. In the research, the lowest total acidity was determined to be from 5.86 to 5.89 g/l and the highest values were determined between 7.92 and 8.25 g/l (Table 2).

They reported in their study about the development of the grapes from the veraison stage to the extreme maturity, the berry weight increased from the date of first sample taking date and the amount of TSS increased from the process of veraison and the total acidity amount increased up to the process of veraison; it started to decrease after this process and the ratio of decline decreased up to the maturity; the amount of tartaric acid has continuously decreased since the beginning of the maturity and the amount of it remained almost fixed up

to the end of the maturity (Agaoglu, 2002; Sen 2008).

In a study carried out with Bogazkere, Cabernet Sauvignon, Chardonnay, Emir, Merlot, Narince and Okuzgozu, Riesling varieties in Kazova region for two years, it is reported that the total acidity decreased rapidly from the period of veraison, the highest total acidity was detected in Bogazkere and Okuzgozu varieties and the lowest one was detected in the variety of Emir in the harvest period (Sen, 2008).

When pH values of grape varieties determined during veraison and maturity periods were examined (Table 3), the highest pH values were determined in Merlot variety (respectively; 2.87 and 2.86) and the lowest values were determined in Malbec variety (respectively; 2.44 and 2.45) in veraison periods in the years of 2011 and 2012.

Table 3. pH during the blooming at veraison stages 2011 and 2012 year for varieties evaluated

Varieties	Veraison			Harvest		
	2011	2012	Average	2011	2012	Average
CabernetSauvignon	2.84a	2.84a	2.84b	3.90bc	3.91b	3.90b
Tannat	2.53b	2.51b	2.52c	3.84c	3.82c	3.83b
Merlot	2.87a	2.86a	2.86a	4.01ab	4.03a	4.02a
Malbec	2.44c	2.45c	2.45d	4.05ab	4.06a	4.05a
Shiraz	2.44c	2.46c	2.45d	4.07a	4.09a	4.08a
LSD %5	0.03	0.02	0.02	0.13	0.06	0.06

The difference between the means with different letters in same column was significant (P<0.05)

As approaching to the maturity time of the grape varieties, it was determined that pH values increased. During the maturity process, pH values among the varieties were determined to vary from 3.82 to 4.09.

As a result of being examined of pH value in the berries during the maturation process of Gewurtztraminer, Pinot Noir, Narince and Shiraz that are among the wine grape varieties grown in Kazova (Tokat) region, pH was determined to vary from 3.27 (Pinot Noir) to 4.20 (Shiraz). When the values obtained from our study are examined, the pH values among the varieties at the time of maturity are seen to vary from 3.82 to 4.09.

Winkler et al., (1974) have reported that pH significantly increased until the grapes matured; with this change in pH, the unsuitable tastes in flavor and eating quality are covered and changed. It is reported that the pH during the grape increased in parallel with the increase in TSS during maturation and used as a

decisive criteria in determining optimum harvest time. As the maturity criteria of grapes especially grown in warmer areas, pH can be used as maturity criteria (Fanizza, 1982). Indeed, Amerine et al., (1972) reported the lowest acceptable acid amount must be 0.65 g/100 ml, also pH in red table wine should be lower than 3.4.

When the maturity index values of grape varieties grown in Diyarbakir province in 2011 and 2012 are analyzed (Table 4), it is seen that the highest value belongs to Merlot grape variety (40.63) and the lowest value (30.37) belongs to Tannat grape variety in 2011.

In 2012, similar to the previous year, the highest value was found to be in the Merlot grape variety (39.92) and the lowest value was in the Tannat grape variety (29.18). The maturity index values of grape varieties are determined to vary from 29.77 to 40.27 on average.

Table 4. Maturity index (Total soluble solids (TSS) / Acidity)
during the veraison at harvest stages 2011 and 2012 year for varieties evaluated

Varieties	2011	2012	Average
CabernetSauvignon	39.81a	39.60a	39.71a
Tannat	30.37b	29.18c	29.77b
Merlot	40.63a	39.92a	40.27a
Malbec	32.34b	33.23bc	32.79a
Shiraz	39.15a	37.15ab	38.15a
LSD %5	4.41	5.62	3.08

The difference between the means with different letters in same column was significant (P<0.05)

In 2007, TSS was found to be 21.7 (%) during maturation period in Shiraz grape variety grown in Kazova; Total Acidity to be (g/l) 6.45 and maturity index to be 33.64, while in 2008 TSS was found to be 22.3 (%), Total Acidity to be (g/l) 7.05 and maturity index to be 31.63 (Uluocak, 2010). In our study, the average maturity index value in Shiraz variety was found to be slightly higher amount as 38.15. As a result of the studies performed in Kazova region, the maturity index values of Bogazkere, Cabernet Sauvignon, Chardonnay, Emir, Merlot, Narince, Okuzgozu and Riesling varieties during harvest periods vary from 18.18 (Bogazkere) to 33.90 (Emir) according to the varieties (Sen, 2008).

When the total phenolic compound amounts in pulp, peel and seeds of grape varieties grown in Diyarbakir province are examined, it has been determined that the differences among the varieties are statistically significant (Table 5).

Table 5.Total phenolic content (μg GAE/mg) in grape berry skin, flesh and seed

Varieties	Peel	Pulp	Seed	Total
CabernetSauvignon	300.58a	107.03c	68.33c	475.94
Tannat	167.06b	697.86b	445.76a	1310.68
Merlot	92.50c	657.46b	447.01a	1196.97
Malbec	108.16c	974.23a	390.23b	1472.62
Shiraz	89.05c	667.13b	411.16ab	1167.34
LSD %5	34.931	129.89	43.96	

The difference between the means with different letters in same column was significant (P<0.05)

When the amount of phenolic compounds found in many different parts of the berries of grape varieties are examined, in the highest values in the peel are found in Cabernet Sauvignon variety, in pulp in Malbec variety and in seed in Tannat variety. When the phenolic compound amounts found in the different parts of the berries of grape varieties are examined, the highest values in the peel are found in Cabernet Sauvignon variety, in the pulp, they are found to be in Malbec variety and in the seed they are found in Tannat variety. The maximum amount of phenolic compounds found in Cabernet Sauvignon variety (300.58 μg GAE/mg), and the least amount is found in the Shiraz grape variety (89.05 μg GAE/mg). Considering the amount of phenolic compounds in the pulp, being different from the peel and the seed, the highest value was found to be in the Malbec grape varieties (974.23 μg GAE/mg) and the least was found to be in the Cabernet Sauvignon variety. When the total amounts of phenolic compounds in the seed are compared, it has been determined that the highest value is found in the Merlot grape variety and the lowest value is found in Cabernet Sauvignon grape variety. The total phenolic compound identified in the peel, pulp and seeds of the varieties has been identified to vary from 475.94 to 1472.62 μg GAE/mg (Table 5).

It has been identified in different research that total phenolic amounts vary according to the variety and year and decreased during maturation period (Doshi et al., 2006; Navarro et al., 2008; Jin et al., 2009). Saidani Tounsia et al., (2009) examined the total phenolic amount in the methanol extract of three types of red grape seeds and found the equivalent of respectively 427.00 mg/100g, 218.00 mg/100g und 112.81 mg/100g of gallic acid for dry weights of Muscat, Shiraz and Carignan varieties. A similar study was carried out by Hogan et al., (2009) and it was examined the total phenolic content of the three types of Virginia black wine grapes in various regions of northern France by made and Cabernet in Virginia black wine grapes in northern France as Cabernet Franc 1, Cabernet Franc 2 and Cabernet Franc 3 and, as a result, they were identified to be equivalent of respectively 1.82 ± 0.07mg/g, 1.47 ± 0.05, mg/g 0.63 ± 0.02 mg /g of gallic acid.

As a result of their study, Ozden and Vardin (2009) have found that the total phenolic compound concentration of some grape varieties grown in Sanliurfa conditions such as Merlot, Chardonnay, Cabernet Sauvignon and Shiraz *(V. vinifera* L.) grape varieties vary from 1805 mg/kg to 3170 mg/kg in terms of total antioxidant activity and certain phytochemical properties. While the highest concentration of phenolic compounds was being found in Chardonnay variety, the lowest concentration was found in Shiraz variety.

In their study, Gokturk Baydar et al., (2011) have identified grape seeds and peel extracts belonging to Cabernet Sauvignon, Kalecik Karasi and Narince grape varieties, antioxidant properties of wine and the content of phenolic compounds. Total phenolic content was determined to vary from 522.49 to 546.50 mg GAE g^{-1} in seed extracts and from 22.73 to 43.75 mg GAE g^{-1} in peel extracts and from 217.06 to 1336.21 mg L^{-1} in wine. The radical scavenging effects of the samples and reducing capacities have varied depending on grape varieties, the parts of the grape and wine type.

Kanner et al., (1994) analyzed total phenolic compound amounts by harvesting the grapes in optimal harvest ripeness in their study conducted with seven different table (Miabell Concord, Flame Seedless, Emperor, Thomson Seedless, Red Globe and Red Malaga) and seven different wine (Calzin Petite Shiraz, Merlot, Cabernet Sauvignon, Cabernet Franc, Sauvignon Blanc and Chardonnay) grapes. They reported that phenolic compounds in wine grapes vary from 230 to 1236 mg/l and Calzin and Petit Shiraz grape varieties have the highest phenolic content.

It has been identified in different researches that total phenolic types vary according to the year and variety and they decrease during the maturity process (Doshi et al., 2006; Navarro et al., 2008; Jin et al., 2009). Saidani Tounsia et al., (2009) examined the total phenolic amount in methanol extract of three types of red grape seeds and they found equivalent to respectively 427.00 mg/100g, 218.00 mg/100g and 112.81 mg/100g of gallic acid for Muscat, Shiraz and Carignan varieties in their dry weights. A similar study was carried out by Hogan et al., (2009) and it was examined the total phenolic content of the three types of Virginia black wine grapes in various regions of northern France by made and Cabernet in Virginia black wine grapes in northern France as Cabernet Franc 1, Cabernet Franc 2 and Cabernet Franc 3 and, as a result, they were identified to be equivalent of respectively 1.82 ± 0.07mg/g, 1.47 ± 0.05, mg/g 0.63 ± 0.02 mg /g of gallic acid. As a result of the analysis made in grape varieties examined in Diyarbakir conditions, the total amount of flavonoids in the pulp, peel and seed of the berry were found to vary greatly among varieties (Table 6).

Table 6.Total flavonoid content (µg QUE/mg) in grape berry skin, flesh and seed

Varieties	Peel	Pulp	Seed	Total
CabernetSauvignon	5.18e	60.99b	9.72ab	75.89
Tannat	13.28b	53.62c	7.61bc	74.51
Merlot	5.85d	29.46d	11.64a	46.95
Malbec	10.17c	58.37bc	7.28bc	75.82
Shiraz	19.65a	122.25a	6.11c	148.01
LSD %5	0.65	5.34	3.06	

The difference between the means with different letters in same column was significant (P<0.05)

When the total amount of flavonoid compounds is examined, the total flavonoid amount found in the peel, pulp and seed was detected to vary from 46.95 to 148.01 µg QUE/mg. It has been detected that the flavonoid content in the peel is found mostly in Shiraz grape variety (19.65 µg QUE/mg) and the lowest one is found in Cabernet Sauvignon (5.18 mg QUE/mg) grape variety. Considering the flavonoid substance content in the pulp, it has been detected that the highest value is found again in Shiraz grape variety (122.25 µg QUE/mg) and the lowest one is found in the Merlot grape variety (29.46 µg QUE / mg).

When flavonoids substance amounts found in the seed were analyzed, the highest value was found to be in Merlot grape varieties (11.64 µg QUE / mg), and the lowest value was found to be in Shiraz grape varieties (6.11 µg QUE/mg) (Table 6).

CONCLUSIONS

In the chemical analysis performed with one-week intervals from the period of veraison to the maturity, it has been determined that while the grapes are ripening, there is an decrease in tartaric acid and an increase in TSS and pH and these parameters vary in accordance with the years and varieties.

As a result of the examination of the grape-juice features of the varieties used in the research, it has been seen that the amount of TSS rapidly increased in all varieties since the period of veraison and it reached to the harvest period of the grapes after 3 to 6-week maturation period according to the varieties. In both hears, the amount of TSS that was low in the period of veraison reached to the 20-25% of desired values during the harvest period.

When the values obtained as a result of tartaric acid analysis of the grape samples received in the periods of veraison and maturity are examined, the acidity value in the period of veraison in trial years was determined to vary from 14.58 g/l (Merlot) to 18.84 g/l (Tannat) and in the period of maturity, it was determined to vary from 5.86 g/l (Cabernet Sauvignon) to 7.92 g/l (Tannat).

When the pH values of grape varieties determined during the periods of veraison and maturity are examined; the highest pH values during veraison periods in 2011 and 2012 were determined to be in Merlot varieties (respectively; 2.87 and 2.86), the lowest values in the Malbec variety (respectively, 2.44 and 2.45). As approaching to the maturation time of the grape varieties, an increase was determined in the pH values.

According to the results of maturity index that is another feature examined in the research, in both trial years it has been determined that the highest value belongs to Merlot grapes variety while the maturity index value varied between 29.77 and 40.27 on average.

As a result of the research, the highest values in terms of the amounts of phenolic compounds in the research have been determined to be in the peel in Cabernet Sauvignon variety (300.58 µg GAE/mg), in the pulp they are found to be in Malbec variety (974.23 µg GAE/mg) and in the seed they are found in Merlot variety (447.01 µg GAE/mg). When the flavonoid amounts are compared, it has been determined that the total flavonoid amount varied from 6.95 µg QUE/mg to 148.01 µg QUE/mg. the flavonoid content found in the peel is found mostly in the Shiraz (19.65 µg QUE/mg) grape variety and the lowest value was found in Cabernet Sauvignon (5.18 µg QUE/mg) grape variety.

Considering the flavonoid substance content in the pulp, the highest value was found again in Shiraz (122.25 µg QUE/mg) grape variety like in the peel and the lowest value was found in Merlot (29.46 µg QUE/mg) grape variety. When the flavonoid amounts in the seed are compared, it has been determined that the highest value belongs to Merlot grape variety (11.64 µg QUE/mg) and the lowest value belongs to Shiraz grape variety (6.11 µg QUE/mg).

Among plant-derived foods, fruits and vegetables are natural sources that are rich in phenolic substances. Today, it is clear that the increase in escaping from the artificial substances will increase the significance of the natural phenolic substances. Besides the use opportunities in the fields of food, ladder and pharmacology, it is seen that understanding the mechanism of action of phenolic substances with significant effects on human health and it is important to investigate the paths to be able to use technologically.

ACKNOWLEDGEMENTS

The authors thanks Dicle University Scientific Research Project Coordinatory for its funding of this research.

REFERENCES

Agaoglu Y.S., 2002. Scientific and Applied Viticulture (Volume 2: Grape Physiology). Kavaklidere Boks: 5, 444p, Ankara, Turkey.

Amerine M.A., Berg H.W., Crue W.V., 1972. The Technology of Wine Making. The Avl Publishing Company. Inc.Vestport, Connecticut, 802p.

Aras O., 2006. Determination of the Total Carbonhydrate, Protein, Mineral Substances and Phenolic Compounds of Grape and Grape Products. Suleyman Demirel University, Graduate School of Natural and Applied Sciences M.Sc. Thesis, 59p.

Babalik Z. Cetin S. HallaC Turk F., Gokturk Baydar N., 2009. Determination of Phenolic Compounds of Cavus Grape Cultivars under Different Training Systems. 7th Symposium on Viticulture and Technologies, 5-9 October, Turkey.

Baydar G. N., Ozkan G., Yasar S., 2007. Evaluation of the Antiradical and Antioxidant Potential of Grape Extracts. Food Control, 18, 1131-1136.

Bayir A., 2011. Investigation of Phenoloic Content and Antiradical Activity of Grape, Mulberry and Myrtle. Akdeniz University, Graduate School of Natural and Applied Sciences PhD Thesis, 147.

Bravdo B.A., Hepner Y., Loigner C., Cohen S., Tabacman H., 1985. Effect of irrigation and crop level on growth, yield, and wine quality of Cabernet Sauvignon. American Journal of Enology and Viticulture. 36: 132-139.

Cangi R., Saracoglu O., Uluocak E., Kilic, D., Sen A., 2011. The Chemical Changes of Some Wine Grape Varieties During Ripening Period in Kazova (Tokat) Ecology. Igdir Univ. J. Inst. Sci. & Tech. 1(3): 9-14

Cangi R., Şen A., Kılıç D., 2008, Determination of Phenological Characters and Effective Heat Summations Required for Maturation of Some Grapes Cultivars Grown in Kazova Region (Tokat-Turhal). Research Journal of Agricultural Sciences (TABAD), 1 (2):45-48.

Chandler S. F., Dodds J. H., 1983. The effect of phosphate, nitrogen and sucrose on the production of phenolics and solasodine in callus cultures of Solanum laciniatum. Plant Cell Reports, 2, 105.

De La Hera Orts M.L., Martinez-Cutillas A., Opez-Roca J.M., Gomez-Plaza E., 2005. Effect of moderate irrigation on grape composition during riperning. Spanish Journal of Agricultural Research. 3: 352-361.

Deryaoglu A., Canbas A., 2003. Physical and Chemical Changes Occured During Maturation of Okuzgozu Grape Variety Grown in Elazig Region. The Journal of Food. 28(2):131-140.

Doshi P., Adsule P., Banerjee K., 2006. Phenolic Composition and Antioxidant Activity In Grapevine Parts and Berries (Vitis vinifera L.) cv. Kishmish Chornyi (Sharad Seedless) During Maturation. International Journal of Food Science and Technology, 41 (Supplement 1), 1–9.

Fanizza G., 1982. Factor Analyses For The Choice of A Criterion of Wine Grape (V.V.) Maturity İn Warm Regions. Vitis, 21: (4), 334-336.

Gokturk Baydar N., Babalik Z., Hallac Turk F., Cetin E.S., 2011. Phenolic Composition and Antioxidant Activities of Wines and Extracts of Some Grape Varieties Grown in Turkey. Journal Of Agricultural Sciences, 17(2011), 67-76.

Hogan S., Zhang L., Li J., Zoecklein B., Zhou K., 2009. Antioxidant properties and bioactive components of Norton (Vitis aestivalis). Food Science And Technology, 42, 1269-1274.

Iland P., 1989. Grape berry composition-the influence of environmental and viticultural factors, Australian Grape grower & Winemaker. 302: 13-15.

Jin Z.M., He J.J., Bi H.Q., Cui X.Y., Duan C.Q., 2009. Phenolic Compound Profiles in Berry Skins From Nine Red Wine Grape Cultivars in Northwest China. Molecules, 14(12), 4922-4935.

Kanner J., Frankel E., Granit R., German B., Kinsella J.E., 1994. Natural Antioxidants in Grape And Wines. Ibid. 42, 64-69.

Kaplama P., 2012. Antioxidant Activities Anthocyanin Profiles and some Physical and Chemical Properties of Grape Cultivars Grown in Erzincan Ataturk University, Graduate School of Natural and Applied Sciences M.Sc. Thesis, 87p.

Karadeniz F., Burdurlu, H.S., Kocan ,N., Soyder, Y., 2005. Antioxidant Activity of Selected Fruits and Vegetables Grown in Turkey, Tübitak Turk J. Agric. For. 29: pp.297-303.

Kelebek H., 2009. Researches on the Phenolic Compounds Profile of Okuzgozu, Bogazkere and Kalecik Karasi Cultivars Grown in Different Regions and Their Wines. Cukurova University, Graduate School of Natural and Applied Sciences PhD Thesis, 251p.

Landrault N, Poucheret P, Ravel P, Gasc, Cros G, Teissedre PL., 2001. Antioxidant capacities and phenolic l evels of French wines from different varieties and vintages. J. Agr. Food Chem, 49: 3341-3348.

Macheix J., Fleuriet A., Billot J., 1990. Fruit Phenolics. CRC, Boca Raton, FL, 1-25.

Mateus N., Proença S., Ribeiro P., Machado J.M., De Freitas V., 2001. Grape and wine polyphenolic composition of red Vitis vinifera varieties concerning vineyard altitude, Cienc. Technol., Aliment., 3(2): pp.102-110.

Matthews M., Anderson, M., 1988. Fruit ripening in Vitis vinifera L responses to seasonal water deficits, American Journal of Enology and Viticulture, 39:313-320.

Morris J.R., Cawthon D.L., 1982. Effect of irrigation, fruit load, and potassium fertilization on yield, quality, and petiole analysis of concord (Vitis vinifera L) grapes, American Journal of Enology and Viticulture, 33: 145-148.

Nadal M., Arola L., 1995. Effects of limited irrigation on the composition of must and wine of Cabernet Sauvignon under semi-arid conditions, Vitis, 34: 151-154.

Navarro S., Leo′N M., Roca-Pe′Rez L., Boluda R., Garcı′A-Ferriz L., Pe′Rez-Bermu′Dez P., Gavidia I., 2008. Characterisation Of Bobal And Crujidera Grape Cultivars, In Comparison With Tempranillo And Cabernet Sauvignon: Evolution Of Leaf Macronutrients And Berry Composition During Grape Ripening Food Chemistry 108, 182–190.

Orak H.H., 2007. Total Antioxidant Activities, Phenolics, Anthocyanins, Polyphenoloxidase

Activities of Selected Red Grape Cultivars and Their Correlations Scientia Hort. Vol. 111, Issue 3, 5 February 2007, Pages 235-241.

Ozdemir G., Tangolar S. 2005. Determination pf Phenological Periods with Heat Summation Values and some Quality Characteristics in some Table Grape Cultivars Grown in Diyarbakir and Adana Conditions.   6th Symposium on Viticulture, 5-9 September, Volume 2, 446-453, Turkey.

Ozdemir G., Tangolar S., Bilir H., 2006. Determination of Cluster and Berry Characteristics with Phenological Stages of some Table Grape Cultivars. Alatarim, 5(2): 37-42.

Ozden M., Vardin H., 2009. Quality and Phytochemical Properties of some Grapevine Cultivars Grown in Sanliurfa Conditions. Harran Journal of Agriculture and Food Science 13(2): 21-27.

Palomino O., Gomez-Serranillos M.P., Slowing K., Carretero E., Villar A., 2000. Studyof polyphenols in grape berries by reversed-phase highperfor mance liquid chromatography. Journal of Chromatography, A. 870: 449-451.

Pehlivan E.C., Uzun H.İ., 2015. Effects of Cluster Thinning on Yield and Quality Characteristics in Shiraz Grape Cultivar. Yuzuncu Yil University Journal of Agricultural Sciences 25 (2):119-126.

Pirinccioglu M., Kızıl G., M Kızıl., Ozdemir G., Kanay Z., Ketani M.A., 2012. Protective effect of öküzgözü grape (*Vitis vinifera* L.) juice against carbon tetrachloride induced oxidative stress in rats. Journal of Food Funct., 3, 668-673.

Revilla I., Luisa M., Gonzalez-Sanjoze L., 2001. Evolution during the storage of red wines treated with pectolytic enzymes: New Anthocyanin Pigment Formation, Journal of Wine Research, 12 (3): pp.183-197.

Rice-Evans C.A., Miller N.J., Paganda G., 1996. Structure antioxidant activity relationship of flavonoids and phenolic acids, Free Radical Biology & Medicine, 20: 933-956.

Saidani Tounsia, M., Ouerghemmi, I., Wannes, W.A., 2009. Ksouri, H.Z.; Marzouk, B.; Kchouk, M.E. Industrial Crops And Products, 30, 292–296.

Singleton V.L., 1982. Grape and wine phenolics: background and prospects. In: Webb, A.D. (ed.),

Proceedings of Grape Wine Centennial Symposium University of California, Davis 215-222.

Slinkard K., Singleton V.L., 1977. Total Phenol Analysis: Automation And Comparison With Manual Methods. American Journal Of Enology And Viticulture 28: 49-55.

Spayd S.E., Tarara J.M., Mee D.L., 2002. Ferguson, J.C., Separation of sunlight and temperature effects on the composition of Vitis vinifera cv. Merlot berries. Am. J. Enol. Vitic., 53(3): pp.171-182.

Sen A., 2008. Determination of the Effect Heat Summation and Optimum Harvest Time of some Grape Cultivars Grown in Kazova (Tokat) Ecology. Gaziosmanpasa University, Graduate School of Natural and Applied Sciences M.Sc. Thesis, 74p.

Tangolar S., Ozdemir G., Tangolar S.G., Ekbic H.B., Rehber Y., 2010. Grape Growing. Caglar Puplishing, 47p, Adana.

Tangolar S., Eymirli S., Ozdemir G., Bilir H., Tangolar S.G., 2002. Determination of Phenological Properties and Cluster and Berry Characteristics of some Grape Varieties Grown in Pozanti Adana. 5th Symposium on Viticulture, 372-380, Turkey.

Tangolar S., Ozdemir G., Bilir H., Sabir A., 2005. Determination of Cluster and Verry Characteristics with Phenologies of some Wine Grape Cultivars Grown in Pozanti/Adana Ecological Regions. 6th Symposium on Viticulture, 5-9   September, Volume 1, 58-64, Turkey.

Toprak F.E., 2011. Phytochemical Characteristics in Kalecik Karasi Grape Cultivar (Vitis vinifera L.) Grown in Ankara and Nevsehir. Ankara University, Graduate School of Natural and Applied Sciences M.Sc. Thesis, 64p.

Uluocak E., 2010. The physical and chemical changes during ripening period of some wine grape varieties grown in Kazova (Tokat) Ecology. Gaziosmanpasa University, Graduate School of Natural and Applied Sciences M.Sc. Thesis, 89p.

Winkler A. J., Cook J.A., Kliewer W.M., Lider L.A., 1974. General Viticulture. 633 P., Univo f California. Pres/ Berkeley

12

DETERMINATION OF POMOLOGICAL AND BIOCHEMICAL COMPOSITIONS ON BERRIES IN DIFFERENT PARTS OF CLUSTERS IN SOME TABLE GRAPE VARIETIES

Mehmet Ali GUNDOGDU[1], Alper DARDENIZ[1], Baboo ALI[2], Ahmet Faruk PEKMEZCI[1]

[1]Canakkale Onsekiz Mart University, Faculty of Agriculture, Department of Horticulture, Terzioglu Campus, 17100, Canakkale, Turkey
[2]Canakkale Onsekiz Mart University, Faculty of Agriculture, Department of Agricultural Biotechnology, Terzioglu Campus, 17100, Canakkale, Turkey
Corresponding author email: gundogdu85@hotmail.com

Abstract

This research work has been conducted in the trail areas of the 'Research and Application Vineyard of Table Grape Varieties' situated in 'COMU Dardanos Campus' during the years 2013 and 2014 in order to determine pomological and biochemical compositions on berries in different parts of clusters of some table grape cultivars namely, 'Cardinal', 'Yalova Cekirdeksizi' and 'Yalova Incisi'. In the research, the samplings have been done from the berries randomly on top of clusters (TC), middle outer side of clusters (MOC), middle inner side of clusters (MIC) and tip of cluster (TIC) of grape varieties (initial clusters). The heaviest amount of berries has been obtained from the MOC of the 'Cardinal', 'Yalova Cekirdeksizi' and 'Yalova Incisi' cultivars. The ripest berries (SSC TA⁻¹) had been taken in TC in 'Cardinal', TC, MOC and MIC in 'Yalova Cekirdeksizi' and TC, MOC and MIC in 'Yalova Incisi' cultivars. In particular, berries of TIC were found more small–light, and a little bit unmatured than berries of other parts of clusters in 'Cardinal' and 'Yalova İncisi' grape cultivars. Pomological and biochemical compositions of berries on different parts of cluster may vary considerably, and also it has been changed at different grape cultivars. For this purpose, the regular monitoring of the maturity level of table and wine grape cultivars should be done by following the precautionary measures. It has also been determined that the samplings should have to be done in equal number from at least 3 different parts such as top, middle and tip sides of clusters. Nevertheless, the tip portions, having the latest blooming on flower clusters to be cut in certain proportions just after the berry formation, have been projected that they will provide an increase in that of the volume and maturity in those berries remaining on clusters.

Key words: Vitis vinifera L., table grape, position of grape on cluster, pomological and biochemical composition of grape berries, grape quality.

INTRODUCTION

According to FAO statistical data; there were 67,067,128 tons fresh grape produced in 6,969,373 ha of vineyards in 2012 in the world. In the same year, the total fresh grape production was recorded as 4,275,659 tons in 462,296 ha grapevine area in Turkey (FAO, 2014). Clusters, uniform in size and colour, have bigger berries is an important factor for increasing the market values of table grapes. Therefore, many researches have been carried out related to the training system and development of new hybrid grape types and varieties aimed to improve the grape quality.

Every berry fertilized on flower cluster, has occurred from pericarp that has juicy pulp and skin (Agaoglu, 1999). Size of berries may vary by variety, growing power of vine stock, water uptake, berry set, berry count and maturity (Celik, 2011). Balance between direct sun exposure on leaf area and size of berries is significantly affected the yield and quality of product (Reynolds et al., 1994). Stoev (1974) reported that leaves found in same direction have been supported by roots in same direction. Similarly, nutrition of bunch on shoot was affected by shoot that have bunches. Todorov (1970) explained that there is a positive correlation between sizes of cluster and shoot having the cluster. The same author also reported that the clusters and berries on strong and productive shoots are more heavy clusters and berries than that of others. However, there are many factors that control the development and composition of grape berries. Therefore, there are significant differences on pomological and biochemical composition between clusters on a vine stock or first and second clusters on same summer shoot or berries of same clusters (Smart et al., 1985; Yılmaz and Dardeniz, 2009). Yılmaz and Dardeniz (2009) mentioned that the best developing clusters in terms of fruit width, length and weight were first clusters on summer shoots. Moreover, maturity indexes of first clusters have been found higher than second clusters in case of 'Amasya' and 'Cardinal' grape cultivars. Besides, clusters located on summer shoots from growing on second bud showed a better development in terms of cluster width, length and weight than those that located on summer shoots growing on first bud. Though there were not any significant differences found on maturity index by the pruning of 2 buds.

Nowadays, different application methods are being used for the improvement of grape quality. For example, Ilgin (1997) reported that the thinning of cluster at the level of 25% in 'Yalova Cekirdeksizi' grape cultivar decreased the yield of grafted vineyards but had not effect on ungrafted vineyards. Although the grape yields decreased 50% of thinning of 75% of flower clusters in ungrafted vineyard while the grape quality has been increased by 25% level of thinning found more of flower clusters in ungrafted vineyard. Dardeniz and Kismali (2002) determined the effect of cluster thinning, a week before blooming at 0%, 30% and 60% levels, on the yield and quality of grapes, and also on the vegetative growth of 'Amasya' and 'Cardinal' cultivars. But in this study, authors have not recommended 30% levels of cluster thinning of 'Amasya' cultivar, but 60% level of cluster thinning has been recommended only in southern latitudes. In case of 'Cardinal' variety, 30% level of cluster thinning at was enough for grape quality.

Florescence starts from top to bottom on the flower clusters. After berry set, berries on the bottom of the cluster impair the quality and image of whole of the cluster because of smaller and late maturing berries. Therefore bunch thinning is recommended for these berries on the bottom of cluster in 2 or 3 weeks after berry set at 25% or 33% level. Thus either enlargement for cluster or increase in berry size was provided. Also it is possible to harvest at 3–7 days early to clusters that have good coloured, allured and uniform size. When the

berries were 5–7 mm, the clusters were tipped at 1/3rd, 1/6th and 1/12th of the cluster length on 'Cardinal' and Uslu table grape cultivars by Dardeniz (2014). In Uslu, cluster length (cm), cluster width (cm), cluster compactness (1–9), number of berries/cluster (n), berry weight (g) and titratable acidity (TA) (%) parameters were affected by the applications. In 'Cardinal', cluster length (cm), cluster compactness (1–9), number of berries/cluster (n), berry weight (g), total soluble solid (TSS) (%), titratable acidity (TA) (%) and maturity index parameters were affected by the applications. Yield was not affected by cluster tipping in both grape cultivars. It was concluded that the cluster tipping applied to the Uslu in a proportion of one–third and to the 'Cardinal' in a proportion of one–sixth of the cluster length would be positively sufficient in terms of increasing the grape quality.

Dardeniz et al. (2012a) compared the growth and productivity of primary and secondary summer shoots, which primary summer shoots were cut at the base following 10–15 cm of growth and secondary buds were forced to sprout giving rise to new summer shoots, of two different table grape varieties, 'Yalova İncisi' ('Honusu' x 'Siyah Gemre') and 'Yalova Cekirdeksizi' ('Beyrut Hurması' x 'Perlette'). In both years (2010 and 2011), vinestock that has secondary shoots showed significant decreases in the levels of fresh grape yield, with especially small grape bunches obtained. In terms of maturity of the grapes, significant differences were not observed among the applications due to low sprout growth on secondary shoots. The results of this study show that some grape products still may be harvested even if primary buds from summer sprout are damaged in cases of late spring frost.

Another research has also been carried out by Dardeniz et al. (2012b) evaluating the changes in chlorophyll content on 8 different table grape varieties leaves in 3 different (5th node, 10th node and 15th node) branch nodes, at 4 periods (15th of June, 1st of August, 15th of September and 1st of November) by SPAD digital chlorophyll meter. Although chlorophyll contents were low on the leaves of 10th and 15th nodes, chlorophyll contents were observed to be equal during the 2nd measurement period in leaves at the 5th and 10th nodes, and also during the 3rd measurement period in leaves at the 10th and 15th nodes. Additionally during the last study period, a reduction of chlorophyll content was observed in leaves at the 5th and 10th nodes. 3–5 leaves per lateral shoot, the terminal 3 leaves at the tip of the shoot contained low chlorophyll content during the 1st study period; however a gradual increment in chlorophyll content was observed in subsequent periods. High total chlorophyll content was observed in lateral shoot leaves throughout all periods of study when compared to leaves at the tip of the shoot. Turker and Dardeniz (2014) aimed to determine the effects of 3 different levels of axillary shoot removal applications (High Level Axillary Shoot Removal (HLASR); Normal Level Axillary Shoot Removal (NLASR); None Axillary Shoot Removal (NASR) on the yield and quality characteristics of the 6 different varieties of *Vitis vinifera* L. As far as the HLASR application is concerned, it reduced the yield of table grape by causing a decrease in the potential of vine stock in all grape varieties especially in the second year of research. Consequently, this application in which all of the axillary shoots are taken from the bottom of vine stock is not recommended for any variety of table grapes. In the case of well–organized spraying program, NASR method of application will contribute to high yield, good quality and early grape production in all table grape varieties.

A research that carried out on 'Horoz Karasi' and 'Gok uzum' grape varieties, was to determine the effects of 1 3[-1] cluster reduction (CR), 1 3[-1] CR + herbagreen (HG) and 1 3[-1] CR + humic acid (HA) applications on grape yield and quality of cultivars were examined. It was suggested that 1 3[-1] CR + HA application increased grape yield, berry weight, berry red and blue colour intensity values of Horoz Karasi grape variety and 1 3[-1] CR application increased grape yield and maturity index values of Gok üzüm grape variety (Akin, 2011).

Akin et al. (2012) reported that the combined leaf fertilizer (TARIS–ZF) significantly increased quality parameters such as berry length, berry weight, maturity index, juice yield and drying index of grapevine cv. 'Gok uzum'. Increasing crop load values (16, 21, and 26 buds/vine) increased fresh grape yield and juice yield; however, maturity index and drying index decreased in comparison to the control. As a result of this study, it was suggested that produce a high yield and to increase quality parameters 16 buds/vine pruning, and fertilization by TARİŞ–ZF may be applied on grapevine cultivation especially on 'Gok uzum' cultivar.

A research that in order to determine the effects of 9 different winter and summer pruning practices in 'Yalova İncisi' grape cultivar is suggested that the thinning practices such as EP + CT (early pruning + cluster thinning) and EP + CT + BT (early pruning + cluster thinning + bunch thinning) have been found recommenddable for acquiring early and high–quality crop yields in those regions of our country where early spring frosts are not considered as dominant. By applying pruning on normal date with high level axillary shoot removal practice; the increases both in the average yield and quality, and ripeness of grapes were quite satisfactory resulting to the increase in leaf size and vine stock potential. On the other hand, overall ripening of grapes resulting by the increase found in leaf size and vine stock potential after the application of GB+SUB practice (Sezen and Dardeniz, 2015).

However, it is known that different treatments affected quality of table grapes, for determinate to effect of these treatment it is required to sample the right way on every cluster. Because it is seen that some grape cultivars' berries on different parts of clusters have significantly pomological and biochemical differences. Therefore, this research in order to determine pomological and biochemical compositions on berries in different parts of clusters in some table grape cultivars such as 'Cardinal', 'Yalova Cekirdeksizi' and 'Yalova İncisi'.

MATERIALS AND METHODS

This research has been conducted in the trail area of the 'Research Vineyard of Table Grape Varieties' situated in 'COMU Dardanos Campus' in the years 2013 and 2014 in 'Yalova İncisi' which has white colour berries with seeds on short cluster and very early cultivar, 'Cardinal' which has red colour berries with seeds on middle–long cluster and early cultivar and 'Yalova Cekirdeksizi' which has white colour seedless berries on middle–short cluster and mid early cultivar. Plant materials, used in this research, have been grafted onto 41B American grape rootstock ('Yalova Incisi' cultivar) and 5BB American grape rootstock ('Cardinal' and 'Yalova Cekirdeksizi' cultivars) in 11 years old. Vines were trained to unilateral cordon system. The spacing in between rows and within rows was 3.0 m and 1.5 m; respectively. The soil was loamy clay, slightly alkaline, medium calcareous and unsalted. Summer shoots that reached sufficient lengths were bended between first and second wire and tipped above 20–30 cm on second wire. 2nd offshoot tipping treatments did 1 month later after 1st treatments. Tipping of offshoots was cut above 1st or 2nd leaves on bottom of the offshoots by secateurs.

All grape cultivars were thinned first and second leaves of summer shoots, lateral, offshoot, water sprouts, and secondary and tertiary shoots just before full blooming in both 2 years (2013–2014). While harvest time was coming, first clusters of grape cultivars packed with plastic bags and labelled and brought to Pomology Laboratory in Canakkale Onsekiz Mart University, Faculty of Agriculture, and Department of Horticulture.

In this experiment grape berries were sampling randomly on Top of Clusters (TC), Middle Outside of Clusters (MOC), Middle Inside of Clusters (MIC) and Tip of Clusters (TIC); and berry width (mm), berry length (mm), berry weight (g of 1 berry), Chroma value of berry, Hue value of berry, soluble solid content (SSC) (Brix%), Titratable acidity (TA) (mg tartaric acid 100 g^{-1}), maturity index (SSC TA^{-1}) and total phenolic compounds (TPC) (mg GAE 100 g^{-1}) were investigated with these berries.

This research, was settled in randomized plot factorial design with 3 replications and each replication was had 100 berries which took 1st clusters on 4 vine stock. LSD multiple

comparison test was used determining the differences among treatments. All of the data analyses were done with SAS system for Windows (ver. 9) statistical package program.

RESULTS AND DISCUSSIONS

Statistical data, in order to determine the pomological and biochemical compositions found into the berries in different parts of clusters of some table grape cultivars, are given in Table 1–6.

In the light of the results of this research, the values regarding to the highest berry width and length have been determined on the TC (20.98 mm and 21.68 mm, respectively) and MOC (21.61 mm and 21.61 mm, respectively) in 'Cardinal' cultivar, when MOC (6.46 g) and TC (6.01 g) had the heaviest level of berries. However; the shortest, narrowest and the lightest berries have been found at the TIC and MIC, respectively. The lowest values of chroma were measured into the MOC (8.67) and TC (9.57). Berries on the TC were found more red tones and good coloured as compared to the other parts of clusters due to their highest values of Hue, shown in Table 1.

The highest SSC have been recorded in those berries which were found on the TC (14.00%) while the lowest SSC were calculated on the berries located at the TIC (12.62%) and MIC (13.16%). The lowest TA was determined in TIC (0.817%) for 'Cardinal' grape cultivar. The highest pH values have been found on the TC (3.47) and MOC (3.42). The ripest berries (SSC TA^{-1}) have been obtained from the TC in case of 'Cardinal' cultivar. Berries of TIC (2029 mg GAE 100 ml^{-1}) and MOC (2056 mg GAE 100 ml^{-1}) had significantly lower TPC in comparison to the other parts of clusters. TC (2112 mg GAE 100 ml^{-1}) having the highest content of phenolic compound, shown in Table 2.

Table 1. Some pomological compositions on berries of different parts of clusters in 'Cardinal' cultivar[*]

Parts of clusters	Berry width (mm) 2013–2014	Berry length (mm) 2013–2014	Berry weight (g of 1 berry) 2013–2014	Chroma value 2013–2014	Hue value 2013–2014
TC	20.98 a	21.68 a	6.01 ab	8.67 c	49.32 b
MOC	20.88 a	21.61 a	6.46 a	9.57 bc	62.30 a
MIC	19.97 b	20.91 b	5.60 b	10.15 b	57.82 a
TIC	19.99 b	20.27 c	5.46 b	11.18 a	59.95 a
LSD	0.3833	0.3418	0.6523	0.9553	5.6721

[*]: Means of 2 years data. TC: Top of clusters, MOC: Middle outer side of clusters, MIC: Middle inner side of clusters, TIC: Tip of clusters.

Table 2. Some biochemical compositions on berries of different parts of clusters in 'Cardinal' cultivar[*]

Parts of clusters	SSC (%)	TA (%)	pH	Maturity index (SSC TA^{-1})	TPC (mg GAE 100 ml^{-1})
	2013–2014	2013–2014	2013–2014	2013–2014	2013–2014
TC	14.00 a	0.663 b	3.47 a	22.24 a	2112 a
MOC	13.21 ab	0.678 b	3.42 a	20.43 b	2056 b
MIC	13.16 b	0.671 b	3.39 ab	20.88 ab	2068 ab
TIC	12.62 b	0.817 a	3.30 b	16.84 c	2029 b
LSD	0.7965	0.0429	0.0972	1.5628	58.9

[*]: Means of 2 years data. TC: Top of clusters, MOC: Middle outer side of clusters, MIC: Middle inner side of clusters, TIC: Tip of clusters.

Although, the highest values of the width of berry have been calculated from the MOC (16.03 mm), MIC (15.14 mm) having the narrowest berries. The largest berries in their sizes have been observed into the MOC (19.21 mm) and TC (18.79 mm). While the berries that located onto the MOC (3.34 g) bearing the heaviest number of berries. On the other hand, the berries found on the MIC (2.77 g) possessing the lowest values of weight. There were no any significant differences found in the berries located on different parts of clusters in accordance to the values of Chroma and Hue, given in Table 3.

Table 3. Some pomological compositions on berries of different parts of clusters in 'Yalova Cekirdeksizi' cultivar[*]

Parts of clusters	Berry width (mm)	Berry length (mm)	Berry weight (g of 1 berry)	Chroma value	Hue value
	2013–2014	2013–2014	2013–2014	2013–2014	2013–2014
TC	15.74 ab	18.79 a	3.06 ab	13.85	106.59
MOC	16.03 a	19.21 a	3.34 a	13.95	106.61
MIC	15.14 b	17.63 b	2.77 b	14.24	106.79
TIC	15.68 ab	17.60 b	3.04 ab	14.72	106.40
LSD	0.6022	1.1285	0.4548	NS[1]	NS

[*]: Means of 2 years data. TC: Top of clusters, MOC: Middle outer side of clusters, MIC: Middle inner side of clusters, TIC: Tip of clusters.
[1]NS= Non–significant.

Differences between SSC and TA berries, found on different parts of clusters, have not been significantly different when the lowest pH values, maturity index values and TPC values have been determined into the berries located at the TIC which were 3.12, 13.52 and 1890 mg GAE 100 ml^{-1}, respectively.

The highest pH values were measured in berries that found on the TC and MOC. Statistical analyses showed that the berries while taking place on the TC, MOC and MIC had the highest maturity (15.32, 15.03 and 14.80, respectively) while the berries found on the MIC (1963 mg GAE 100 ml^{-1}) and TC (mg GAE 100 ml^{-1}) having the highest numbers of the total phenolic compounds. Monagas and Bartolomé (2005), reported that the maturation and sun exposure factors affected synthesis of phenolic compounds especially flavones are found in the skins of berries for the berries against sun burn (Table 4.).

Table 4. Some biochemical compositions on berries of different parts of clusters in 'Yalova Cekirdeksizi' cultivar[*]

Parts of clusters	SSC (%)	TA (%)	pH	Maturity index (SSC TA^{-1})	TPC (mg GAE 100 ml^{-1})
	2013–2014	2013–2014	2013–2014	2013–2014	2013–2014
TC	14.53	0.951	3.20 a	15.32 a	1959 a
MOC	14.18	0.943	3.20 a	15.03 a	1907 ab
MIC	14.55	0.984	3.18 ab	14.80 a	1963 a
TIC	13.89	1.029	3.12 b	13.52 b	1890 b
LSD	NS[1]	NS	0.0716	0.9129	45.41

[*]: Means of 2 years data. TC: Top of clusters, MOC: Middle outer side of clusters, MIC: Middle inner side of clusters, TIC: Tip of clusters.
[1]NS=Non–significant.

According to the results of this research work, the highest berry width have been determined on MOC (19.29 mm) and the lowest berry width have been calculated on MIC (18.00 mm) in 'Yalova Incisi' cultivar.
Berries on MOC and TC (24.41 mm and 23.94 mm, respectively) had the longest berries. When MOC (5.79 g) had the heaviest berries,

berries on TIC (4.67 mm) and MIC (4.68 mm) had the lightest grapes.
The lowest chroma values were measured on MIC (13.04), MOC (13.58) and TC (13.65). Berries on TC had more yellow tones and good coloured than the other parts of clusters due to the highest values of Hue (Table 5).

Table 5. Some pomological compositions on berries of different parts of clusters in 'Yalova Incisi' cultivar[*]

| Parts of clusters | Berry width (mm) | Berry length (mm) | Berry weight (g of 1 berry) | Chroma value | Hue value |
	2013–2014	2013–2014	2013–2014	2013–2014	2013–2014
TC	19.01 ab	23.94 a	5.54 b	13.65 b	106.74 c
MOC	19.29 a	24.41 a	5.79 a	13.58 b	108.16 b
MIC	18.00 c	23.07 b	4.68 c	13.04 b	109.58 a
TIC	18.54 bc	23.01 b	4.67 c	15.46 a	107.46 bc
LSD	0.5548	0.8148	0.1656	0.8094	1.3988

[*]: Means of 2 years data. TC: Top of clusters, MOC: Middle outer side of clusters, MIC: Middle inner side of clusters, TIC: Tip of clusters.

The highest SSC have been recorded in berries on TC (13.47%) while the lowest SSC were calculated on the berries of TIC (12.35%).
The lowest TA was determined in MIC (0.472%) for 'Yalova Incisi' grape cultivar. The highest pH values were found on MOC (3.84) and TC (3.81). The ripest berries (SSC TA^{-1}) have been obtained from MIC (28.24),

TC (27.59) and MOC (26.95) in 'Yalova Incisi' cultivar.
Berries of TIC (1900 mg GAE 100ml^{-1}) had significantly lower TPC in comparison to other parts of clusters. MIC (1924 mg GAE 100ml^{-1}) was highest content of phenolic compound (Table 6).

Table 6. Some biochemical compositions on berries of different parts of clusters in 'Yalova Incisi' cultivar[*]

| Parts of Clusters | SSC (%) | TA (%) | pH | Maturity Index (SSC TA^{-1}) | TPC (mg GAE 100 ml^{-1}) |
	2013–2014	2013–2014	2013–2014	2013–2014	2013–2014
TC	13.47 a	0.488 ab	3.81 a	27.59 a	1911 ab
MOC	12.99 b	0.483 ab	3.84 a	26.95 a	1904 ab
MIC	13.29 ab	0.472 b	3.73 b	28.24 a	1924 a
TIC	12.35 c	0.528 a	3.67 b	23.55 b	1900 b
LSD	0.3795	0.0460	0.0534	2.4693	21.101

[*]: Means of 2 years data. TC: Top of clusters, MOC: Middle outer side of clusters, MIC: Middle inner side of clusters, TIC: Tip of clusters.

CONCLUSIONS

According to research results, the largest berries were determined on TC and MOC in 'Cardinal' cultivar and on MOC in 'Yalova Cekirdeksizi' and 'Yalova Incisi' grape cultivars. The largest berries were taken from TC and MOC in case of all cultivars, though grape berries of MOC had the heaviest berries in all cultivars.
The ripest berries (SSC TA^{-1}) have been obtained from the TC in 'Cardinal'; TC, MOC

and MIC in 'Yalova Cekirdeksizi' and the TC, MOC and MIC in 'Yalova Incisi' cultivars. Particularly, the berries of TIC have been found tinier, lighter and lesser mature as compared to the berries of other parts of the clusters in case of 'Cardinal' and 'Yalova Incisi' grape cultivars.
Pomological and biochemical compositions of berries on different parts of clusters may vary considerably and also they have changed in different grape cultivars. Therefore, the tip reduction for monitoring maturity on table and

wine grape cultivars should be treated with caution and it was determined that samplings have to be done equally at least 3 different parts (top, middle and tip) of clusters. Nevertheless, the tip portions, having the latest blooming on flower clusters to be cut in certain proportions just after the berry formation, have been projected that they will provide an increase in that of the volume and maturity in those berries remaining on clusters.

ACKNOWLEDGEMENTS

Authors would like to thank to Mr. Furkan Yildirim (Agricultural Engineer) for his constant supports and regular cooperation throughout the experimental period.

REFERENCES

Ağaoğlu S., 1999. Bilimsel ve Uygulamalı Bağcılık, Vol. I Asma Biyolojisi, Kavaklıdere Eğitim Yayınları, Ankara, No: 1, 205.

Akin A., 2011. Effects of cluster reduction, herbagreen and humic acid applications on grape yield and quality of Horoz Karasi and Gök üzüm grape cultivars. African Journal of Biotechnology. 10 (29): 5593–5600.

Akin A., Dardeniz A., Ates F., Celik M., 2012. Effects of various crop loads and leaf fertilizer on grapevine yield and quality. Journal of Plant Nutrition. 35: 1949–1957.

Çelik S., 2011. Bağcılık (Ampeloloji). Cilt I, 3. Baskı. Namık Kemal Üniversitesi. Ziraat Fakültesi Bahçe Bitkileri Bölümü. 428 s. Tekirdağ.

Dardeniz A., Akçal A., Gündoğdu M.A., Killi D., Kahraman K.A., Özkaynak C., Erdem E., 2012a. Yalova İncisi ve Yalova Çekirdeksizi üzüm çeşitlerinde primer ve sekonder yazlık sürgünlerin gelişim ve verimlilik durumlarının karşılaştırılması. Uluslararası Tarım Gıda ve Gastronomi Kongresi, Antalya, 1–8.

Dardeniz A., Şeker M., Killi D., Gündoğdu M.A., Sakaldaş M., Dinç S., 2012b. Sofralık üzüm çeşitlerinin yapraklarındaki klorofil miktarının boğumlar bazındaki dönemsel değişiminin belirlenmesi. Uluslararası Tarım Gıda ve Gastronomi Kongresi Antalya, 9–14.

Dardeniz A., 2014. Effects of cluster tipping on yield and quality of Uslu and Cardinal table grape cultivars. ÇOMÜ Zir. Fak. Derg. 2 (1): 21–26.

FAO, 2014. Agricultural Statistical Database. http://faostat.fao.org (access date: 20 Eylül, 2013).

Ilgın C., 1997. Yuvarlak Çekirdeksiz üzüm çeşidinde farklı ürün yükünün üzüm verim ve kalitesi ile vegetatif gelişmeye etkileri üzerine araştırmalar. Ege Üniversitesi Fen Bilimleri Enstitüsü. Doktora tezi. Bornova/İzmir, 63–65.

Kısmalı İ., Dardeniz A., 2002. Cardinal ve Amasya üzüm çeşitlerinde iki farklı yeşil budama uygulamasının gelişme, üzüm verimi ve kalitesine etkileri üzerinde araştırmalar. V. Ulusal Bağcılık Sempozyumu, Nevşehir, 221–227.

Monagas M., Bartolomé B., 2005. Updated knowledge about the presence of phenolic compounds in wine. Crit. Rev. Food Sci. 45: 85–118.

Reynolds A.G., Price S., Wardle D.A., Watson B., 1994. Fruit environment and crop level effects on Pinot Noir. Vine Performance and Fruit Composition in the British Columbia. Amer. J. Enol. Vitic. 45: 452–459.

Sezen E., Dardeniz A., 2015. Farklı kış budama dönemleri ve yaz budaması uygulamalarının Yalova İncisi üzüm çeşidinin verim ve kalitesine olan etkilerinin belirlenmesi. ÇOMÜ Zir. Fak. Derg. 3 (1): 15–27.

Smart R.E., Robinson J.B., Due G.R., Brien C.J., 1985. Canoph microclimate modification for the cultivar Shiraz. II. Effects on Must and Wine Composition. Vitis 24. 119–128.

Stoev K., 1974. Asma bünyesinde asimilantların hareketliligi. (Bulgarian). Bitki Fizyolojisi Dergisi. 21.

Todorov,H., 1970. Somakların salkıma dönüşme safhasında çiçek tomurcuklarını silkmesi ve diğer değişimler üzerine araştırmalar. (Bulgarca). Gradnarska i Lazarska Nauka, 1.

Türker L., Dardeniz A., 2014. Sofralık üzüm çeşitlerinde farklı düzeylerdeki koltuk alma uygulamalarının verim ve kalite özellikleri üzerindeki etkileri. ÇOMÜ Zir. Fak. Derg. 2 (2): 73–82.

Yılmaz E., Dardeniz A., 2009. Bazı üzüm çeşitlerindeki salkım ve sürgün pozisyonunun üzüm verim ve kalitesi ile vejetatif gelişime etkileri. Süleyman Demirel Üniversitesi Ziraat Fakültesi Dergisi. 4(2): 1–7.

13

KNOWLEDGE OF QUALITY PERFORMANCE OF SOME TABLE GRAPE VARIETIES GROWN AND OBTAINED IN THE EXPERIMENTAL FIELD FROM U.A.S.V.M. BUCHAREST

Marinela Vicuţa STROE[1]

[1]Department of Bioengineering of Horticulture - Viticulture Systems
University of Agronomical Sciences and Veterinary Medicine, 59 Marasti Blvd, District 1,
011464, Bucharest, Romania
Corresponding author email: marinelastroe@yahoo.com

Abstract

In our country, the first varieties of grape-vines with known genetic origin, have been created since the sixth decade of the last century, and the outstanding achievements obtained in improving varieties were presented and published over the years, through many treatises. In general, the main objectives of the unitary genetic improvement of grapevine program were sought in obtaining varieties (table grapes, wine, seedless), that would be characterized by a higher production potential than the genitors and that would show greater resistance to pests and diseases specific to grapevines. Unfortunately, these new obtained varieties mostly are known neither nationally nor externally, only a few managed to get in and pass through this transition period, the border of the area where they were created. Although some of these are very valuable, both in terms of productivity and quality, in these circumstances, they will be doomed to anonymity. In this paper, we will refer to the five varieties of table grapes produced in our university - Muscat Timpuriu de Bucuresti, Augusta, Chasselas de Băneasa, Triumf and Select varieties classified in three different eras of maturation (early, middle, tardive). The productive and qualitative performances achieved by these varieties, expressed through carpometric values and organoleptic perspectives (yield, gluco-acidometric index, shape, color of skin, firmnes of flesh, particular flavor), can become attractive for vineyard in the south of Romania and can effectively contribute to fill the conveyor varietal of grape varieties for table grapes. Therefore, the promotion of these local creations, through various means, would be a win for both wine growers (producers) and consumers due to very high production potential, but also because of the particular organoleptic qualities that they posses.

Key words: capacity, grape table, performance, quality, varieties

INTRODUCTION

In our country, the first varieties of grape-vines with known genetic origin, they have been created since the sixth decade of the last century, and outstanding achievements obtained in improving varieties were presented and published over the years, through many speciality papers, over the years (Constantinescu et al., 1959, 1960, 1962,1965, 1966; Constantinescu, 1975; Constantinescu and Negreanu, 1960; Dvornic, 1960, 1974; Gorodea et al., 1976; Gorodea, 1983; Neagu et al., 1968; Negreanu and Lepădatu, 1971; Oprea and Gorodea, 1980; Oprea et al., 1983, 1986; Ioniţă et al.,1981; Lepădatu, 1979; Toma and Ispas, 2008). The main objectives of the unit program of genetic improvement of grapevines, coordinated by the Research Institute for Winegrowing and Winemaking Valea Calugareasca, sought is and still soughting in

particular at creating varieties with higher potential of productivity and quality, with installment periods of maturation at table grapesvarieties, with increased resistance to diseases and pests, but also with higher resistance to weather and extreme phenomena (Stroe et al., 2013). Therefore, in our country in the period 1970-2000 were approved 22 varieties of table grapes and three seedless, and after 2000, six more varieties of table grapes. Given these goals, mentioned above, creating a new varieties of grapevines and improving the main varieties of table grapes from the range which makes up the conveyor varietal of our country were the main concerns of the research under taken in within our institution – U.A.S.V.M. Bucharest beginning with the year 1957. Among the achievements obtained in our institution, we mention the creation and approval of two early varieties of table grapes: Muscat Timpuriu de Bucureşti and Augusta, two varieties of table grapes with middle maturity - Chasselas de Baneasa and Triumf and a variety with late maturity- variety Select (Table 1). The main data about these varieties can be found in Vitis International Variety Catalogue (www.vivc.de).

This study aimed to follow the elements that define the quality of the five new varieties in the south area of Romania in order to popularise them.

Varieties come basically from the same wine area where they imposed, but in Romania,

except the first two, were rarely investigated and even much less cultivated.

The first two and last two varieties are distinguished by a slight degree of similarity between them, having in common some genetic lineage (Variety Queen Vineyards as paternal variety, in the first case and the variety Afuz ali in the second case).

This study was addressed on the need to know the quantitative and qualitative performance of these varieties in order to promote them at least at national level.

Table 1. Genetic origin of studied varieties

Prime name	Muscat Timpuriu de Bucuresti	Augusta	Chasselas de Baneasa	Triumf	Select
Variety number VIVC	8256	14781	2480	12655	11471
Country of origin of the variety	Romania	Romania	Romania	Romania	Romania
Species	Vitis vinifera L.	Vitis vinifera L.	Vitis vinifera L.	Vitis vinifera L.	Vitis vinifera L.
Pedigree as given by breeder/bibliography	Coarna albă x Queen vineyards	Italia x Queen vineyards	Chasselas dorè	Lignan x Afuz li	Bicane x Afuz ali
Pedigree confirmed by markers	-	Italia x Queen vineyards	-	-	-
Prime name of pedigree parent 1	Coarna albă	Italia	Chasselas blanc	White Luglienga	Bicane
Prime name of pedigree parent 2	Queen vineyards	Queen vineyards	-	-	Afuz ali
Year of crossing	1970	1984	1978	1970	1970
Last update	15.01.2016	22.12.2015	22.12.2015	22.12.2015	22.12.2015

MATERIALS AND METHODS

They were studied five varieties of table grapes, obtained in U.A.S.V.M. Bucharest-Muscat Timpuriu de Bucuresti, Augusta, Chasselas de Băneasa, Triumf and Select. The varieties were conducted on semi-stem type of pruning Guyot on semi-stem with a load of 42 buds/vine. Were watched mainly the elements of fertility and productivity, currently used in studying varieties of grape-vines, in special on those who have shown interest in the appreciation of carpometric and organoleptic elements covered by this study: average weight of a grape, weight of 100 berries, production/vine (kg/vine), sugars (g/l), total acidity (g/l tartaric acid), gluco-acidometric index, shape, color of skin, firmnes of flesh, particular flavor. Grape harvesting was performed at full maturity of each variety. This study was approached from two perspective: knowledge of these varieties to promote and popularize them at least at national level. For

this, were used data sheets of each individually varienty, from speciality literature and for updating the variety data were pursued the viticultural year 2014-2015 being known that the variety must provide a high intrinsec quality given by the constant productions obtained year after year, regardless of the direction of production (for table, for wine, a raw material for distilled products etc.).

Figure 1. Variety Muscat Timpuriu de Bucuresti

Variety Muscat Timpuriu de Bucuresti - Short presentation. Vigorous variety with a short growing period between 155-165 days, which is well suited to the lead half stem form. Presents poor tolerance to cold (-16°C -...-18°C), medium drought tolerance, manifesting a high sensitivity to mildew and powdery mildew. The variety is characterized by abundant fruit fulness evenon secondary shoots.

Figure 2. Variety Augusta

Variety Augusta - Short presentation. Variety of medium vigor which obtain good results in 20 buds/m^2, divided on long elements. Presents middle frost tolerance (-18^0 C ...- 20^0 C), powdery mildew and gray mold of grapes and manifest greater sensitivity to mildew.

Figure 3. Variety Chasselas de Băneasa

Variety Chasselas de Băneasa - Short presentation. The variety is characterized by a medium growth vigor of the variety from which it was obtained. But it is more sensitive to cold than Chasselas doré and shows good disease resistance.

The best results were obtained at a load of 12 to 16 buds/m^2 spread over long strings with 12-14 buds, (Țârdea and Rotaru, 2003).

Figure 4. Variety Triumf

Variety Triumf - Short presentation. The variety is very vigorous and the author (Dvornic, 1974; Indreas and Visan, 2000; Stroe, 2012), recommends Guyot type of pruning on semi-stem with a load of 14 to 15 buds/m^2.Presents good tolerance to cold (-18°C ... -20°C), drought and oidium, especially susceptible to powdery mildew and gray rot of grapes. It is not attacked by moths.
It yields the same as Muscat Timpuriu de Bucuresti on the secondary shoots and most often, they can be used to recover the production of grapes climate accidents.

Figure 5. Variety Select

Variety Select - Short presentation. Select varieties are distinguished by great vigor, with strong growth and a middle period growth. It manifests a good resistance to frost and oidium and tolerance to powdery mildew and gray mold. Select doesn't makes small berry and doesn't makes small grains. In plantations, the author recommends between 18-20 buds/m^2.

RESULTS AND DISCUSSIONS

The analysis of climatic elements for the wine year 2014-2015 and the values of the four synthetic indexes (Table 2) shows that when the thermal resources are high, the water resources are low and the most fluctuating indicator is the bioclimatic one, whose spectrum is within the 9.9 - 12.76. The observations made show that the area in which the didactic-experimental field of U.A.S.V.M. Bucharest is found is favorable for growing varieties of grape-vines studied (registered in the south of Romania), and the elements of microclimate positively put their mark on the behavior of the studied varieties.

Table 2 Evolution of climatic elements (2001-2015)

	Specification	Average	Year
		2001-2011	2014-2015
Indices agroclimatics	The hydrothermic coefficient CH)	0.75	0.73
	The real heliotermic index (IHr)	1.3	1.23
	The viticultural bioclimatic index (Ibcv)	9.9	12.76
	Index of the oenoclimatic aptitude (IAOe).	5231	5153
	Huglin index	2392	2548.9

In awarding the same charges of 42 buds/vine in the viticultural year 2014-2015 it observes, that there are differencies is their behavior, given by the fertility of varieties and the obtained values keeps the varieties in standard limits, and in some cases, such as the variety Triumf the values recorded are even higher (Table 3). Their qualitative appreciation is based not only by analyzing the elements of fertility (absolute fertility coefficient, absolute productivity index), which in general, are constant, but also in the accumulated sugar levels, amid a total acidities quite balanced. The data show a highlight from this stand point varieties Muscat Timpuriu de Bucuresti (171 g/l), Chasselas de Băneasa (178 g/l), but neither the other varieties are found in imbalance, the minimum being recorded by variety Select (145 g/l). Regarding the appreciation of organoleptic and carpometric elements which makes the subject of this study is observed and maintained a constant distinct, surpassing in terms the viticultural year analyzed average values found in the speciality literature (Dvornic, 1974; Gorodea et al., 1976; Oprea et al., 1983, 1986). The varieties of table grapes can be harvested before full maturity, practically at the maturity of consumption based on gluco-acidometric index. Normally this index is between 2.5 ÷ 4.5. At the tested varieties ranged from 2.2 - 3.06 (Table 4), the highest values recorded Muscat Timpuriu de Bucuresti and Chasselas de Băneasa (respectively 3.05, 3.06).

Although the variety Select, usually reach full maturity later belonging to maturing eras V-VI, in terms of the viticultural year 2015 it reached the optimal harvest in advance (mid-September). In the context of the results presented above, it can be appreciated that the viticultural area in which they were created and are current cultivated these varieties, leaves its mark on their quality potential, at the precocity maturing, as evidenced by productions obtained that are constant from quantity and quality point of view, year after year, no matter the variety and age of maturation.

Table 3. The synthesis of the main fertility elements of varieties study

Varieties	% fertile shoots	Absolute fertility coefficient	Relative fertility coefficient	Absolute productivity index (g/shoot)	Relative productivity index (g/shoot)
Muscat Timpuriu de Bucuresti	41	1.5	0.8	483	258
	51.9*	1.0*	0.6*	380*	228*
Augusta	61	1,7	1.1	552	356
	63	1.07	0.63	466	275
Chasselas de Baneasa	75	1,6	1.1	411	282
	65	1.4	0.8	327	187.2
Triumf	40	1,1	0,8	484	352
	53	1.2	0.9	547.2	410.4
Select	50	1.4	0,6	256	602
	40	1.1	0.4	451	164

*years 2014-2015

Table 4. Physical and chemical characteristics of the grapes belonging to the studied varieties

Varieties	Average weight of a grape (g)	Weight of 100 berries (g)	Production (kg /vine)	Sugar (g/kg)	Total acidity (g tartaric acid/l)	Gluco-acidome-tric index	Berry			
							Shape	Color of skin	Firmnes of flesh	Particular flavor
Muscat Timpuriu de Bucuresti	322 380*	383 402*	3.0 2.6*	190 171*	5.9 5.6*	3.22 3.05*	ovoid	green yellow	very firm	muscat
Augusta	325 436	440 502	3.74 3.04	149 155	5,3 5,9	2.8 2.62	ovoid	green yellow	slightly firm	none
Chasselas de Baneasa	257 234	448 398	3.7 3.2	151 178	6,0 5.81	2.51 3.06	globose	green yellow	soft	none
Triumf	440 456	408 430	3.5 3.7	150 152	6.8 6.73	2.2 2.25	ovoid	green yellow	very firm	none
Select	430 410	470 492	3.1 2.6	135 145	6.42 6.57	2.10 2.2	ovoid	green yellow	slightly firm	none

*years 2014-2015

CONCLUSIONS

Muscat Timpuriu de Bucuresti is maturing immediately after variety Muscat Perla of Csaba and far exceeds the size of berries and flavored taste. In addition, grapes precocity, discreet flavor, pleasant taste and attractive appearance make this new variety to have a good potential for the viticulture from our country. It can be grown allover the country, especially in the southern regions, where ensure the early supply of the market.

Variety Augusta is also a sort of early, maturing in the second decade of August, with 5-6 days after Cardinal variety has large berries, but shows a gradual ripening of the grapes.

Variety Chasselas de Băneasa presents larger berries than those of the variety Chasselas doré, but has its lower organoleptic qualities. It impose with large enough production and can contribute to diversification of varietal conveyor of varieties with middle ripening maturity.

Triumph variety is distinguished by the attractive appearance of the grapes very pleasant taste, refreshing and good resistance to transport.

Select variety has a quite compact grape, very showy and doesn't makes small berry and doesn't makes small grains, it retains their organoleptic qualities even after 2-3 weeks if left on the vine, but accumulate modest sugars.

Quantity and quality performance of the varieties analyzed, can become attractive for the decision taking them in culture, at least for the viticulture in the south of Romania.

Therefore, the promotion of these local creations, through various means, would be a win for both wine growers (producers) and consumers due to very high production potential, but also because of the particular organoleptic qualities they hold.

REFERENCES

Constantinescu Gh., Negreanu E., 1960 - Soiuri noi de viţă de struguri de masă cu coacere timpurie. Lucrări ştiinţifice I.A.N.B, Seria B, 109-116.

Constantinescu Gh. şi colab., 1959, 1960, 1962, 1965, 1966 - Ampelografia R.S.R., Volumele II, III, IV, V, VI,VII, Editura Academiei Bucureşti.

Constantinescu Gh., 1975 - Probleme de genetică teoretică şi aplicată. Institutul de cercetări pentru cereale şi plante tehnice Fundulea. Secţia de ameliorarea plantelor, vol. VII, nr. 4: 213-214.

Dvornic V., 1960 - Un hibrid nou de persectivă pentru struguri de masă. Lucrări ştiinţifice, I.A.N.B, Seria B, 139-143.

Dvornic V.,1974 - Comportarea elitei pentru struguri de masă Chasselas de Băneasa în condiţii de silvostepă. Analele I.C.D.V.V.Valea Călugărească, volumul V: 61-68.

Gorodea Gr. şi colab, 1976 - Studiul soiurilor de vită-de-vie pentru struguri de masă introduse recent în colecţiile ampelografice. Analele I.C.D.V.V., vol. VII: 33-54.

Gorodea Gr., 1983 - Contribuţii la studiul variabilităţii ereditare a rezistenţei la ger a viţei-de-vie. Analele I.C.D.V.V., vol. X: 55-64.

Indreaş A., Vişan L., 2000 - Principalele soiuri de struguri de masă cultivate în România. Editura Ceres, Bucureşti. ISNB 973-40-0469-7, p. 32, 34, 42, 52.

Ioniță I. si colab., 1981 - Soiuri de viță de vie pentru struguri de masă obținute în rețeaua I.C.D.V.V. în perioada 1976-1980, Sesiunea științifică Blaj, Alba.

Lepădatu V., 1979 - Soiuri noi pentru struguri de masă. Cercetarea în sprijinul producției – viticultură și vinificație, București.

Neagu M. I. și colab., 1968 - Studiul însușirilor agrobiologice ale soiurilor de struguri pentru masă recent introduce în cultură. Analele I.C.D.V.V.Valea Călugărească, volumul I: 102-109.

Negreanu E., Lepădatu V., 1971 - Ereditatea unor însușiri și caractere calitative și cantitative la hibrizii F 1 de viță-de-vie. Analele I.C.D.V.V.Valea Călugărească, volumul III: 21-36.

Oprea Șt., Gorodea Gr., 1980 - Soiuri noi pentru struguri de masă. Cercetarea în sprijinul producției viticultură și vinificație, București.

Oprea Șt. și colab., 1983 - Soiuri noi de vita de vie pentru struguri de masă recent create in Romania. Analele I.C.D.V.V.Valea Călugărească, vol. X: 43-54.

Oprea Șt. și colab., 1986 - Soiuri noi de vita de vie pentru struguri de masă și vin create în România în perioada 1982-1985. Analele I.C.D.V.V.Valea Călugărească, vol. XI: 35-48.

Stroe Marinela, 2012 – Ampelografie. Editura Ceres, București, ISBN 978-973-40-0943-5, (COD CNCSIS 236), 182, 184, 187, 194.

Stroe Marinela, Dragoș Matei, Damian Ion, Sofia Ispas, Elena Dumitru, 2013 - Research regarding local vineyards germplasm resource monitoring conditions determined local climate change. 36 [th] World Congress of Vine and Wine, 11 [th] General Assembly of The O.I.V., 2th-7th june, 2013, Bucharest (Romania), Vine and Wine berween Tradition and Modernity, ISNB O.I.V. 979-10-91799-16-4, 194-195.

Țârdea C., Rotaru L., 2003 - Ampelografie, volumul II. Soiuri de viță-de-vie pentru struguri de masă și soiuri apirene. Editura Ion Ionescu de la Brad, Iași, ISNB 978-8014-94-8, 22, 34, 84, 88, 169.

The European Vitis Database, Genetic resources of grapes, website, www.eu-vitis.de/index.php, accessed on February 2016.

Toma Otilia, Ispas Sofia, 2008 - 115 ani de cercetare, învățământ și producție vitivinicolă și pomicolă la Pietroasa și Istrița, București, Istrița, Pietroasele-Buzău, 30-31 Octombrie 2008. Editura INVEL-Multimedia ISBN 978-973-7753-90-8, 77-78.

Vitis International Variety Catalogue, Passport of variety Muscat Timpuriu de Bucuresti, Variety number VIVC 8256, passport of Augusta, VIVC number 14781, passport of Chasselas de Băneasa, VIVC number 2480, passport of Triumf, VIVC number 12655, and passport of Select, VIVC number 11471, website, http://www.vivc.de, accessed on February 2016.

http://www.oiv.int/oiv/files/5%20-20Publications/5%20-%201%20Publications%20OIV/EN/59_Liste_descrip teurs_2ed_EN.pdf.

14

SPECIES COMPOSITION OF PLANT PARASITIC NEMATODES *PRATYLENCHUS* SPP. IN CONVENTIONAL AND ORGANIC PRODUCTION OF RASPBERRIES

Elena TSOLOVA[1], Lilyana KOLEVA[2]

[1]Institute of Agriculture-Kyustendil, Sofjisko shose str., 2500 Kyustendil, Bulgaria,
Email: elena_tsolova@abv.bg
[2]University of Forestry, Sofia, 10 Kliment Ohridski Blvd, 1756, Sofia, Bulgaria,
Email: liljanamarkova@abv.bg
Corresponding author email: liljanamarkova@abv.bg

Abstract

The sustainable organic production involves a lot of difficulties in pest control. Plant-parasitic nematodes of the genera Pratylenchus cause significant economic damage to raspberry production. To study the species composition of root-lesion nematodes of Pratylenchus spp., observations were carried out of raspberry plants in conventional and organic cultivation systems. The species composition of the established nematodes by different farming technologies was similar. In the conventional production, the species composition of nematodes was relatively homogeneous, as the number of established Pratylenchus species was 4, while, in the organic production, the species composition of nematodes is characterized by variety, but their number was 6. The number of individuals of the Pratylenchus species in conventional cultivation was significantly higher.

Key words: root-lesion nematode Pratylenchus spp., raspberry, organic production.

INTRODUCTION

Raspberry-growing is currently occupying an important place in the fruit production in many European countries, which creates new jobs in the agriculture sector and delivers food products and raw materials for the food industry. Maintaining a healthy and productive raspberry plantations for years creates significant difficulties and efforts are directed to a preliminary assessment of the overall process. The economic damage is considerable in the infested soil with plant-pathogenic nematodes. The yield can be completely lost in heavily infested fields, and the presence of certain species of Nematoda restricts the growing of a range of crops in the contaminated areas (Nicol et al., 2011).
The worldwide damage caused by root lesion nematodes of the genus *Pratylenchus* (Filipjev, 1936) increased, and thus the interest in them. In the most regions, the root lesion nematode *Pratylenchus penetrans* continues to be a major limiting factor in the production of raspberries. This species is endoparasitic, destroyes the root

tissue, reduces the flow of water from the soil into the roots and the transport of nutrients. In many cases in the field, there are mixed populations of plant- parasitic nematodes, rather than individually occurring species.
This study of the species composition of plant parasitic nematodes in conventional and organic systems of production of raspberries based on a comparative assessment will be achieved not only to determine the effectiveness of the methods and means of plant protection and will ensure the plant health and the growth potential.
The aim of this work was to investigate the qualitative and quantitative structure of root lesion nematodes in the genus *Pratylenchus* in different growing technologies.

MATERIALS AND METHODS

Characteristics of experimental fields
The study was conducted in two raspberry plantations near the town of Kostinbrod (GPS: 42°49'5.80"N; 23°13'30.54"E). The plants are planted at a distance of 2.50 x 0.50 m or 88

plants/ha. Two growing systems were tested conventional growing and organic growing. The test plot was 25 m^2, four replications of sequences of time.

Sampling methods

The plant health condition of the fields has been satisfactory; there were zones of reduced plant growth and plant damage.

According to the meteorological data, abiotic factors, which could cause these symptoms, are excluded. The period of sampling of plant and soil samples was consistent with that recommended by Knuth et al. (2003). In order to determine the increase or decrease of density of the populations, the samples were taken depending on the season, during no-vegetation and vegetation period. The soil samples were taken randomly from 15 – 25 cm depths and then transferred in plastic bags to the laboratory.

Extraction of the nematodes

The methods for the extraction of the nematodes from the soil and roots and their subsequent mounting on permanent slides for identification are according to the Baermann pan method described by Townshend (1963). Species characterization and identification were based on morphology of various life stages (Loof, 1978; Bongers, 1988; Handoo and Golden 1989).

Statistical analysis

The statistical analysis was carried out using Statistica 99 Edition statistical package.

RESULTS AND DISCUSSIONS

The established species of the genus *Pratylenchus* in different cultivars and variations of fertilization are shown in Tables 1 and 2.

Besides the most commonly found species of *P. crenatus*, *P. neglectus* and *P. pratensis*, two species of this genus: *P. thornei* and *P. sonvallariae* were isolated in organic growing. Although the species of *P. penetrans* was established, its presence in both fields was sporadic.

During the first year of investigations (2011), symptoms of plant damage and the increased density of nematodes of this genus were not observed.

But in 2012, the density of species of the genus *Pratylenchus* was increased during fruiting seasons in the conventional growing systems compared to the organic growing.

During the investigation of the samples (soil and plant), a total number of individual nematodes of the genus *Pratylenchus* was 620; 384 individuals in the conventional framing of the genus *Pratylenchus* and 236 individuals in the organic growing.

The highest density was recorded in September 2012- 74 individuals in conventional growing, and lowest in the organic growing in August 2012-9 individuals. Throughout the research period in 2012, the total number of isolated nematodes of the genus *Pratylenchus* did not confirm the increase of their density in the organic growing.

It should be noted that the number of species of the genus *Pratylenchus* is difficult to predict, since they are located at a specific moment in the soil and the roots.

Figure 1 and 2 present the percentage of the number of species and their density in the conventional and the organic growing during the study.

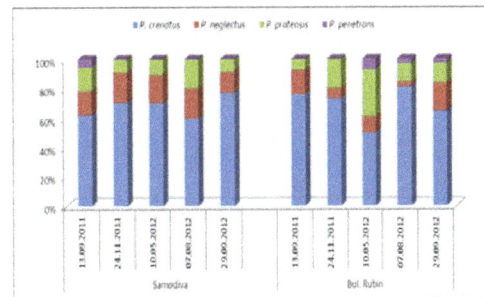

Figure 1. Percentage of species of the genus *Pratylenchus*, extracted from soil and plant samples in conventional growing of raspberry

Throughout the study period from the total number of samples infected with *Pratylenchus* spp., these species were isolated: in the conventional growing: the species *P. crenatus* was 69.7%, *P. neglectus*- 13.4% and *P. pratensis*- 14.7%; in the organic growing: *P. crenatus* was found 48.7%, *P. neglectus*- 29.3%, *P. pratensis* 14.5%, *P. thornei* 3.4% and *P. sonvallariae* 2.5% (Figure 1 and 2).

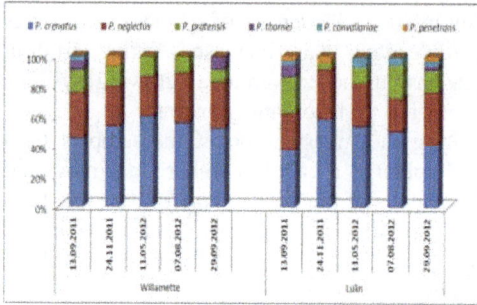

Figure 2. Percentage of species of the genus
Pratylenchus, extracted from soil and plant samples in
organinc growing of raspberry

Postive significant correlations (R=0.15; P
<0.05) between the different variants of
fertilisation and the density of nematodes were
found in the conventional growing. Correlation
between nematodes of the genus *Pratylenchus*

and variants of fertilization (R=-0.06; P>0.05)
was not demonstrated in the organic growing.

The population density of *Pratylenchus* spp.
was negatively correlated with the total density
of the migratory root nematodes (R=-0.26; P
<0.01).

The sum of the number of individuals of the
species *P. crenatus* was 38 individuals, which
is indication of the most common species in the
organic growing. *P. neglectus*, especially in the
autumn of 2012, was 1.5 times more common
species than other species of the genus
Pratylenchus.

Table 1. Species of root lesion nematodes of the genus *Pratylenchus*,
extracted from fields with conventional growing of raspberry

sampling time/cultivar	P. crenatus	P. neglectus	P. pratensis	P. penetrans	undetermined species
13.09.2011					
Samodiva	*V2-9 V3-14	V0-1 V1-1 V2-2 V3-3	V0-2 V3-4	V1	V3-1
Bul. Rubin	V0-12 V1-13 V2-10 V3-26	V0-2 V1-4 V2 - 5	V0-2 V2-1 V3-2		
24.11. 2011					
Samodiva	V0-1 V1-2 V2-10 V3-11	V0-2 V2-1 V3-4	V0-1 V3-3		V0 -1
Bul. Rubin	V0-11 V3-11	V3-2	V0-1 V2-1 V3-4		V1 -1
10.05.2012					
Samodiva	V0-4 V1-9 V3-13	V0-1 V2-3 V3-3	V0-1 V1-1 V3-2		
Bul. Rubin	V0-1 V1-4 V2-2 V3-7	V1-1 V2-1 V3-1	V0-2 V2-4 V3-3	V3-3	
07.08.2012					
Samodiva	V0-2 V1-0 V3-4	V2-1 V3-1	V2-1 V3-1		V3-1
Bul. Rubin	V1-1 V2-13 V3-12	V3-1	V0-2 V2-1 V3-1		
29.09.2012					
Samodiva	V0-1 V1-4 V2-10 V3-12	V0-1 V1-1 V2-1 V3-2	V0-1 V1-1 V3-1		
Bul. Rubin	V0-10 V1-14 V3-24	V0-3 V1-1 V3-7	V0-1 V1-2 V2-2 V3-5	V0-2	V2 -1

*V- variants of experiment (various fertilization) and number of established nematode species

Table 2. Species of root lesion nematodes of the genus *Pratylenchus*,
extracted from fields with organic growing of raspberry

sampling time/cultivar	*P. crenatus*	*P. neglectus*	*P. pratensis*	*P. thornei*	*P. convallariae*	*P. penetrans*	undetermined species
13.09.2011							
Willamette	*V0-5 V1-3 V2-2 V3-5	V3-10	V1-1 V2-1 V3-3	V0-2	V2-1		
Lulin	V0-7 V1-2 V3-5	V2-2 V3-5	V1-2 V2-3 V3-4	V0-3	V3-1	V3-1	
24.11.2011							
Willamette	V2-4 V3-4	V2-1 V3-3	V2-1 V3-1			V3-1	V3-1
Lulin	V0-4 V1-2 V2-2 V3-6	V2-5 V3-3	V2-1			V0-1	
11.05.2012							
Willamette	V0-3 V1-4 V3-4	V0-2 V1-2 V3-5	V2-1 V3-1				V3-1
Lulin	V1-3 V2-5 V3-7	V0-4 V2-2 V3-2	V0-2 V1-1		V0-1 V1-1		V2-1 V3-1
07.08.2012							
Willamette	V1-2 V2-1 V3-2	V0-3	V0-1				
Lulin	V0-1 V1-2 V2-3 V3-3	V0-1 V1-2 V3-1	V0-1 V1-3		V3-1		
29.09.2012							
Willamette	V0-4 V1-3 V3-5	V0-2 V1-2 V2-2 V3-1	V3-2	V3-2	V3-2		
Lulin	V0-1 V1-2 V2-6 V3-5	V0-2 V1-4 V2-2 V3-3	V3-5	V0-1	V0-1	V3-1	

*V- variants of experiment (various fertilization) and number of established nematode species

Many authors reported that the species *P. thornei* and *P. convallariae* were isolated from heavy soils (Whish et al., 2014).

Despite their differences in preferences for the type of soil (Sturhan, 2014), the species *P. crenatus* and *P. neglectus* were found in joint populations in both growing technologies. The frequent occurrence of these two together is reported by Kleynhans et al. (1996) Kruse (2006), Söğüt et al. (2014) and Esteves et al. (2015) in various agricultural ecosystems.

As regards the reaction of the soil it has been found that the population density of *P. crenatus* and *P. neglectus* was significant different. The species *P. crenatus* prevailed in the soil with slightly alkaline reaction, pH ± 8. This species was present 1.5-2.5 times more than in the soil with a neutral to slightly acid, pH 7.5–6.0 (organic growing). The species *P. neglectus* was detected in significantly less samples of the organic growing soil with low pH values.

Larvae of genus *Pratylenchus* were isolated more frequently from the samples, thus so that making it difficult to determine or its determination could not be carried out. The prevalence of *Pratylenchus* spp. is higher in regions with extensive agriculture (Vrain and Dupré, 1982; Vrain and Rousselle, 1980). Kimpinski (1985), Bélair (1991), Zasada and Moore (2010) demonstrate that the significant increase in population density of *Pratylenchus* spp. requires soil treatment in raspberry fields with nematicides.

The root lesion nematode, *Pratylenchus penetrans*, is a production-limiting pest in red raspberry, *Rubus idaeus*. Nowadays authors continue searching for genetic resistance, as a tool to manage *P. penetrans* in raspberries. This would reduce the impact of this nematode on raspberry productivity as well as reduce the plant chemical treatments to keep populations in control (Zasada and Moore, 2014).

These results indicate that more research is needed to learn about the relationship between fertilization, cultivars and root lesion nematodes of genus *Pratylenchus* in different growing technologies.

Given that the culture practices within each field remained almost unchanged for 10 years, this study provides a current representation of the distribution of nematode populations of small areas in the conventional and the organic production of raspberries.

It should be noted that a redistribution of certain species of plant-parasitic nematodes stays possible.

CONCLUSIONS

The species of the genus *Pratylenchus* were dominant in both of growing technologies, as *P. crenatus* and *P. neglectus* are occurred most often.

The number of identified species of the genus *Pratylenchus* in the conventional growing was 4 and 6 in the organic growing.

Although the number of established species of the genus *Pratylenchus* in the conventional production was smaller, the number of individuals of each species was significantly higher.

REFERENCES

Bélair, G., 1991. Effects of preplant soil fumigation on nematode population densities and on growth and yield of raspberry. Phytoprotection, 72:21-25.

Bongers, A.M.T., 1988. De nematoden van Nederland. Pirota Schoorl, Bibliotheek uitgave KNNV.

Esteves, I., Maleita, C., Abrantes, I., 2015. Root-lesion and root-knot nematodes parasitizing potato. European Journal of Plant Pathology, 141(2): 397-406.

Handoo, Z. A., Golden, A. M., 1989. A key and diagnostic compendium to the species of the genus Pratylenchus Filipjev, 1936 (lesion nematodes). Journal of Nematology, 21(2):202-218.

Kimpinski, J., 1985. Nematodes in strawberries on Prince Edward Island., Canada. Plant Disease, 69:105-107.

Kleynhans, K. P. N., Berg, E., Swart, A., Marais, M., Buckley, N. H., 1996. Plant nematodes in South Africa. ARC-Plant Protection Research Institute.

Knuth, P., Lauenstein, G., Ipach, U., Braasch, H., Müller, J., 2003. Untersuchungsmethoden für pflanzenparasitäre Nematodenarten, die in Deutschland von Rechtsvorschriftenbetroffen sind. Braunschweig: Eigenverlag, Ber. Biol. Bundesanst. Land- Forstwirtsch., 121:1-49

Kruse, J., 2006. Untersuchungen zur Schadwirkung und Populationsentwicklung wandernder Wurzelnematoden in getreidebetonten Fruchtfolgen Mecklenburg-Vorpommerns (Doctoral dissertation, Universitätsbibliothek Giessen).

Loof, P. A. ,1978. The genus Pratylenchus Filipjev, 1936 (Nematoda: Pratylenchidae): a review of its anatomy, morphology, distribution, systematics and identification. Swedish University of Agricultural Sciences, Research Information Centre.

Söğüt, M. A., Göze, F. G., Önal, T., Devran, Z., Tonguc, M., 2014. Screening of common bean (Phaseolus vulgaris L.) cultivars against root-lesion nematode species. Turkish Journal of Agriculture and Forestry, 38(4):455-461.

Sturhan, D., 2014. Plant-parasitic nematodes in Germany–an annotated checklist. Soil Organisms 86 (3): 177–198

Townshend, J. L., 1963. A modification and evaluation of the apparatus for the Oostenbrink direct cottonwool filter extraction method. Nematologica, 9:106-110.

Vrain, T. C., Rousselle G. L., 1980. Distribution of plant-parasitic nematodes in Quebec apple orchards. Plant disease, 64(6):582-83.

Vrain, T.C, Dupré M., 1982. Distribution des nematodes phytoparasites dans les sols maraîchers du sud-ouest du Québec. Phytoprotection, 63:79-85.

Whish, J. P. M., Thompson, J. P., Clewett, T. G., Lawrence, J. L., Wood, J., 2014. Pratylenchus thornei populations reduce water uptake in intolerant wheat cultivars. Field Crops Research, 161:1-10.

Zasada, I. A., Walters, T. W., Pinkerton, J. N., 2010. Post-plant nematicides for the control of root lesion nematode in red raspberry. HortTechnology, 20(5):856-862.

Zasada, Inga A., Moore Patrick P., 2014. Host Status of Rubus Species and Hybrids for the Root Lesion Nematode, Pratylenchus penetrans." HortScience, 49 (9):1128-1131

15

STUDY OF THE INFLUENCE OF AGING IN DIFFERENT BARRELS ON SHIRAZ WINES

Mihaela ŞERBULEA[1], Arina Oana ANTOCE[1]

[1]University of Agronomic Sciences and Veterinary Medicine of Bucharest,
Faculty of Horticulture, Department of Bioengineering of Horti-Viticultural Systems,
59, Mărăşti Blvd., District 1, 011464 Bucharest, Romania
Corresponding author email: aantoce@yahoo.com

Abstract

The Syrah/Shiraz grape variety is rather newly introduced in Romania and the producers are still working on establishing optimum winemaking technologies for it. Aging in oak barrels is an important technological step for this high tannic wine, therefore, the selection of the appropriate barrel to improve the structure, colour and flavour of this wine is of great importance. This study aims to compare the influences on the colour and sensory parameters induced on Shiraz wine by aging it for 1 year in barrels made of oak with various origins (French, Romanian, Russian and Hungarian oak) and various toasting levels. Wine kept in stainless steel tank was also used as control. It was observed that the French oak tends to differ compared to Russian and Hungarian oak as regards the colour differences induced in the wines, while the influence of Romanian oak is rather similar to that of French oak. Concerning the sensory quality of the wine aged in barrels, preliminary results point out that barrels from the provider "Transilvania bois" (Romanian and Russian oak) offer constant sensory quality, but some French oak barrels may ensure outstanding sensory quality. Regarding the toasting degree, for Shiraz wines the medium plus toasting offered the best results for structure and aromatic profile after 1 year of aging. These findings need to be confirmed with more precise aromatic profile analyses.

Key words: oak barrel, Shiraz, flavor, maturation, wine.

INTRODUCTION

Aging wine in oak barrels is a well known technique used to improve the quality of red wine by reducing the tannin astringency, increase the intensity of colour through tannin-anthocyians condensation and to improve the structure and taste under the influence of the substances extracted from the wood. The earliest literature on the use of oak containers for wine can be traced back to the Roman Empire (Zhang *et al.*, 2015).

The topic is important for many wineries, therefore plenty of studies have been performed in order to determine the various effects that wood contact and aging in oak barrels induce in the wines. However, selecting the oak type and toasting degree for the oak to match the structure and complexity of a certain varietal wine is a difficult task, and not many studies are found on this subject.

The present study focuses on several oak barrel types destined for the aging of red wine from

the Shiraz grape variety grown in Dealu Mare region of Romania.

The barrels selected for comparison are produced from wood of various origins (French, Russian, Hungarian and Romanian), with various levels of toasting (heavy, medium plus and medium).

The oak wood used in winemaking is mainly from two sources: American oak (Quercus alba - the white oak) and French oak (Quercus robur – the pedunculate oak or sometimes the less common Quercus petraea - sessile oak). American oak induces intense vanilla and coconut flavours, while the French oak induces a more subtle favour. The Eastern European Oak is actually Quercus robur too, as the French oak, but the flavour it imparts is considered in-between American and French Oak, due to a higher content of volatile phenols and phenolic aldehydes than the French oak, even though they belong to the same species (De Simon *et al.*, 2003).

Studies dating back as far as 1974 (Singleton, 1974, Singleton 1995) showed that the chemical composition of oak with various types is quantitatively different, but more recent studies are also available (Alanon *et al.*, 2011). Beside the botanical species, the geographic origin is also important regarding the proportions of various compounds found in the wood (Prida and Puech, 2006; Guchu *et al.*, 2006).

The aging of wines in oak wood barrels leads to the extraction of numerous compounds that have an impact on wine colour, astringency and bitterness, either directly or indirectly. The ellagitannins extracted from wood and the presence of oxygen have a major impact on taste and colour through the process of co-pigmentation of anthocyanins and tannins during red wine aging (Asen *et al*, 1972; Mateus *et al.*, 2001, 2002).

New types of compounds, such as oaklins, have been reported to contribute to changes in the colour and astringency of wines during aging (Sousa *et al.,* 2005).

The grape variety of the wine kept in barrels is also important, as the wine composition too influences the way in which the wine is aging. In this study Shiraz was selected as the grape variety, a rather rare variety for Romania, but for which the interest has been growing steadily during past 10 years.

The wines of Shiraz have a good structure, being quite tannic and requiring aging.

For these reasons, the aim of this work was to study the influences of oak barrels on the Shiraz wine sensory and colour characteristics, in order to determine the most appropriate type of wood in accordance to its origin.

MATERIALS AND METHODS

Raw material

To carry out the study regarding the influence of barrels on the aging of red wines, different types of barrels and Shiraz wines were used.

The Shiraz wine was produced in 2013, from grapes harvested in Dealu Mare Vineyard. The grapes were harvested in plastic boxes of 10 kg each, on October 3, 2013, when they reached the sugar concentration of 260 g/l, which corresponded too with the phenolic maturation phase and a moderate level of total acidity (5.48 g/l tartaric acid). Grapes harvested in this way allowed for both quality and high yield in

the resulted wine. Shiraz is a grape variety that does not benefit from over-maturation, not only because it does not accumulate more sugar as in the case of other varieties, but also loses aroma and acidity.

Winemaking process

The harvested grapes were brought directly to the wine cellar and crushed within maximum four hours after harvesting. For antioxidant protection a dose of 40 mg/l of sulfur dioxide (SO_2) was applied during the crushing process. The winemaking was based on classic red wine maceration fermentation process.

To accomplish the maceration-fermentation phase, 8000 kg of the crushed grapes were introduced in a 10,000 l stainless steel tank. To extract more color from the skins, 1 g/100 kg of enzyme Lallzyme EX-V was added during maceration. The fermentation was started by inoculation of 20 g/hl 15 RP selected yeast from Lallemand. A dose of 20 g/hl OptiRed (yeast membranes) was also added (from Lallemand products catalog).

Maceration-fermentation for this red wine took place at a temperature of 25-29°C, for 15 days. During all this period the homogenisation of the crushed berries was performed 3-4 times a day, by punching down the cap formed at the upper part of the tank.

After the alcoholic fermentation the wine was separated from the solid parts. Malolactic fermentation (MLF) began at the end of alcoholic fermentation, when the wine was still warm (18-20°C) and yeast population was decreasing. To facilitate MLF the wine was slightly aerated and inoculated with 0.5 g/hL malolactic bacteria. The MLF was slow and accomplished in 20-40 days. In the end the wine was again slightly aerated. After the completion of MLF the wine was separated from the lees by racking and then sent to maturation in oak barrels of 225 l volume.

The wine cellar where winemaking was performed is built on the principle of gravity, which prevents excessive pumping and oxidation, allowing for the production of wines with increased quality.

Types of barrels

For the wine maturation different types of barrels were used, obtained from various manufacturers. The oak used for barrels was of different origins and processed to have different toasting levels. Furthermore, the barrels were either new or in their second or third year of use. For this experiment, the types of barrels available were as follows: barrels of Russian and Romanian oak from the Romanian producer Transilvania Bois; barrels of French oak from the French producers – Radoux, Boutes and Francois Freres; barrels of Hungarian oak from a Hungarian producer - Trust Hungary. Toasting levels varied from medium, medium plus to heavy toast.

Analyses

The main grape and wine parameters were determined in accordance to the usual methods: the sugar content by refractometry, total acidity by automatic titration by using a Mettler Toledo titrator, alcohol concentration by ebuliometry and sulphur dioxide by Ripper method.

The evolution of wines from different barrels was monitored every 3 months over a period of 1 year, by tasting with an in-house panel of wine experts. Organoleptic analyses of the wines were done by examining and recording the visual, smell and taste traits on a special evaluation sheet. An OIV score sheet for the evaluation in international contests was used to score the wine on a scale of 100 points (OIV, 2008).

The determination of total polyphenol was performed with an UV-VIS – spectrophotometer, by diluting the wine 100 times and measuring the absorbance at 280 nm in cuvettes of 1 mm path length. The value of the absorbance measured at 280 nm was reported after multiplying it with 100.

The determination of color intensity, hue and chromatic characteristics of wine was also performed with an UV-VIS – spectrophotometer. The absorbance of wine was measured at three wavelengths, 620, 520, 420 nm, representative for the description of wine colour. Young red wines have a maximum absorbance at 520 nm, while for aged wines color absorbance shifts, their maximum getting closer to 420 nm.

The reference method for reporting the color parameters was a spectrophotometric method by which tristimulus values and chromaticity coordinates X, Y, Z, are determined according to the standards of International Commission on Illumination (CIE, 2015).

RESULTS AND DISCUSSIONS

The wines of Shiraz were introduced in barrels and aged for 1 year and then their parameters determined. As control, the Shiraz wine aged in stainless steel tank was used.

In Table 1 the main wine parameters are included. It can be seen that the barrel, irrespective of type of oak or year of production (and usage) does not influence much the main wine parameters. The only noticeable exception is the volatile acidity, which is clearly higher in all barrels (around 1 g/l acetic acid) as compared to the volatile acidity determined from wine kept in the stainless steel tank (0.75 g/l acetic acid).

Table 1. Main physico-chemical parameters of Shiraz wines aged in barrels and in stainless steel tank

Barrel type	Free SO₂ mg/l	Total SO₂ mg/l	pH	Alcohol % v./v.	Total acidity g/l tartaric acid	Volatile acidity g/l acetic acid
Control (Stainless steel tank)	53	54	3.68	14.8	5.05	**0.75**
Radoux R TGS H TH – French oak	48	93	3.51	15.5	5.78	0.99
Radoux R TGS M TH – French oak	32	43	3.52	15.4	5.63	0.97
Radoux R TGS Rev TH – French oak	39	74	3.53	15.4	5.93	0.98
Radoux Evol R TGS M+ TH – French oak	31	81	3.52	15.3	5.63	1.08
Francois Freres 2011 TG M TH – French oak	27	65	3.53	15.5	5.66	0.95
Boutes Tradition M – French oak	49	71	3.53	15.3	5.73	1.00
Transilvania bois - Romanian oak	39	55	3.52	15.4	5.76	0.98
Transilvania bois 2013 M – Russian oak	23	52	3.53	15.4	5.78	0.98
Transilvania bois M+ 2013 – Russian oak	30	43	3.52	15.3	5.78	0.99
Transilvania bois M 2011 – Russian oak	33	74	3.53	15.3	5.70	0.92
Transilvania bois M+ 2011 – Russian oak	39	77	3.53	15.3	5.70	0.92
Trust M+ – Hungarian oak	26	61	3.53	15.3	5.76	0.97

The organoleptic analyses revealed that the barrels induced significant differences in the aspect, smell and taste of the wines, and also that these characteristics evolved during the maturation.

In Table 2 the sensory evolution of wines during their maturation is presented, while Table 3 presents the sensory descriptions and scores obtained by the wines evaluated after 1 year of aging in barrels.

Table 2. Average scores awarded to Shiraz wines matured in barrels,
evaluated at various time intervals during 4 wine tasting sessions

Barrel type and year of usage	Score March 2014	Score July 2014	Score October 2014	Score March 2015
FrFreres M - II	79.0	81.8	76.8	78.0
Trans B Rus M - I	82.2 Silver	80.0	84.2 Silver	78.0
Radoux H - I	78.5	81.0	82.2 Silver	82.0 Silver
Trust Hung M+ - III	76.6	79.4	76.2	83.0 Silver
Trans B Rus M - II	76.6	79.6	78.6	83.0 Silver
Trans B Rus M+ - I	85.0 Gold	80.2	81.4	83.0 Silver
Trans B Rom M -III	76.6	81.4	76.6	84.0 Silver
Radoux Rev M - I	81.8	82.4 Silver	77.4	84.0 Silver
Radoux Evol M+ - I	82.8 Silver	78.6	76	85.0 Gold
Radoux M - I	77.6	82.8 Silver	81.8	86.0 Gold
Trans B Rus M+ - I	76.6	77.8	77.6	86.0 Gold
Boutes Trad M - I	76.4	73.2	78.0	86.0 Gold

As it can be seen in Table 2, most of the wines aged in new barrels (first year of usage) were highly appreciated by the winetasters after 1 year of maturation. Among the them, those aged in Transilvania Bois – Russian oak, Radoux (both classic and evolution style of toasting) – French oak and Boutes – French oak received top marks, irrespective of the toasting level.

The scores for all the wines ranged from 78-86, which, in accordance to the OIV rules for wine contests signify: good wines (78-81.9 points), silver medal wines (82-84.9), gold medal wines (85-91.9), and great gold medal wines (92-100). Generally, keeping the wines in contact with oak for 1 year improved the perceived quality, but in some cases the quality was not much improved, remaining around 78-82 during all the evaluated period. The Romanian oak only improved the wine up to a silver medal level, but this should be interpreted with caution as the barrel employed was in its third year of usage. It can be noticed that the greatest

improvement is generally achieved in new barrels, followed by barrels used in their second and third year.

Aside of the effect on wines' aroma, the barrels are expected to influence the polyphenolic composition of wines (Table 4).

The total polyphenol index of the wine stored in barrels for a year was, on average 37.95 ± 3.46, with only one wine exceeding the range 35.5-38.0 and reaching TPI = 48.57.

This higher polyphenol wine was kept in a French oak new barrel from Radoux, obtained with a special toasting technique called "Evolution" (Radoux, 2009), in which the maximum toasting temperature was lowered compared to the classical toasting process, while the length of the toasting operation extended through a "re-cooking" phase.

Colour has also undergone some changes during barrel aging. The samples were analyzed after one year of storage in tank or in barrel and the CIELab colour parameters determined are included in Table 4.

Table 3. Wine sensory description and average scores awarded to Shiraz wines matured in barrels after 1 year of aging

Type barrel	Score	Description
FrFreres M - II	78	The medium toast induces less structure and complexity in the taste, a volatile acidity covering the wine fragrance
Trans B Rus M - I	78	Intensely colored, pigments remaining on the tasting glass surfaces, due to a perceptible volatile acidity, which, through acetaldehyde favors copolymerisation
Radoux H - I	82	Intensely colored, with a dominant aroma of vanilla, but still with harsh tannins which dry the mouth; some sweetness is also perceivable
Trust Hung M+ - III	83	Fruity aroma of cherry and sour cherry, quince and fig jam, with a slight hint of volatile acidity, still aggressive tannins
Trans B Rus M - II	83	Powerful, astringent, with aggressive tannins in taste, but with a well balanced acidity
Trans B Rus M+ - I	83	Floral aroma and vanilla notes, still rough, but long, with a burning aftertaste due to alcohol and tannins
Trans B Rom M -III	84	Fine olfactory quality, well integrated medium toast oak and long taste
Radoux Rev M - I	84	More classic style wine, giving the feeling of a shorter term contact with wood, with notes of vanilla and spices, complex in taste, yet harsh, but ready for the market
Radoux Evol M+ - I	85	Long, lingering, peppery aftertaste with persistent, complex flavour of vanilla, tobacco, toast, coffee and caramel
Radoux M - I	86	Fruity aroma of cherries and bitter- sweet cherries, vanilla and truffles notes, long, complex, drinkable, but still a bit harsh in aftertaste
Trans B Rus M+ - I	86	Very well balanced wine, round, lingering, complex, elegant, with notes of over-ripen fruits and spices
Boutes Trad M - I	86	Elegant wine, intense and round in the same time, well balanced acidity and bitterness, well integrated oak flavor

Table 4. Total polyphenolic index and CIE*Lab* parameters of Shiraz wines aged in barrels and in stainless steel tank

Barrel type	TPI	a	b	c	h	L	ΔE*
Control (Stainless steel tank)	46.14	49.38	6.11	49.76	0.12	52.86	0
Radoux R TGS H TH – French oak	36.51	49.56	15.78	52.01	0.31	46.07	11.82
Radoux R TGS M TH – French oak	37.86	54.22	11.87	55.50	0.22	41.39	13.72
Radoux R TGS Rev TH – French oak	37.91	51.35	13.12	53.00	0.25	44.40	11.16
Radoux Evol R TGS M+ TH – French oak	48.57	52.57	13.89	54.37	0.26	40.78	14.72
Francois Freres 2011 TG M TH – French oak	36.91	55.56	13.22	57.11	0.23	40.66	15.41
Boutes Tradition M – French oak	37.86	48.78	11.83	50.20	0.24	50.68	6.15
Transilvania bois - Romanian oak	35.54	52.67	11.06	53.82	0.21	47.23	8.19
Transilvania bois 2013 M – Russian oak	36.28	51.72	14.13	53.62	0.27	34.87	19.83
Transilvania bois M+ 2013 – Russian oak	37.28	54.63	12.11	55.96	0.22	39.58	15.49
Transilvania bois M 2011 – Russian oak	37.76	54.85	13.85	56.57	0.25	42.07	14.36
Transilvania bois M+ 2011 – Russian oak	37.43	52.95	10.86	54.05	0.21	46.92	8.41
Trust M+ – Hungarian oak	35.49	54.55	15.03	56.58	0.27	40.83	15.85

* ΔE is the colour difference calculated against the colour of wine in tank

All the wines have an intensely red colour, with some differences in the colour tone. Some samples aged in barrels, in accordance to parameter a, have their colour shifted on thee green-red axis toward more red tones (a=50-55), as compared to the wine stored in stainless steel tank (a=49.4). Not all the wines kept in barrels evolved toward a higher value of parameter a, showing that the type of barrel and the amount of oxygen which enters through it is important, although no clear tendency regarding the origin of oak or type of toasting was identified. As regards the parameter b, the position on the blue-yellow axis, this one was clearly influenced by the barrel. The values of parameter b are more or less double (b=10.9-15.8) for the wines in barrels than the value of the wine kept in tank (b=6.1), clearly showing a shift toward yellow components, which means that they lost their blue-violet tones and acquired more yellow-brown tones, typical for wine oxidative maturation. This tendency is easier to observe when we place the samples in the *ab* colour space (Fig. 1).

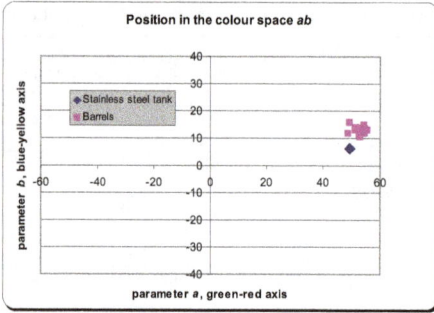

Figure 1. The position of wine the tank (blue) and in barrels (magenta) in the *ab* colour space

As expected, the evolution of wine in barrels is faster than in tank, the hue increasing in all the wines kept in barrels, irrespective of the oak origin and type of toasting (Fig. 2).

However, although for the separate colour parameters changes related to a certain type of oak could no be clearly demonstrated, the overall colour difference, ΔE, which takes into account the variations of a, b and L parameters ($\Delta E = ((\Delta L^*)^2 + (\Delta a^*)^2 + (\Delta b^*)^2)^{(1/2)}$) seems to show that some types of oak induce more colour changes than others (Fig. 3).

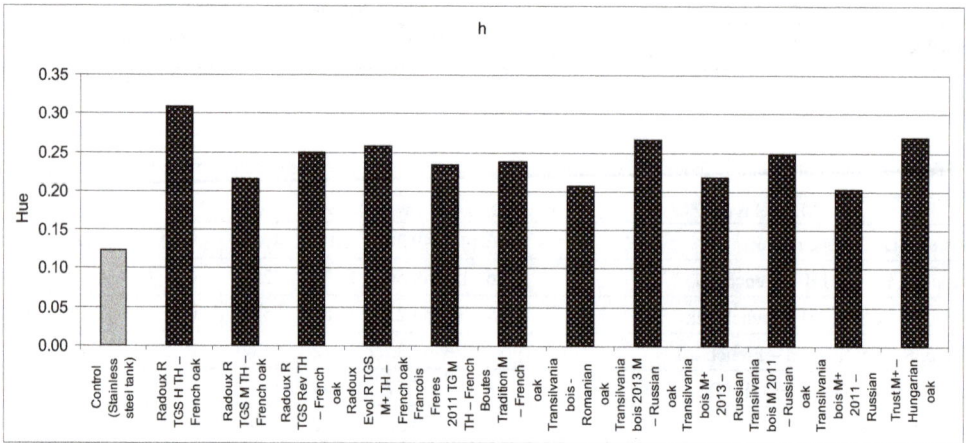

Figure 2. The wine hue in the tank (left sample) and in barrels

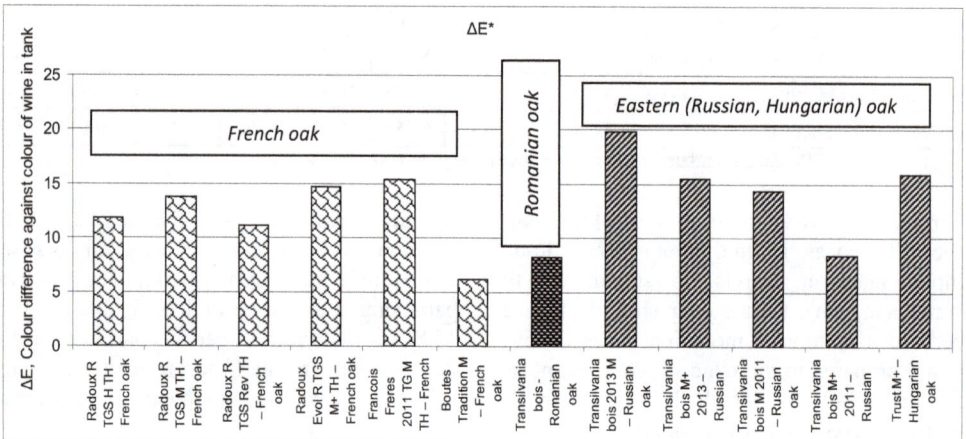

Figure 3. The difference in color ΔE wines depending on the type of barrel compared with the control sample kept in tank (groups of barrels made of French oak, Romanian Oak and Russian or Hungarian oak were outlined with different patterns)

It was observed that Eastern oak (Russian and Hungarian oak) tends to induce more colour differences in the wines (Fig 3 – the right group, average $\Delta E=14.8\pm1.8$) as compared to French oak (Fig 3 – the left group, average $\Delta E=12.2\pm1.4$). In this respect, the Romanian oak (Fig. 3, the middle sample, $\Delta E=8.19$) tends to behave more like the French oak, although, by applying statistical analysis for the three groups, due to the small number of barrels analyzed and the inherent barrel-to-barrel variation (Towey and Waterhouse, 1996), no statistically significant difference was found at the 0.05 level of certainty. Moreover, it must be reminded that the barrel made of Romanian oak is in its 3rd year of usage and that the colour is not only influenced by the porosity and oxygen allowed to dissolve in wine, but also by the tannins transferred by the barrel to wine and their condensation with the existing polyphenols of wine. However, also in the third year of usage are the barrel Boutes Tradition-French oak ($\Delta E=6.68$) and barrel Trust M+ – Hungarian oak ($\Delta E=15.85$), showing that the Romanian oak behaves rather like the French oak than the others, suggesting that the Eastern oak is allowing more oxygen in the wine than the French oak.

CONCLUSIONS

In order to determine the evolution of Shiraz wine in barrels and select the most appropriate barrels for the wine style the simplest analyses to perform are colour determination by CIE*Lab* method and sensory analysis.

Some colour differences were induced by aging in barrels, and based on these changes we observed that the French oak effects tend to differ compared to Russian and Hungarian oak, while the Romanian oak places itself closer to the French oak in this respect.

Transilvanian bois barrels (Romanian and Russian oak) offer constant sensory quality (silver and gold medal scores), but some French oak barrels led to outstanding sensory qualities in wines (gold medal scores).

Among the tested barrels, the Radoux Evolution barrel, produced by a special toasting technique, stood out by producing the wine with highest total polyphenol index, one of the highest colour differences compared to the control wine and the gold medal score obtained in sensory evaluation.

Based on the studied parameters, no clear correlations were found regarding the colour of the final wine and the toasting degree and the barrel year of usage. Apparently, the medium plus toasting degree is more suitable for the structure and aromatic profile of Shiraz wine, thus they were described as being more complex and round in the same time, with the classical notes of coffee and spices. However, more precise analyses are needed to draw accurate conclusions regarding the changes in aromatic profile induced by certain types of barrels or toasting levels.

REFERENCES

Alañón M.E., Castro-Vázquez L., Díaz-Maroto M. C., Hermosín-Gutiérrez I., Gordon M. H., Pérez-Coello M. S., 2011. Antioxidant capacity and phenolic composition of different woods used in cooperage. *Food Chem.*, 129, 1584–1590.

Asen S., Stewart R.N. Norris K.H., 1972, Co-pigmentation of anthocyanins in plant tissue and its effect on colour, *Phytochemistry*, 11, 1139–1144.

Bo Zhang, Jian Cai, Chang-Qing Duan, Malcolm J. Reeves, Fei He, 2015, A Review of Polyphenolics in Oak Woods, *Int. J. Mol. Sci.*, 16, 6978-7014.

De Simón B.F., Hernández T., Cadahía E., Dueñas M., Estrella I., 2003. Phenolic compounds in a Spanish red wine aged in barrels made of Spanish, French and American oak wood. *Eur. Food Res. Technol.* 216, 150–156.

Guchu E., Díaz-Maroto M.C., Díaz-Maroto I.J., Vila-Lameiro P., Pérez-Coello M.S., 2006. Influence of the species and geographical location on volatile composition of Spanish oak wood (Quercus petraea Liebl. and Quercus robur L.), *J. Agric. Food Chem.*, 54, 3062–3066.

Mateus N., Silva A.M.S., Santos-Buelga C., Rivas-Gonzalo J.C. and Freitas V., 2002. Identification of anthocyanin-flavanol pigments in red wines by NMR and mass spectrometry. *J. Agric. Food Chem.*, 50, 2110–2116.

Mateus N., Silva A.M.S., Vercauteren J., Freitas V., 2001. Occurrence of anthocyanin-derived pigments in red wines, *J. Agric. Food Chem.*, 49, 4836–4840.

Prida A., Puech J. L., 2006. Influence of geographical origin and botanical species on the content of extractives in American, French, and East European oak woods. *J. Agric. Food Chem.*, 54, 8115–8126.

Singleton V.L., 1995. Maturation of wines and spirits: Comparisons, facts, and hypotheses, *Am. I. Enol. Vitic.*, 46, 98-115.

Singleton V.L., 1974. Some aspects of wooden container as a factor in wine maturation. In Chemistry of Winemaking. A.D. Webb (Ed.) American Chemical society, Washington D. C.

Sousa C., Mateus N., Perez-Alonso J., Santos-Buelga C., De Freitas V., 2005. Preliminary study of oaklins, a new class of brick-red catechinpyrylium pigments from the reaction between catechin and wood aldehydes. *J. Agric. Food Chem.*, 53, 9249–9256.

Towey J.P., Waterhouse A.L., 1996, Barrel-to-barrel variation of volatile oak extractives in barrel-fermented chardonnay. *Am. J. Enol. Vitic.*, 47, 17–20.

***2009. Radoux, Summary of technical documents from the Radoux Cooperage research and Development department, www.tonnellerieradoux.com/pdf/2009_TRTechnicalDocument.pdf, accessed February 2016.

***2015. Lallemand products catalog, http://www.lallemandwine.com/ products/catalogue/

***2015. International Commission on Illumination,. Standards, http://www.cie.co.at/index.php/Publications/Standards), accessed in 2015.

***2009. Resolution OIV/Concours 332a/2009. OIV standard for international wine and spirituous beverages of vitivinicultural origin competitions, p. 15.

16

THE QUANTITY AND QUALITY OF GRAPES OF 'PREZENTABIL' TABLE GRAPES VARIETY BY THE INFLUENCE OF BIOLOGICALLY ACTIVE SUBSTANCES

Gheorghe NICOLAESCU[1], Antonina DERENDOVSKAIA[1], Silvia SECRIERU[1], Dumitru MIHOV[2], Valeria PROCOPENCO[1], Mariana GODOROJA[1], Cornelia LUNGU[3]

[1]State Agrarian University of Moldova, 44 Mircesti str., MD-2049, Chişinău, Moldova
[2]Tera Vitis Ltd., Burlacu vil., Cahul dis., Moldova,
E-mail: dmytrii@gmail.com
[3]Dionysos Mereni Joint-stock Company, 44 Mircesti str., MD-2049, Chişinău, Moldova,
E-mail: cornelusha10@yahoo.com
Corresponding author email: gh.nicolaescu@gmail.com

Abstract

The table grape production is an effective branch in Moldova. Increasing the quality and quantity of table grape is a necessity for society. The purpose of the research from this article is to study the influence of Gobbi Gib 2LG on the quantity and quality of grapes of 'Prezentabil' variety. The research was conducted in the vineyards of the „Terra Vitis" LTd, from Southern wine region in Moldova. Research results have shown that the dose of 0,98 l / ha is most useful for conditions south of Moldova, for 'Prezentabil' variety.

***Key words**: Prezentabil, table grape, growth stimulators.*

INTRODUCTION

There are a lot of agricultural branches, including viticulture, that use a new efficient proceeding – which is the use of growth regulators or biological active substances.

All growth regulators (natural and synthetic ones) are organic substances which, if used in low concentrations, are able to cause essential modifications within the growth and development processes of the plant body and incite their regulation. A specific particularity of the regulators action is their capacity to influence on the processes that are not responding on the influence of normal agricultural proceeding (Winkler A.J., 1966 (Уинклер А.Дж., 1966); Wear R.J.,1976; Winkler A.J., 1997; Smirnov K.V. and others, 1987 (Смирнов К.В. и др., 1987).

The use of gibberellins within table grape variety technology in most of the countries around the world (Japan, USA, Russia, Italy, Ukraine, Bulgaria etc.) is an obligatory agricultural process. Treating the inflorescences (within the blooming period, within the postfecundation period) leads to considerable modifications of the morphological and mechanical bunch particularities and to productivity increase as well as grain quality modifications.

MATERIALS AND METHODS

The purpose of the research was to study the influence of the Gobbi Gib 2LG, produced by „L Gobbi" Ltd., Italy on the table grapes varieties productivity.

To achieve the final purpose it was necessary to track down the following objectives:
- the action of Gobbi Gib 2LG on the table grapes variety Presentabil, on its berry morphological parameters and mechanical properties;
- the action of Gobbi Gib 2LG on the productivity and quality of the grapes;
- finding out the optimal concentration of the Gobbi Gib 2LG which has a more efficient action within the table grapes seedless varieties;

The research in the field of studying the action of Gobbi Gib 2LG as growth regulator needed to increase the productivity and quality of the grapes was effectuated by "Terra-Vitis" Ltd. located in the Cahul district.

As the object of study it was taken the Presentabil a table grapes variety grafted on the Berlandieri x Riparia SO4 rootstock.

The GG2LG was used by means of treating the vines within different stages of its development: the technology used in Italy (3.6; 4.6 1/ha) – on 8 cm shoots length; one week before the blossom; while blossoming 30% of the bloom; while blossoming 50% of the bloom; while blossoming 80% of the bloom; the treatment of Ø 3-4 mm grains; 8-10 days after the last treating; the technology suggested for Moldova was the treatment of Ø 3-6 mm grains (2,0 and 2,4 1/ha).

RESULTS AND DISCUSSIONS

The reaction of the Prezentabil table grapes variety with seeds to the Gobbi Gib 2LG treatment.

Within the control variant the average bunch weight is 503,3; the weight of the berries in the bunch – 494,4; the weight of the cluster 8,9g. The bunch characteristic value (berries weight/cluster weight) - 55,6. The bunch is big, conically-shaped, dense.

The number of berries in the bunch – 167,0 pcs, including- 21 psc. of the undeveloped berries. The berries have an oblong shape with a length of - 24,1; width - 14,7 mm. A 100 berries weight – 395,8 g., The berry characteristic value (pulp weight/berries skin) – 10,1. As a rule, one big seed forms in each berry. The characteristic value of the seeds (pulp weight / seed weight) is rather high and reaches 71,28.

Within Prezentabil the mechanical properties of the berries, especially the crushing strength, is higher than seedless varieties and comes up to 2476 g/cm^2.

The harvest constitutes 3,8 kg/vine. There is s high sugar content in the juice of the berries – 22,6 %, titratable acidity content – 8,9 g/dm^3.

The usage of Gobbi Gib 2LG following the Italian technology.

While this type of treating Gobbi Gib 2LG was used in the postfecundary period (treating Ø grains of 3-4 mm -13.06.2013), (treating after 8-10 days before the last one - 23.06.2013) the bunch weight increased up to 16,4% (GG2LG-0,65 1/ha) and 27,2% (GG2LG-0,82 1/ha); the weight of the berries in the bunch accordingly up to 16,1 and 26,7 %. The cluster weight increased up to 1,4-1,6 times which leaded to a bunch characteristic value decrease (Table 1, Figure 1).

Under the influence of Gobbi Gib 2LG the number of berries in the bunch increased up to 1,2-1,3 times, but at the same time the quantity of undeveloped berries almost double if compared to the control. The dimensions of the berries, a 100 berries weight is the same as the one of the control, or a bit lower, as far the increase of the number of berries in the bunch brought to a decrease of their dimensions. The characteristic value of the berries is as that of the control or a bit higher.

The crushing strength of the berries is the same as that of the control. The harvest increases up to 1,2 (GG2LG-0,65 1/ha) -1,3 (GG2LG-0,82 1/ha) times. The sugar content is the same as that of the control, or a little higher. At the same time the concentration of the titratable acidity in the berries decreases.

It is important to notice that the biggest differences while using the specimen, if compared to the Witness, were observed in the dose of GG2LG-0,82 1/ha. Within this variant rise the bunch weight and parameters as well as the weight of grains in the bunch which leads to an increase of harvest vine up to 1,3 times. A small increase of seeds characteristic value takes place.

The usage of Gobbi Gib 2LG following the Moldova technology.

Using Gobbi Gib 2 LG within the post fecundation period (treating Ø grains of *3-6 mm* -13.06.2013) leads to the increase of bunch weight and the weight of the grains in the bunch up to 1,3 times, whatever the concentration of the specimen. It should be mentioned that took place the increase of the bunch, its length and width, especially in the middle part of it.

Figure 1. The Gobbi Gib 2LG influence on the external appearance of the bunch and berries.
The Prezentabil variety , "Terra vitis" Ltd., 2013, (Italian technology). The variant of experiment:
1-Control – H_2O; 2-GG2LG-0,65 l/ha; 3-GG2LG-0,82 l/ha

Table 1. The reaction of the Prezentabil variety to the Gobbi Gib 2LG treatment within postfecundary period.
„Terra vitis" Ltd., 2013, (Italian technology)

Value	The variant of experience					DL 0,95
	Control-H_2O	GG2LG-0,65 l/ha		GG2LG-0,82 l/ha		
	$\overline{x}$	$\overline{x}$	% to the control	$\overline{x}$	% to the control	
Bunch weight, g	503,3	586,0	116,4	640,0	127,2	
including - berry	494,4	573,9	116,1	626,2	126,7	
- cluster	8,9	12,1	136,0	13,8	155,1	
Bunch characteristic value (berries weight / cluster weight)	55,6	47,4	-	45,4	-	
Bunch dimensions, cm						
- length	22,0	25,0	113,6	28,0	127,3	
- width / on the top	16,0	20,0	125,0	23,0	143,8	
at the middle	11,0	12,0	109,1	15,0	136,4	
at the bottom	6,5	6,3	98,9	7,0	107,7	
Pedicle dimensions, mm	3,4	6,8	200,0	6,0	176,5	
The quantity of berries per bunch, pcs, total	167,0	208,0	124,6	214,0	128,1	
including undeveloped berries	21,0	43,3	-	47,0	-	
Berry size, mm						
- length	24,1	22,3	92,5	23,6	97,9	
- width	14,7	15,5	105,4	16,3	110,9	
100 berries' weight, g	395,8	368,4	93,1	403,2	101,9	
Berry characteristic value (pulp weight/ skin weight)	10,1	12,3	-	10,3	-	
Seeds characteristic value (pulp weight/ seeds weight)	71,3	73,3	-	77,5	-	
Crushing strength of berries, g/cm^2	2476	2146	86,7	2516	101,6	
Harvest, kg/vine	3,8	4,5	118,4	4,9	129,0	0,49
The content of:						
- sugar, %	22,6	22,0	-	23,3	-	
- titratable acidity, g/dm^3	8,9	8,5	-	8,3	-	

Under the influence of Gobbi Gib 2LG the number of grains in the bunch increased up to 1,5 (*GG2LG-0,98l/ha*) -1,2 (*GG2LG-1,3l/ha*) times, and at the same time grain dimensions, in most of the cases, are not bigger than those of the Witness. A 100 grains weight is the same as that of the Witness or a little lower.

The characteristic value of the grain (pulp weight/skin weight) rises up to 1,1-1,2 times;

the characteristic value of the seminal index (pulp weight/seeds weight) is the same as that of the Witness or a little higher (Table 2, Figure 2).

Under the influence of Gobbi Gib 2 LG the harvest increases up to 1,3 times. The sugar content in the grains increases and the the concentration of the titratable acidity decreases.

Table 2. The reaction of the Prezentabil variety to the Gobbi Gib 2LG treatment within postfecundary period. „Terra vitis" Ltd.,2013 (Moldova technology)

Value	The variant of experiment					DL 0,95
	Control -H$_2$O	GG2LG-0,98l/ha		GG2LG-1,3l/ha		
	$\overline{x}$	$\overline{x}$	% to the control	$\overline{x}$	% to the control	
Bunch weight, g	503,3	649,0	129,0	651,5	129,4	
including - berry	494,4	637,9	129,0	640,5	129,6	
- cluster	8,9	11,1	124,7	11,0	123,6	
Bunch characteristic value (berries weight / cluster weight)	55,6	57,5	-	58,2	-	
Bunch dimensions, cm						
- length	22,0	23,3	105,9	22,7	103,2	
- width / on the top	16,0	21,3	133,1	19,5	121,9	
at the middle	11,0	14,7	133,6	11,8	107,3	
at the bottom	6,5	7,0	107,7	7,5	115,4	
Pedicle dimensions, mm	3,4	4,0	117,6	6,6	194,1	
The quantity of berries per bunch, pcs, total	167,0	255,0	152,7	196,5	117,7	
including undeveloped berries	21,0	50,0	-	18,0	-	
Berry size, mm						
- length	24,1	23,5	97,5	24,0	99,6	
- width	14,7	16,0	108,8	16,0	108,8	
100 berries' weight, g	395,8	376,2	95,1	404,6	102,2	
Berry characteristic value (pulp weight/ skin weight)	10,1	11,5	-	12,1	-	
Seeds characteristic value (pulp weight/ seeds weight)	71,3	105,2	-	83,9	-	
Crushing strength of berries, g/cm^2	2476	2445	98,7	2442	98,6	
Harvest, kg/vine	3,8	4,9	129,0	5,0	131,6	0,49
The content of:						
- sugar, %	22,6	23,6	-	21,6	-	
- titratable acidity, g/dm^3	8,9	8,6	-	8,4	-	

Figure 2. The Gobbi Gib 2LG influence on the external appearance of the bunch and berries.
The Prezentabil variety, "Terra vitis" Ltd., 2013, (Moldova technology).
The variant of experiment:1-Control – H$_2$O;2-GG2LG-0,98l/ha;3-GG2LG-1,3l/ha

The results of testing *Gobbi Gib 2LG* within the *Prezentabil* variety with seeds show that its action on the vines depends on the soil's biological particularities, treatment concentration and the duration of its usage. Taking into consideration the characteristic values sum (bunch weight and its parameters, the quality and weight of the grains in the bunch) it is necessary to note the variant Gobbi Gib 2 LG - 0,98 l/ha. While this very variant takes place an essential increase of the seed characteristic value up to 1.5 times which shows the rise of the seedless degree of grains. Using Gobbi Gib 2 LG within the postfecundation period for 3-6 mm diameter grains leads to a harvest increase up to 1,3 times, rise of the seedless berries quantity which allows sugar quantity growth in the juice of the grain and the early ripening. The last one is very important, especially for early maturation sorts.

CONCLUSIONS

In the issue of the received data after using *Gobbi Gib 2LG* within the vitis it may be said that its action depends on the biological particularities of the sorts notwithstanding the method of using it. Within the Prezentabil variety the efficiency of the specimen showed through:

1. The rise of the bunch, grains, cluster parameters; also modifies bunch characteristic value (grain weight/cluster weight);

2. Rises the quantity of grans in the bunch up to 1,2-1,7 times, at the same time grain parameters decrease (length, width);

3. The harvest rises up to 1,2-1,3 times if compared to the Witness. Grows the seedless grains quantity; rises the seminal characteristic value which heightens sugar accumulation in the grain ant fastens the maturity.

Taking into consideration the obtained results on Gobbi Gib 2LG one can confirm that the specimen may be included into the table grape sorts levelling technological system as growth regulator aiming to increase the productivity and the quality of the production based on the 2 schedules:

Within the varieties with seeds (Prezentabil):

I schedule (Italian method), through vine spraying, using the specimen within periods:

✓ post fecundation period

II schedule (Moldova method), through spraying the zone of the bunch placement using the specimen only within one single period:

✓ post fecundation period

REFERENCES

Nicolaescu, Gh., Cazac, F., 2012. Producerea strugurilor de masă soiuri cu bobul roze şi negru (ghid practic) / Gheorghe, Nicolaescu, Fiodor, Cazac. Chişinău. Elan Poligraf. p. 248

Nicolaescu, Gh., Cazac, T., Vacarciuc, L., Cebotari,V., Cumpanici, A., Nicolaescu Ana, Hioară Veronica, 2010. Filiera vitivinicola della Repubblica Moldova – situazione attuale e prospettive di sviluppo. Istituto Nazionale per il Commercio Estero – Ufficio di Bucarest, Univ. Agraria di Stato di Moldova; Chişinău. Print-Caro SRL. p. 142

Nicolaescu Gh., 2013. Particularităţile culturii soiurilor de struguri pentru masa in Italia, Spania şi Ucraina. Raport la Forumul Naţional al Producătorilor şi Exportatorilor strugurilor de masă din 08.09.2013

Wear R.J., 1976. Grape growing. Awilez-interscience publication:New Zork, Chichester, Brisbane, Toronto p. 371

Winkler A.J., Cook J.A., Kliwer W.M., Lider L.A., 1997. General viticulture. University of California press. Berkeley, Los Angeles, London p. 710.

Смирнов К.В., Калмыкова Т.И., Морозова Г.С., 1987. Виноградарство. М.: Агропромиздат. 367 с.

Уинклер А.Дж., 1966. Виноградарство США. Перевод с англ. Москва: Колос. 638 с.

17

OCCURRENCE OF *ISARIOPSIS* LEAF SPOT OR BLIGHT OF *VITIS RUPESTRIS* CAUSED BY *PSEUDOCERCOSPORA VITIS* IN TURKEY

Fatih Mehmet TOK, Sibel DERVIŞ

Mustafa Kemal University, Faculty of Agriculture, Department
of Plant Protection, 31040 Antakya, Hatay, Turkey
Corresponding author email: ftok@mku.edu.tr

Abstract

In mid-autumn 2013, a disease of Vitis rupestris affecting 30 to 50% of the plants was observed in an orchard of the Kırıkhan district of Hatay province in the east Mediterranean region of Turkey. Symptoms included presence of initially tiny and angular chlorotic halos but latterly large black spots. On the spots, cercosporoid structures were present. Based on cultural and morphological characteristics of the consistently isolated fungus and pathogenicity tests, the causal agent of this leaf spot or blight was identified as Pseudocercospora vitis. This appears to be the first report of P. vitis on V. rupestris in Turkey. Since Isariopsis leaf spot caused by P. vitis can also affect grapevines (V. vinifera), it might negatively impact orchards and vineyards in the future and might be considered a potential threat to Turkish grapevine production.

Key words: leaf blight, Vitis rupestris, Pseudocercospora vitis, Mycosphaerella personata.

INTRODUCTION

Grapevine is a widely planted and economically important crop in Turkey, hosting valuable grape germplasm resources with a total area under production of 467,093 hectares, producing 4,175,356 tonnes of grapes (FAO, 2014). In Kırıkhan and Hassa districts of Hatay, a coastal southeastern province located in the east Mediterranean region having about 20% of the grapevine growing areas of the country, some family farmers cultivate *Vitis rupestris*, a grape species commonly known as an American grapevine rootstock.

In mid-autumn 2013, a disease of *V. rupestris* plants was observed in an orchard of the Kırıkhan district of Hatay province in the east Mediterranean region of Turkey.

This study represents the first attempt to identify and characterize a cercosporoid species causing *Pseudocercospora* leaf spot, a disease of the aerial part of the *V. rupestris* vines, in Turkey using morphological and pathogenicity approaches.

MATERIALS AND METHODS

In October 2013, symptoms consisting of necrotic leaf spots were observed on *V. rupestris* cv. du Lot grown in an orchard in the Kırıkhan District and affected leaves sampled. Leaf tissues bordering these lesions surface disinfested with NaOCl (1%) for 2 min, rinsed in sterile water, blotted dry, and plated onto potato dextrose agar (PDA).

Isolated fungus was transferred to V-8 juice agar and malt extract agar (MEA). Plates were incubated for 30 days at 25°C under NUV light and a 12-h light/dark photoperiod for morphological examination. Morphological and colony characteristics of the fungus were examined.

Pathogenicity tests were performed on 10 1-year-old potted plants of grapevine (*V. rupestris*) cv. du Lot. Inoculations were performed by spraying a conidial suspension (3.0×10^4 conidia/ml) prepared in sterile water by harvesting conidia from 2-week-old cultures on V8 agar until runoff onto the leaves of healthy seedlings using manual pressure sprayer. Three control plants were sprayed with sterile water. Inoculated and control plants were covered with plastic bags to maintain a relative humidity of 100% for 48 h and then transferred to a greenhouse.

RESULTS AND DISCUSSIONS

Leaf spots were initially tiny, irregular to angular chlorotic halos or patches on both leaf surfaces - abaxial (Figure 1) and adaxial (Figure 2), and as the disease progressed, they were progressively turned purplish brown to black lesions. These spots later coalesced (amphigenous or confluent) reaching to 2 cm in diameter mostly with a serpentine and well defined outline encircled with dark borders or chlorotic halos on the upper leaf surfaces and became brittle with age (Figure 3). On the corresponding lower leaf surfaces, only the narrow centers of these lesions were brown to black but surrounding tissue was a large necrotic area (Figure 4). These coalesced spots caused leaf blight and premature defoliation. The disease incidence approached 30 to 50% on vines (cv. du Lot). The leaf spots were more severe on the leaves near the ground and progressed to the upper leaves. When infected tissue was examined under a stereomicroscope, typical cercosporoid hyphomycete structures were observed within the lesions on both leaf sides but mostly on the upper side.

A slow-growing fungus was consistently isolated from the affected tissues after 5 days of incubation at 25°C.

Figure 1. Irregular to angular chlorotic halos on the abaxial leaf surface of *Vitis rupestris* du Lot

Colonies were gray with black stromatic structures in the centers on the upper side (Figure 5) and dark green on the underside. The fruiting structures were slender, black, bristlelike synnemata (200-500 μm long) bearing pale olivaceous to pale brown, elongate conidia (25 to 100 × 4 to 7 μm) 3 to 7 transverse septa and no longisepta (Figure 6). Morphological characteristics of the fungus were consistent with previous descriptions of

Pseudocercospora vitis (Lév.) Speg. (Ascomycetes, Mycosphaerellales), the anamorph of *Mycosphaerella personata* Higgins as described by Ellis (1971) and Harvey and Wenham (1972).

Figure 2. Initial symptoms of Isariopsis leaf spot on the adaxial leaf surface of *Vitis rupestris* du Lot

Figure 3. Leaves of *Vitis rupestris* du Lot, with leaf spots on abaxial surface, late in the season

Pathogenicity was confirmed by fulfilling Koch's postulates. Necrotic spots appeared on the inoculated leaves 20 days after inoculation, and were identical to the ones observed in the field. Disease incidence on inoculated leaves varied from 30% to 91% and severity from 2 to 3 to 3 to 6 lesions per leaf. No symptoms were observed on control plants. Fungal colonies morphologically identified as *P. vitis* were reisolated from lesions on inoculated leaf tissues, fulfilling Koch's postulates. Control plants remained symptomless.

Figure 4. Leaves of *Vitis rupestris* du Lot, with leaf spots on adaxial surface, late in the season

Based on these results, the disease was identified as *Pseudocercospora* or *Isariopsis* leaf spot of *V. rupestris* caused by *P. vitis*. Since the pathogen was named *Isariopsis clavispora* formerly, the disease is still often referred to as *Isariopsis* leaf spot. Its occurrence primarily reported on *V. vinifea* and vines of wild species in the United States and throughout the warmer grape-growing areas of the world under one or another of its several synonyms (Pearson, 1998).

On *V. rupestris*, it was only recorded in Italy (Chupp, 1953) and Kansas (Anonymous, 1960) based on Farr and Rossman (2017).

To our knowledge, this is the first report of *P. vitis* on this host plant species in Turkey. Therefore, this present report is considered one of the few reports of this grapevine disease on *V. rupestris* in the world.

The appearance of the disease was more frequent at the end of the plant's vegetative cycle, in this American cultivar.

The main damage resulting from the attack of the pathogen was the premature fall of leaves, which caused a weakening of the plant and reduction of production in the following year, in 2014.

Cultural practices that increase air circulation such as shoot positioning and thinning may aid in management of the disease.

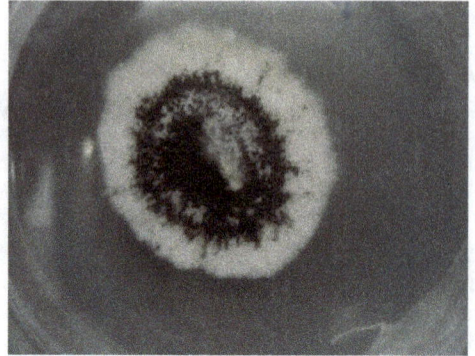

Figure 5. Colony of *Pseudocercospora vitis* after 30 days growth on potato dextrose agar

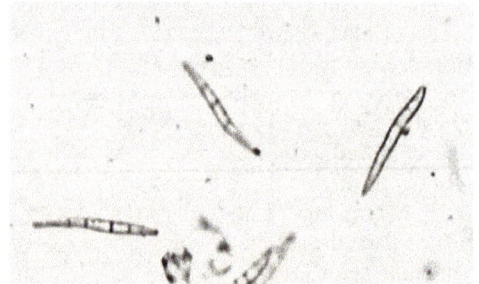

Figure 6. Conidia of *Pseudocercospora vitis*

CONCLUSIONS

A prominent leaf spot disease was found to occur on *V. rupestris* in Kırıkhan district of Hatay province in Turkey. A leaf spotting hyphomycete, *P. vitis,* was isolated, identified and proved to be pathogenic on grapevine. Associated symptoms were described and illustrated. Keeping vines healthy, destroying crop residues and spraying with standard fungicides (mid to late season) are recommended to perform in the case of severe attacks of the disease.

ACKNOWLEDGEMENTS

We thank Prof. Dr. Gülşen Sertkaya from the Department of Plant Protection at the University of Mustafa Kemal, for her participation in the expeditions and Assoc. Prof. Dr. Önder Kamiloğlu from the Department of Horticultural Science at the University of Mustafa Kemal, for the identification of the plant species.

REFERENCES

Anonymous, 1960. Index of Plant Diseases in the United States. U.S.D.A. Agric. Handb. 165, 1-531.

Chupp C., 1953. Monograph of the fungus genus *Cercospora*. Published by the Author, Ithaca, New York, 667 pages.

Ellis M.B., 1971. Dematiaceous Hyphomycetes. Kew, UK: Commonwealth Mycological Institute.

FAO, 2014. FAOSTAT. Food and Agriculture Organization of the United Nations, Rome, Italy Web. http:// www.fao.org/faostat/en/#data/QC

Farr D.F., Rossman A.Y., 2017. Fungal Databases. Syst. Mycol. Microbiol. Lab., Online publication, ARS, USDA, Retrieved 16 February, 2017.

Harvey I.C., Wenham H.T., 197.) A Fungal Leaf Spot Disease of Grapes *Cercospora vitis* (Lév) Sacco , New Zealand Journal of Botany, 10 (1), 87-96.

Pearson R., 1998. Compendium of Grape Diseases. (A.C. Goheen ed.) St. Paul, Minnesota: American Phytopathological Society.

18

INFLUENCE OF TEMPERATURE AND HUMIDITY IN BLOOMING PHENOPHASE CONCERNING ON FRUIT SET IN SOME TABLE GRAPES (*VITIS VINIFERA* L.)

Marinela Vicuţa STROE, Toniţa Valentina DUNUŢĂ,
Daniel Nicolae COJANU

University of Agronomic Sciences and Veterinary Medicine of Bucharest,
Department of Bioengineering of Horticulture-Viticulture Systems,
59 Marasti Blvd., District 1, 011464, Bucharest, Romania
Corresponding author email: marinelastroe@yahoo.com

Abstract

The aim of this study was to determine the influence of temperature and humidity on blooming phenophase and also on the percentage of fruit set of three table grapes varieties such as Muscat Hamburg cv., Afuz Ali and Victoria cv. (Vitis vinifera L). For assessing the optimal timing analysis it was used a grading system called BBCH which is a universal scale for describing monocots and dicots numbered from 00-97, a special focus was made on the principal growth stage blooming and includes stages 61 and 67 and growth stage of fruit development stages 71 and 77. Results shows that there was a direct correlation between the average values of two analyzed factors and percent of fruit set, and also the size of fruits. Average temperatures below + 20°C during blooming and a relative humidity of 65% has determined a good percentage binding at Muscat Hamburg (63%), followed by Afuz Ali variety (80%) and Victoria variety (86%). Surprisingly, even though the last two variety have formed more berries during the process of fecundation, proportion of very small berries (2-4 mm) and small berries(6-7mm) was high situated between 51.07% and 55.24%.

Key words: blooming, grows, phenology, table grape, variety.

INTRODUCTION

In the last decade, world production and consumption of fresh table grapes registered a noticeable increase Moreover, the consumer market, both globally and nationally, had a significantly increased every year, drawing attention to ensure competitive products in terms of sensory characteristics but also a high (OIV, 2013, Rolle et.al., 2015).
To achieve a competitive production, which means a good quality and high yield, a number of factors are related mainly environmental (Shinomiya et al., 2015) and climate change (Jones, 2005), and also genetic varieties and applied technology Among the environmental factors involved in plant development cycles, phenological temperature is considered to be a key factor that can profoundly alter the timing entire spectrum phenology for different plant species but also within the same species in different varieties (Parker et al., 2013). Phenology of varieties is changeable not only in terms of temperature impact but also by a complex interaction between temperature, humidity, precipitation, hour of insolation, registered at the beginning of each phenophases (Chuine et. Al., 2003, Cleland et al., 2007). From the factors mentioned above, blooming phenophases and fruit set are the most influenced. Blooming phenophase - after Baggiolini is stage I, and stage 65 after scoring system BBCH, (www.diprove.unimi.it/GRAPENET/index.php) or stage 23 after scoring system made by Eichorn and H. Lorenz, (1977), (Pierot and Rochard, 2013), starts about 55- 65 days after bud stage, when flowers and inflorescences are fully developed (Stroe, 2014). Blooming period can last, depending on variety and climatic conditions, between 7-16 days. In northern areas this phenophase can lasts 16 days and it starts when average temperature is approximate 17,5°C, in the south vineyards can lasts only 12 days, and in the centre of the country where the thermal regime is characterized by a warm temperate

climate (Savu and Stroe, 2005) phenological phase lasts 7-8 days. In practice, it was observed that in the first 2-3 days after onset of blooming there was a percent of 20-30% open flowers, in the coming 3-4 days, 60-70% bloom and only a small percentage of flower bloom at the end phenophase, but in the inflorescence, the first flowers whom are opening are the one in the middle of it, followed by those from the base and, in the end, the top inflorescence (Mustea, 2004). Fruit set is a process of pollination and fertilization, a phenomenon influenced by biological, physiological, climate and technological factors.

Among the biological factors the most important is type of flowers, morphologic and functional, and also ability of pollen germination. Depending on the unfavorable conditions and fecundation during blooming the percentage of fruit set will be different, and berries will have a different evolution: very small berries is the lack of fecundation of flowers, berries remain small as a grain of millet after reaching 2-4 mm in diameter they do not increase, acquires a yellow-green base and forms a isolator layer, and therefore it occurs shaking berries (Dobrei et. al., 2005).

Millerandage is a common thing among grape varieties that require foreign pollen, and in this case, the berries stop growing the size of a peppercorn because of the lack of hormonal substances. Later, they mature earlier, and also can accumulate higher amounts of sugars and bring a good influence in production of quality wine grape varieties, but for quality of grape table varieties is a disadvantage. Even so, varieties studied in this paper are adapted to temperate continental climate in Romania they have a different response reaction in terms of the percentage of fruit set (Table 1).

Basically, adverse climatic conditions which means low temperatures (below 20°C), high humidity 55-65%, heavy rainfall - manifested in this phenophase can trigger a massive shaking of flowers, resulting in the formation of rare berry cluster, which occurs most often the millerandage phenomenon.

Studies have shown that percentage of fruit set is an average of 36% to Rhine Riesling, Gewurztraminer from 35%, 66% at Italian Riesling, 37% to Chasselas and 65% to Oporto (Irimia, 2012).

This phenomenon must not be mistaken with pollination and fertilization disfunctions, due to the fact that, in the first case, it represents a sugar redistribution disorder, with different effects depending on the period when it is initiated (Bernaz, 2003).

Under this circumstances, May occur other physiological diseases caused by either lack or excess of nutrients during some phases of the anual vegetative cycle.

Based on this consideration, the paper aim was to analyzed the influence and effect of temperature and humidity on fruit set development in the climatic conditions of the year 2016.

Results of this experience were analyzed under the quality parametres of a high yield/commodity production.

MATERIALS AND METHODS

Studied varieties and growth conditions

In this study were analyzed three varieties of table grapes: Muscat Hamburg cv., Afuz Ali cv., from the world's assortment and Victoria cv., a romanian variety obtained in 1978 by Victoria Lepădatu and Gh. Condei. The main data about these varieties can be found in Vitis International Variety Catalogue (www.vivc.de) and The European Vitis Database. In vineyard, all these three varieties have proven over the years, requiring a specific temperature and humidity during blooming phenophase. Table grape varieties are located in the experimental field of the ampelographic collection from the University of Agronomic Sciences and Veterinary Medicine of Bucharest. They have been conducted on the semi-stalk; the type of pruning in the prior year was Guyot on semi-stem, with a load of 30 buds/vine. During experiment observations and measurements were currently made for determining the elements of fertility and productivity, with a special focus was made on carpometry indices that define the productive potential: % fertile shoots, absolute fertility coefficient, relative fertility coefficient, absolute productivity index (g/shoot), relative productivity index (g/shoot), number of grape/vine, average weight of a grape (g), yield (kg/vine).

Short presentation of Muscat Hamburg variety.
Is a variety with medium vigor growth holding up relatively well at lower temperature during

winter and has a good fertility manifested by 70% fertile shoots. In normal years, the flowers have the capacity to fruit set up to 50% because the capacity of pollen germination is small, and that is why indicated to cultivate in vineyard with alongside varieties like Chasselas doré and Afuz Ali (Stroe, 2012).

Short presentation of Afuz Ali variety

Hardy variety, but sensitive to low winter temperatures (-16 °C, - 18 °C) with a long growing season 180-210 days, that is why fails to mature wood well. It has a great sensitivity to frost and this leads to poor resistance to diseases like bacterial cancer cancer and anthracnose.

Short presentation of Victoria variety

Variety with a medium to high vigor and has a good fertility, manifested by a percentage of 63-73% fertile shoots. Variety is characterized by good resistance to frost (-20 °C) and drought, it behaves well towards spring frosts because of late budbreak. It has an average resistance to mildew and very poor resistance to powdery mildew (Stroe and Veliu, 2010).

Phenological and temperature data

In this study it was used a update version of BBCH, an universal scale used for describing monocots and dicots numbered from 00-97, with special focus on the principal growth stage 6: Flowering, stages 61 to 67 and principal growth stage 7: Development of fruits, 71 to 77 (Pierot and Rochard, 2013). Three observations were made on all inflorescence of 6 vine for each variety separately in different stages of development as shown in table 2.

Meteorological parameters were analyzed in the period 20 May-20 June 2016 using daily averages of 6 points hourly results of the day (5:00, 08:00 11:00 14:00 17:00 20:00 23:00) it was recorded by weather station Bucharest - Baneasa, Romania. The calculation was conducted on a higher timeframe, to ensure better accuracy of results, even so phenophase of blooming at studied varieties lasted 12 days (20 - 31 May 2016).

RESULTS AND DISCUSSIONS

Although well adapted to the temperate continental climate from Romania, the varieties

Muscat Hamburg cv., Afuz Ali and Victoria cv. responded differently in vineyard in what concerns the percent of fruit set, the pergentage of formed flowers shaken, and also the millerandage phenomen, which in particular years can put their marks on the obtained crops. Climatic particularities of the wine year 2016 resulted in earliness in flowering terms Cleland (2007), noticing that the varieties analysed bloomed in the third decade of May (Figure 1) within a period of 12 days (20 - 31 May 2016).

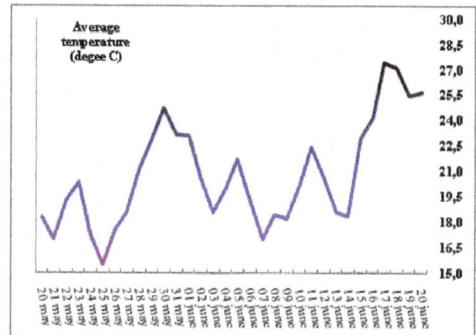

Figure 1. Evolution of the average temperature (°C) for period 20 May - 20 June 2016

In this amount of time, the average daily temperatures recorded presented in figure 1 were of 19.69°C with an inferior limit of 15.5°C recorded in 25th of May and a maximum of de 20.4°C in 23th May, bassically, when the blooming phenophase was at its best, beeing well known that in the first days from the flowering 20-30% of the total flowers, in the following days are flowering 60-70% and only a small percentage open at the end of the phenophase Mustea (2004).

Regarding values of relative humidity (%) figure 2 gives an indication of their impact on analyzed varieties, because in the same period, the average was 70.65% with a peak of 83.12% in May 25th, values far exceeding the normal limit of blooming knowing that optimal conditions is in range 55-65%.

Following the recordings in figure 3, we notice that the percent of the fruit set is different from a variety to another, the Muscat Hamburg variety records the smallest percent (63%), followed by the Afuz ali variety (80%) and then Victoria variety (86%).

Table 1. The genetic origin of studied varieties

Prime name	Muscat de Hamburg	Afuz Ali	Victoria
Variety number *VI*VC	8226	122	13031
Country of origin of the variety	U. K.	Liban	România
Species	*Vitis vinifera* L.	*Vitis vinifera* L.	*Vitis vinifera* L.
Pedigree as given by breeder/bibliography	Trollinger x Muscat de Alexandria	-	-
Pedigree confirmed by markers	Schiava grossa x Muscat Alexandria	-	Cardinal x Afuz Ali
Prime name of pedigree parent 1	Schiava grossa	-	Cardinal
Prime name of pedigree parent 2	Muscat Alexandria	-	Afuz Ali
Year of crossing	1850	-	1964
Last update	18.01.2017	18.01.2017	18.01.2017

Table 2. Phenological study of vine during blooming and development of fruits

Principal growth stage 6: Flowering (BBCH MODIFIED PHENOLOGICAL SCALE FOR COST ACTION FA1003)		
6.1: Beginning of flowering: 10% of flowerhoods fallen		
6.7: 70% of flowerhoods fallen		
Principal growth stage 7: Development of fruits		
7.1: Fruit set: young fruits begin to swell, remains of flowers lost		
7.7: Berries beginning to touch (if bunch are tight)		
I.Observation (30.05.2016) 70% of flowerhoods fallen	II. Observation (05.06.2016) Small-berry grape only formats	III. Observation (01.07.2016) Berries beginning to touch

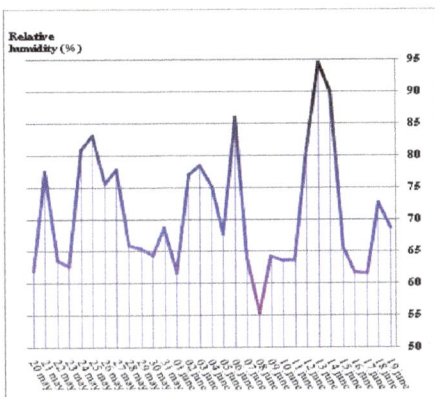

Figure 2. Evolution of the average relative humidity (%) for the period 20 May-20 June 2016

Analysing figure 4 which give details about size categories in which the formed berries belong, expressed as a percentage, it is observed that the Muscat Hamburg variety stands out with a great percentage of normal berries (65.9%), variety specific, followed by Victoria variety with 49.10% and then we find the Afuz Ali variety with a lower percent of 44.69%. It is surprisingly that even if the last varietes formed more berries in the process of fecundation, the proportion of berries remained very small (millet dimensions of 2-4 mm) and small (peppercorn dimensions 6-7 mm) was highly pronounced (55.24% and 51.07%).

All this results are affecting in a negative way the crops obtained (Table 3), literally, these shows a low productive potential, under the limit of these variety potential, which is given by the high percentage of very small and small berries remained.

A similar evolution was recorded in what concerns the average weight of a grape (220 g/grape - Muscat Hamburg cv., 270 g/grape - Afuz Ali cv. and 281g/grape Victoria cv.), determining a low yield, between 2.24-4.05 kg/vine, and the merchandise production in this case is being the lowest Stroe et al. (2016).

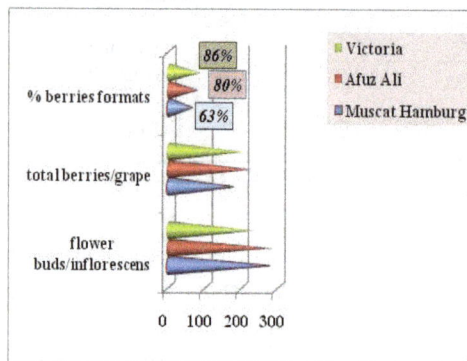

Figure 3. Evolution of the berries formats (%) for the period 20 May -20 June 2016

CONCLUSIONS

The average temperature and humidity values recorded in the blooming phenophase have left their mark on the three varieties studied, demonstrating through the obtained results that they are very pretentious and requires in this phenophase higher temperatures (over 20°C) and relative moist values of 65%.

Regarding the categories of the formed berries was recorded that Afuz Ali and Victoria varieties although formed much more berries, about half of them remained small (6-7 mm) and very small (2-4 mm).

Although Muscat Hamburg is known as a variety with millerandage problems in the given conditions of 2016, this was exceeded in this particular study of Afuz Ali and Victoria varieties, Victoria and Afuz Ali varieties are having in its pedigree Cardinal variety.

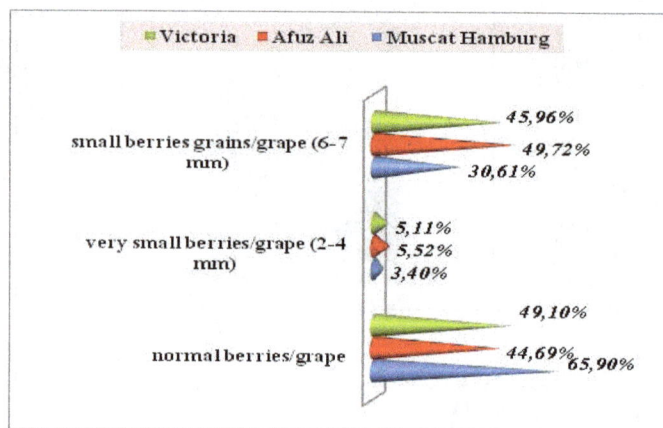

Figure 4. Evolution of the categories of grains formed of studied varieties

Table 3. The synthesis of the main fertility elements and physical characteristics of varieties study

Varieties	% fertile shoots	Absolute fertility coefficient	Relative fertility coefficient	Absolute productivity index (g/shoot)	Relative productivity index (g/shoot)	No. of grapes/vine	Average weight of a grape (g)	Yield (kg / vine)
Muscat Hamburg	58.33	0.91	1.57	200.2	345.4	11	220	2.42
Afuz Ali	61.11	0.83	1.36	348.6	571.2	15	270	4.05
Victoria	50.00	0.80	1.33	240	399	8	281	2.24

For an improvement, as a tehnological intervention, it can be applied several actions intervening on the sugar redistribution system: pinching fertile shoots a few days before blooming, annular incision below the inflorescence, application of growth retardant treatments, to briefly stop vegetative growth and redirect nutrients to inflorescences, balanced fertilization with potassium, which favors the sugars migration and accumulation of reserves, and also reducing nitrogen fertilizer rates.

REFERENCES

Bernaz Gh., 2003. Refacerea viilor vătămate de accidente climatice și boli fiziologice-Gh. Editura M.A.S.T 2003, 26-27.

Chuine, I., Kramer, K., Hanninen, H., 2003. Plant development models. Phenology:An Integrative Environmental Sciences, vol. 39, 217–235.

Cleland, E.E., Chuine, I., Menzel, A., Mooney, H.A., Schwartz, M.D., 2007. Shifting plant phenology in response to global change. Trends Ecol. Evol. 22 (7), 357–365.

Dobrei Alin, Liliana Rotaru, Mihai Mustea, 2005. Cultura viţei de vie. Editura Solness, Timişoara, 62.

Irimia L. M., 2012. Biologia, ecologia și fiziologia viței-de-vie. Editura "Ion Ionescu de la Brad", Iași, 79, 87.

Isabelle Pierot, Joël Rochard, 2013. Adaptation aux changements climatiques - projet europeen Leonardo Da Vinci: E-viticlimate. The XXXVI-th World Congress of Vine and Wine June 2-7-th 2013, Bucharest. ISBN O.I.V. 979-10-91799-16-4.

Jones G. V., White M. A., Cooper O. R., Storchmann K., 2005. Climate change and global wine quality. Climatic Change 73, 319-343.

Luca Rolle, Fabrizio Torchio, Simone Giacosa, Susana Río Segade, 2015. Berry density and size as factors related to the hysicochemical characteristics of Muscat Hamburg table grapes (Vitis vinifera L.). Journal of Food Chemistry, 173 (2015), 105-113.

Mustea M., 2004. Viticultură - Bazele biologice, înfiinţarea și întreţinerea plantaţiilor tinere de vii roditoare. Editura "Ion Ionescu de la Brad", Iași, 151-153, 157.

O.I.V., 2013. OIV vine and wine outlook 2008–2009. Paris: Organisation Internationale de la Vigne et du Vin.

Parker Amber et al., 2013 - Classification of varieties for their timing of flowering and veraison using a modelling approach: A case study for the grapevine species Vitis vinifera L. Agricultural and Forest Meteorology 180 (2013) 249 – 264.

Ryo Shinomiya et al., 2015. Impact of temperature and sunlight on the skin coloration of the 'Kyoho' table grape. Scientia Horticulturae 193 (2015) 77–83.

Savu Georgeta Mihaela, Stroe Marinela Vicuţa, 2005. Evaluarea condiţiilor ecoclimatice, cu ajutorul unor indicatori sintetici, în regiunea viticolă a Dealurilor Munteniei și Olteniei. Simpozion ştiinţific anual, "Horticultura – ştiinţă, calitate, diversitate-armonie", U.S.A.M.V. „Ion Ionescu de la Brad", Iași, 27-28 mai, XLVIII Vol 1 (48), Seria Horticultură, CD-ROM I.S.S.N. 1454-7376, 303-308.

Stroe Marinela, 2014. Lucrări practice Ampelografie. Editura Invel-Multimedia, Ediţie revăzută și adăugită, 2014, ISBN 978-973-1886-80-0, 7, 15.

Stroe Marinela Vicuţa, Raluca Veliu, 2010. The agrobiological and technological evaluation of some table grape variety with different maturation periods in vineyard Ostrov. Sesiunea Ştiinţifică anuală „Horticultură - Ştiinţă, Calitate, Diversitate și Armonie", U.S.A.M.V. „Ion Ionescu de la Brad" Iași, Facultatea de Horticultura, Anul III, Vol 53, Seria Horticultură, CD ROM ISSN 2069-847X, 437- 443.

Stroe Marinela, 2012. Ampelografie. Editura Ceres, Bucureşti, ISBN 978-973-40-0943-5, (COD CNCSIS 236), 187, 193, 203.

Stroe Marinela, Bejan Carmen, Carmen Florentina Popescu, 2016 - Research regarding of the technological parameters of some table grape varieties in the experimental field from U.A.S.V.M. Bucharest - Bulletin of University of Agricultural Sciences and Veterinary Medicine Cluj-Napoca. Horticulture, vol. 73, no 1 (2016), ISSN 1843-5254.

The European Vitis Database, Genetic resources of grapes, website, www.eu-vitis.de/index.php, accessed on February 2017.

Vitis International Variety Catalogue, Passport of variety Muscat Hamburg, Variety number VIVC 8226, passport of Afuz Ali, VIVC number 122, passport of Victoria cv, VIVC number 13031, website, http://www.vivc.de, accessed on ianuary 2017.

http://www.diprove.unimi.it/GRAPENET/index.php, Cost action FA1003: East-West collaboration for Grapevine Diversity Exploration and Mobilization of Adaptive Traits for Breeding, PHENOTYPING TRIAL 2012, First circular 12th March 2012.

19

SOME NUTRIENT CHARACTERISTICS OF GOLDENBERRY (*PHYSALIS PERUVIANA* L.) CULTIVAR CANDIDATE FROM TURKEY

Aysun OZTURK[1], Yasin ÖZDEMİR[1], Barış ALBAYRAK[2],
Mehmet SİMŞEK[3], Kutay Coşkun YILDIRIM[3]

[1]Atatürk Horticultural Central Research Institute, Department of Food Technologies,
Yalova, Turkey
[2]Atatürk Horticultural Central Research Institute, Land and Water Resources, Yalova, Turkey
[3]Atatürk Horticultural Central Research Institute, Vegetables Production Department,
Yalova, Turkey
Corresponding author email: ozturkaysun@hotmail.com

Abstract

This study was the first record on nutrient characteristics of goldenberry types in Turkey. In this study; 6 types of goldenberry were used as material which were collected from different side of Turkey and cultivated in observation parcel of Atatürk Horticultural Central Research Institute Yalova (Turkey). Obtained data will be used for future selection and registration steps of these types. For this purpose carotene, vitamin C, crude fiber, total phenolic and mineral matter content and antioxidant activity of goldenberries were determined. The results showed that goldenberry types are rich in vitamin C, total phenolic content and minerals (especially phosphorus) content. Also they have been found as a rich source of crude fiber and carotene. 6th type has the highest total phenol (48.60 mg GAE/100g), crude fiber (4.10%) and vitamin C (34.86 mg AAE/100g) content and 3rd type has the lowest total phenol (42.16 mg GAE/100g), vitamin C (31.40 mg AAE/100g) and carotene (1.98 mg/100g) content. According to evaluation of these results 6th type was determined as featured types for registration as new cultivar.

Key words: *Physalis peruviana L., carotene, crude fiber, total phenolic content, mineral matter*

INTRODUCTION

Physalis peruviana L., belongs to the *Solanaceae* family and *Physalis* genus, commonly known as goldenberry or cape gooseberry in English speaking countries. Different varieties of *P. peruviana* are recognized in many countries and are cultivated commercially in tropical and subtropical countries such as South Africa, America, New Zealand and Spain (Morton, 1987).

They are grown in warm climates region of Turkey, especially Black Sea, Marmara, Aegean and Mediterranean regions (Besirli, 1998).

Generally, the fruit of goldenberry is consumed fresh or dried like raisin. It is also used in sauces, glazes for meats, seafood (NRC, 1989). Currently, there are different products processed from goldenberry fruits such as juice, pomace jams, chocolate-covered candies, other products sweetened with sugar as a snack (Ramadan and Moersel, 2009). In European markets, it is used as ornaments in meals, salads, desserts and cakes (Puente et al., 2010). The fruits of goldenberry were reported to have many medicinal properties such as antispasmodic, diuretic, antiseptic, sedative, analgesic, helping to fortify the optic nerve, throat trouble relief, elimination of intestinal parasites and amoeba (CCI, 1994).

Studies indicated that eating the fruit of goldenberry reduces blood glucose after 90 min postprandial in young adults, causing a greater hypoglycaemic effect after this period (Rodríguez & Rodríguez, 2007).

The fruit of goldenberry has been stated as a rich source of carotene and its major components are its high amounts of polyunsaturated fatty acids, vitamins A, B and C and phytosterols, as well as the presence of essential minerals, vitamins such as E and K₁, with anolides and physalins, which together would give them medicinal properties described above (Ombwara, 2004; Puente et al., 2010).

The aim of this study was to investigate some chemical properties and nutrient composition of goldenberry fruit which were collected from different side of Turkey and cultivated in Atatürk Horticultural Central Research Institute Yalova (Turkey). Obtained data was used for future selection and registration steps of these types. With this study; main features of goldenberry which new known for the food industry and consumers of Turkey were determined.

MATERIALS AND METHODS

Fruits of 6 types of goldenberry were used as material. They were collected from different side of Turkey, numbered from 1 to 6 and cultivated at same conditions in Atatürk Horticultural Central Research Institute.

Figure 1. Goldenberry fruits

Mineral Matter Analysis. Golderberry samples were dried, milled and analyzed according to method of wet decomposition by using sulfuric acid + hydrogen peroxide solution according to Anonymous (1980). Amounts of K, Ca, Mg, Fe, Cu, Zn and Mn elements were determined with atomic absorption spectrophotometer (Perkin-Elmer AAnalyst 700, USA). P element was calculated as colorimetric with vanadomolibdofosforic acid method (Lott et al., 1956).

Ascorbic Acid Analysis. Official spectrophotometric method was used according to AOAC 1970.

Crude Fiber Analysis. Samples were boiled with acid (1.25% H_2SO_4), then alkaline (1.25% NaOH). Later, the residues were burned in an ash furnace (AOAC, 1970).

Extraction of samples for Amount of Total Phenolic Compounds and Antioxidant

Activity Analysis: 3 g samples were weighed fresh goldenberry fruits and mixed with 25ml methanol then homogenized for 2 min. and homogenates were kept at +4 °C for 12 h. Samples were centrifuged at 15,000 rpm for 20 min. and liquid phases in methanol were collected with Pasteur pipettes into amber bottles then were stored at -20 °C until analysis. These extracts prepared were used both antioxidant activity and amount of total phenolic compounds (Thaipong et al., 2006).

Amount of Total Phenolic Compounds Analysis. Amount of total phenolic compounds was determined by the Folin - Ciocalteu method. The 150 μl of extract, 2400 μl of distilled water, and 150 μl of 0.25 N Folin - Ciocalteu reagent were combined in a plastic vial and then mixed well using a Vortex. The mixture was allowed to react for 3 min then 300 μl of 1N Na_2CO_3 solution was added and mixed well. The solution was incubated at room temperature (23°C) in the dark for 2 h. The absorbance was measured at 725nm using a Hitachi U-2900 brand spectrophotometer (Tokyo, Japan) and the results were expressed in gallic acid equivalents (GAE mg/100 g) using a gallic acid (0–0.1 mg/ml) standard curve (Thaipong et al., 2006).

Antioxidant Activity Analysis. The DPPH (2,2-difenil-1-pikrilhidrazil) assay was done according to the method of Thaipong et al. (2006). The stock solution was prepared by dissolving 24 mg DPPH with 100 ml methanol and then stored at -20°C until needed. The working solution was obtained by mixing 10mL stock solution with 45 ml methanol to obtain an absorbance of 1.1±0.02 units at 515 nm using the spectrophotometer. Mushroom extracts (150 μl) were allowed to react with 2850 μl of the DPPH solution for 24 h in the dark. Then the absorbance was taken at 515 nm. The standard curve was linear between 25 and 800 μM Trolox. Results were expressed in μM TE/g.

RESULTS AND DISCUSSIONS

Reducing sugar rates of goldenberry types were determined between 6.55 and 7.80 % (Table 1). In a similar study; Sharoba and Ramadan (2010) reported that reducing sugar rates 8.23% in goldenberry fruit.

In our study, the carotene contents of the goldenberry types were found very close to each other. These values were between 1.98-2.30 mg / 100 g (Table 1). Ombrawa (2004) and Sharoba and Ramadan (2010) reported carotene contents respectively as 1.61 and 2.38 mg/100g.

According to Table 1, ascorbic acid amounts of goldenberry samples have showed variability between 35.10-31.40 mg/100 g. Ascorbic acid amounts were given 43.0 mg/100 g in study done by NRC (National Research Council, 1989) and 26,00 mg/100 g in study done by CCI (Corporación Colombia Internacional, 1994) and also 20,00 mg/100 g by Fischer et al.(2000).

The highest crude fiber ratio was calculated as 4.10 % in 6th type goldenberry sample, while the lowest crude fiber ratio was calculated as 3.18 % as in 1th type goldenberry sample (Table 1). Other researchers reported crude fiber ratios as 4.9 % (NRC, 1989); 4,8 % (CCl,1994) and 3.6 % (Repo Depo de Carrasco and Zelada, 2008).

Table 1. Reducing sugar, carotene, ascorbic acid and crude fiber content of goldenberry types

Types	Reducing Sugar (%)	Carotene (mg/100g)	Ascorbic Acid (mg/100g)	Crude Fiber (%)
1	6.55	2.10	33.90	3.18
2	6.79	2.26	33.15	3.24
3	7.80	1.98	31.40	3.85
4	7.20	2.24	35.10	3.94
5	6.21	2.20	32.54	4.02
6	7.75	2.30	34.86	4.10

Amounts of total phenolic compound (TPC) for goldenberry are showed in Table 2. As a result of the analysis, 6th type has the highest TPC amount (48.60 mg GAE/100 g) and 3rd type the lowest TPC (42.16 mg GAE/100 g). In a study conducted by Restrepo (2008), amount of TPC has been specified 40,45±0,93 mg GAE/100 g; in another study has been reported as 39,15±5,43 mg GAE/100 g (Botero, 2008). In terms of TPC amounts, our results have been showed compatibility with literature values.

Antioxidant activity of goldenberry types were determined between 604.69 - 514.06 TEµM/100 g (Table 2). Both Restrepo (2008) and Botero (2008) carried out DPPH method in their studies and they stated antioxidant activity of goldenberry fruit as 210.82±9.45 and 192.51±30.13 TEµM/100 g, respectively. When Table 2 is examined, it was seen that our results are higher than results of Restrepo (2008) and Botero (2008).

Table 2. Amounts of total phenolic compound and antioxidant activities of goldenberry types

Types	Amount of Total Phenolic Compounds (mg GAE/100g)	Antioxidant Activity (TEµM/100g)
1	46.91	566.27
2	47.67	554.69
3	42.16	514.06
4	48.39	604.69
5	48.53	552.08
6	48.60	569.24

Mineral matter amounts of types were found as N 1160-1460 mg/100 g, P 178,46-233,49 mg/100 g, K 1794,98-19,68 mg/100 g, Ca 34,12-43,65 mg/100 g, Mg 102,50-122,51 mg/100 g, Fe 3,68-4,09 mg/100 g, Mn 0,59-0,78 mg/100 g, Zn 1,78-2,32 mg/100 g, Cu 2,19-3,28 mg/100 g, calculated on the basis of total dry matter (Table 3).

In a study carried out in Colombia, mineral matter amounts of goldenberry fruits (on the basis of fresh weight) were informed as: P 27 mg/100 g, K 467 mg/100 g, Ca 23 mg/100 g, Mg 19 mg/100 g, Fe 0,09 mg/100 g, Mn 0,20 mg/100 g, Zn 0,28 mg/100 g, Cu 64 mg/100 g, Na 6 mg/100 g, Cl 1 mg/100 g, S 10 mg/100 g, Ni 0,02 mg/100 g (Leterme et al., 2006). In others studies also were reported: P 55,3 mg/100 g, Ca 8,0 mg/100 g, Fe 1,23 mg/100 g (Ombwara, 2004); P 34 mg/100 g, K 210 mg/100 g, Na 2 mg/100 g, Ca 28 mg/100 g, Fe 0,3 mg/100 g, Mg 7 mg/100 g (Musinguzi et al., 2007).

Amounts differences of mineral matter in between our research and literature have been caused by different climate and growing conditions of the goldenberries.

Table 3. Mineral matter of goldenberry types (mg/100 g dry matter)

Types	N	P	K	Ca	Mg	Fe	Mn	Zn	Cu
1	1160	214.91	1911.99	37.37	102.92	3.76	0.59	1.88	2.19
2	1200	217.61	1968.77	40.55	103.13	3.68	0.60	1.78	2.44
3	1220	178.46	1801.85	34.12	102.50	3.72	0.65	1.99	2.88
4	1460	233.49	1794.98	43.65	122.51	4.09	0.78	2.32	2.97
5	1325	222.31	1903.15	41.39	107.82	4.04	0.65	2.14	3.28
6	1455	220.62	1954.12	42.54	115.36	4.02	0.62	2.18	3.08

CONCLUSIONS

Plant phytochemicals especially phenolic compounds are thought to protective cells against oxidative damage and aging effects etc. In this study fruits of goldenberry types are detected as rich source for vitamin C, carotene and phenolic compounds. This study was the first record on nutrient characteristics of goldenberry types in Turkey. According to the analysis results, it was seen remarkably that goldenberry fruit is a good source crude fiber and carotene, furthermore it have low sugar content. 6th type has the highest total phenol, crude fiber and vitamin C content. According to evaluation of results 6th type was considered proper to select and stand out for registration and certification as a new cultivar.

ACKNOWLEDGEMENTS

This research work was carried out Food Technology Department with the support of Atatürk Horticultural Central Research Institute. It also was financed Republic of Turkey Ministry of Food, Agriculture and Livestock. (The project number is TAGEM/HSGYAD/12/A05/P03/14).

REFERENCES

Anonymous, 1970. Association of official analytical chemists (AOAC), Horwitz W, Chichilo P, Reynolds H. Washington, syf 129-131.

Anonymous, 1980. Soil and Plant Testing and Analysis as a Basis of Fertilizer Recommendations. F.A.O. Soils Bulletin 38/2, 95.

Beşirli G., Sürmeli N., 1998. Agriculture Village Journal.

Boter, A., 2008. Aplicación de la Ingeniería de Matrices en el desarrollo da la uchuva mínimamente procesada fortificada con calcio y vitaminas C y E. Facultad de química farmacéutica, vol. Magíster en ciencias farmacéuticas énfasis en alimentos. p185, Medellín: Universidad de Antioquía.

Corporaciòn Colombia International (CCI), Universidad De Los Andes & Departamento De Planeaciòn Nacional, 1994. Análisis internacional del sector hortofrutícola para Colombia. Bogotá: El Diseño

Fischer G., Ebert G., Ludders P., 2000. Provitamin A carotenoids, organic acids and ascorbic acid content of cape gooseberry (Physalis peruviana L.) ecotypes grown at two tropical altitudes. Acta Horticulturae, 531, 263−268.

Leterme P., Buldgen A., Estrada F., London A.M., 2006. Mineral content of tropical fruits unconventianal foods of the Andes and the rain forest of Colombia. Food Chemistry 95, 644-652.

Lott W.L., Gallo J.P., Medaff J.C., 1956. Leaf Analysis Technic in Coffee Research. Ibec. Research Institute II: 21-24.

National Research Council (NRC), 1989. Goldenberry (Cape Gooseberry). Lost crops of the incas: Little-known plants of the andes with promise for worldwide cultivation. Washington D.C., National Academy Press, 240−251.

Morton, J. F. Cape gooseberry. In Fruits of Warm Climates; Morton, J. F., Ed.; Creative Resource Systems: Winterville, NC, 1987; pp 430-434.

Musinguzi E., Kikafunda J., Kiremire B., 2007. Promoting indigenouswild edible fruitsto complement roots and tuber crops in alleviating vitamin A deficiencies in Uganda, Arusha, Tanzania: Proceedings of the 13th ISTRC Symposium, 763−769.

Ombwara F.K., Wamocho L.S., Mugai E.N., 2004. The Effect of Nutrient Solution of golStrength and Mycorrhizal Inoculation on Anthesis in Physalis peruviana. Proceedings of the fourth workshop on sustainabla horticultural production in the tropics. 24-27 Nowember, 117-122.

Puente L. A., Claudia A P., Eduardo S.C., Misael C., 2010. Physalis peruviana Linnaeus, the multiple properties of a highly functional fruit: A review. Food Research International.

Ramadan M.F., Moersel J.T., 2007. Impact of enzymatic treatment on chemical composition, physicochemical properties and radical scavenging activity of goldenberry (Physalis peruviana L.) juice. J. Sci. Food Agric. 87, 452–460.

Ramadan M., Moersel J.T., 2009. Oil extractability from enzymatically treated goldenberry (*Physalis peruviana* L.) pomace: Range of operational variables. International Journal of Food Science & Technology, 44(3), 435–444.

Repo Depo De Carrasso R., Zelada C., 2008. Determinación de la capacidad antioxidante y compuestos bioactivos de frutas nativas peruanas. Revista de la Sociedad Química Perú, 74(2), 108–124.

Restrepo A., 2008. Nuevas perspectivas de consumo de frutas: Uchuva (*Physalis peruviana* L.) y Fresa (*Fragaria vesca* L.) mínimamente procesadas fortificadas con vitamina E. Facultad de Ciencias Agropecuarias, vol. Magíster en ciencia y tecnología de alimentos, Medellín: Universidad Nacional de Colombia, 107.

Rodríguez S., Rodríguez E., 2007. Efecto de la ingesta de Physalis peruviana (aguaymanto) sobre la glicemia postprandial en adultos jóvenes. Revista Médica Vallejiana, 4(1), 43–52.

Ross A.F., 1959. Dinitrophenol method for reducing sugar Patato Processing. (Ed: W.F. Talburt). The AVI Publishing Com. İnc., Wesport, Connecticut. 469–470.

Sharoba A.M., Ramadan M.F., 2010. Rheological behavior and physicochemical characteristics of goldenberry (*Physalis peruviana*) juice as affected by enzimatic treatment. Journal of Food Processing and Preservation.

Thaipong K., Boonnprakob U., Crosby K., Cisneros-Zevallos L., Byrne D. H., 2006. Comparison of ABTS, DPPH, FRAP, and ORAC assays for estimating antioxidant activity from guava fruit extracts. Journal of Food Composition and Analysis 19, 669–675.

20

DETERMINATION OF STRENGTH PROPERTIES FOR MECHANICAL HARVEST OF ROCKET (*ERUCA VESICARIA*)

Deniz YILMAZ, Mehmet Emin GOKDUMAN

Süleyman Demirel University, Agriculture Faculty,
Agricultural Machinery and Technologies Engineering Department,
Doğu Campus, Isparta, Turkey
Corresponding author email: denizyilmaz@sdu.edu.tr

Abstract

Rocket (Eruca vesicaria) is a vegetable from the family Turpentaria (Brassicaceae) that eats leaves as a salad. Although rocket vegetable to produce small areas our country, it has started to make production in large and larger areas in recent years. This study aimed to determine the strength of Rocket (Eruca vesicaria) specifications for mechanical harvesting. For this purpose, properties as the maximum force, stress in the maximum force point, work at maximum force point, shearing force, deformation at maximum force, bioyield force, and shearing stress of rocket (Eruca vesicaria) stalk, leaf and root have determined. Average values for maximum force, stress, work to maximum force and deformation in maximum force were determined as 4.820 N, 0.474 MPa, 0.015 J and 22.149 mm, respectively. The shearing force and shearing stress were found to be as 2.150 N and 0.219 MPa, respectively. Average values for bioyield force were determined to be 3.856 N. These features can be used in determining the design and operating conditions for the mechanical harvester cutting blade.

Key words: Rocket (Eruca vesicaria), strenght properties, mechanical harvesting.

INTRODUCTION

'Rocket' is a common name used for some species in the family Brassicaceae that have a pungent aroma and a sharp taste (Figure 1). They are native to the Mediterranean and Near East, and they possibly acquired their original common name from the Lat-in-speaking Roman citizens who in-habited this area. The common name and many of its derivatives, including rughetta, rucola, roquette and others, most likely descended from the Latin word roc meaning harsh or rough (Pignone, 1997). Common names currently used to describe these species include roquette, rucola, arugula and rocket. As with all common names, the choice of common name varies with ethnicity, location and language group.

Rocket is traditionally grown in Italy, Portugal, Egypt, and Turkey (Bianco et al., 1997; Mohamedien, 1995; Pimpini et al., 1997; Silva Dias, 1997; Tuzel, 1995), it has also been successfully investigated as a new crop for Indiana and US Midwest (Morales et al., 2002), where it can be cultivated in open field and protected areas. In the past years, rocket has increasingly become popular also in the Central Europe.

Figure 1. Rocket (*Eruca vesicaria*) plant

It has 23,000 (1000 ha) of farmland in Turkey. 3.4 percent of this area (809,000 ha) used for vegetable production. Vegetable production has been increasing in recent years. According to 2015 data; the rocket production is 9110 tons, while tomato takes place on top with 12,615,000 tons in production volume of about 30 million tons.

The vegetable mechanization is conducted by hand in Turkey. Mechanization is needed due to the increase in production area.

The recent studies focused on chemical, herb and oil properties of rocket (*Eruca vesicaria*)

(Doležalová et al., 2013; Nurzyńska-Wierdak R., 2006). However, studies on strength properties of rocket are limited. This study covers determination of maximum force, stress in the maximum force point, work at maximum force point, shearing force, deformation at rupture force, bioyield force, shearing stress of rocket (*Eruca vesicaria*) stalk, leaf and root.

MATERIALS AND METHODS

For this study, rocket (*Eruca vesicaria*) plants were harvested by hand from the experimental field in Suleyman Demirel University, Isparta, Turkey.

Diameter and cross-sectional area of the experimental samples were measured before the shearing tests. Moisture content of the plants was determined at harvest time. Specimens were weighed and dried in an oven at 102°C for 24 h and then reweighed (ASABE, 2006). It was provided concise but complete information about the materials and the analytical and statistical procedures used.

A universal testing machine (LF Plus, UK) with a 500 N load cell and a computer-aided cutting and picking apparatus (Figures 2, 3) was used to measure the strength properties of the rocket plant. Knife material was hardened iron. All the tests were carried out at a speed 0.8 mm s^{-1}, and data were recorded at 10 Hz. All data were analysed by nexygen software program.

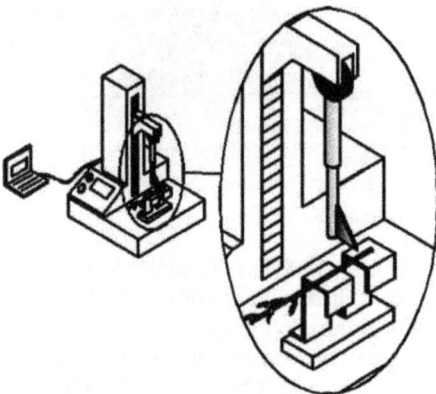

Figure 3. Picking system

The shearing forces on the load cell with respect to knife penetration were recorded by computer (Ozbek et al., 2009).

The shearing stress in N/mm^2 was calculated using the equation of Shahbazi et al., 2012:

$$\tau = \frac{F_{s\,max}}{A}$$

Where F_{smax} is the maximum shearing force of the curve in N, and A is the area of the stalk at the deformation cross-section in mm^2.

The rocket plants were attached to the apparatus from its stalks (Figure 4). The shearing tests were conducted with 0.8 mm.s^{-1} knife speed progress (Simonton, 1992).

Figure 2. Cutting system

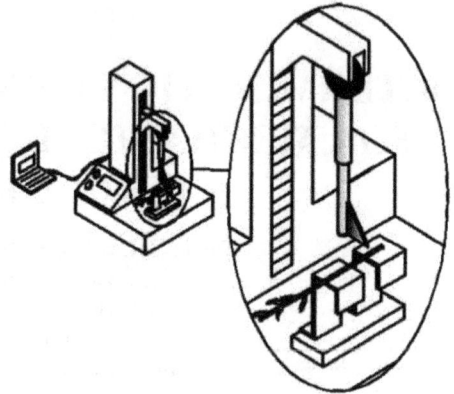

Figure 4. Measuring the cutting of rocket (*Eruca vesicaria*) plant

Picking force can be defined as the force required to separate leaf from ovary point (picking force of leafs). The load cell of the machine was then pulled upward to determine the picking force of the rocket leafs (Figure 5).

Figure 5. Measuring the picking force of rocket leaf

Bioyield force, shearing force, bending stress, shearing stress, and shearing deformation were calculated from the force-deformation curves at the inflection point as defined by ASAE Standard (1985). S368.1 (ASAE Standards, 1985) was obtained from all curves (Figure 6).

The energy of shearing was determined as the area under these curves (Chen et al., 2004; Srivastava, 2006).

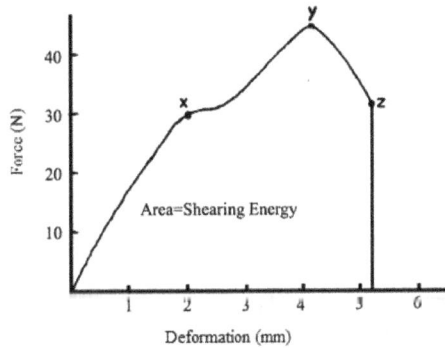

Note. Labels on the graph indicate the following points:

x – bioyield force, y – maximum force, z – shearing force (Liu, 2012).

Figure 6. Typical force-deformation curve of rocket (*Eruca vesicaria*) stalk during shearing loading

RESULTS AND DISCUSSIONS

Moisture content of the plants was determined as 89% at harvest time and all tests were conducted at harvest moisture. The strength measurements of rocket stalks are given in Table 1.

Table 1. Average strength properties of rocket stalk

	Maximum force (N)	Bioyield force (N)	Shearing force (N)	Stress in maximum force (MPa)	Energy in maximum force (J)	Shearing stress (MPa)	Shearing deformation (mm)	Area (mm²)
Stalk	4.820	3.856	2.510	0.474	0.015	0.219	22.149	10.534
Standard deviation	2.940	1.256	2.544	0.154	0.012	0.133	2.509	6.124

The maximum force was observed as 4.820 N at rocket stalk. The bioyield force of 3.856 N was observed at Stalk. Shearing force is one of the most important plant characteristics affecting plant harvesting. If the weight of the plant is known, the shearing force and the shearing height can be used to determine the speed of the blade to be used in harvesting (Igathinathane et al., 2010; Taghijarah et al., 2011). The maximum shearing force was observed as 2.510 N at Stalk. The maximum

stress value (0.474 MPa) was observed at Stalk. The energy at maximum force was found to be as 0.015 J. Deformation has an important place among the strength characteristics of the plant. The maximum shearing deformation (22.149 mm) was observed at Stalk. The average cross-sectional area of rocket was determined as 10.534 mm² at harvest moisture (89.9 %). The strength measurements of rocket leaf are given in Table 2.

Table 2. Average strength properties of rocket leaf

	Maximum force (N)	Bioyield force (N)	Shearing force (N)	Stress in maximum force (MPa)	Energy in maximum force (J)	Shearing stress (MPa)	Shearing deformation (mm)
Leaf	6.233	4.986	3.659	0.235	0.024	0.114	7.477
Standard deviation	3.044	1.563	2.088	0.102	0.017	0.085	2.623

The maximum force required to separate leafs from stalk was determined as 6.233 N. As a function of the maximum force the bioyield force was found to be 4.986 N. Lower shearing forces required for mechanical harvesting leads to savings in power and energy usage. Leaf shearing force of rocket observed 3.659 N is higher than stalk shearing force. The maximum stress in maximum force value (0.235 MPa) was observed at leaf. The energy at maximum force was found to be as 0.024 J.
Average shearing deformation value for rocket leaf was determined as 7.447 mm. The strength measurements of rocket root are given in Table 3.

Table 3. Average strength properties of rocket root

	Maximum force (N)	Bioyield force (N)	Shearing force (N)	Stress in maximum force (MPa)	Energy in maximum force (J)	Shearing stress (MPa)	Shearing deformation (mm)	Diameter (mm)
Root	21.798	17.439	18.912	1.097	0.084	1.017	19.785	3.924
Standard deviation	8.207	6.566	11.642	0.652	0.047	0.713	3.708	0.845

The maximum force and shearing force are important design parameters for harvesters and they should be known for power requirement. Therefore, the maximum force and shearing force were determined as 21.798 and 18.912 N, respectively. The bioyield force of (17.438 N) was observed at root. The stress value in maximum force was found to be as 1.097 MPa. The energy at maximum force (0.084 MPa) was observed at root. Average shearing deformation value of rocket root was observed as 19.785 N. The average root diameter of rocket *Eruca vesicaria* was determined as 3.924 mm at harvest moisture (89.9 %)

CONCLUSIONS

This study was carried out to determine the strength properties of rocket plant (*Eruca vesicaria*) at leaf, stalk and root sections in the harvest moisture. Properties as the maximum force, bioyield force, shearing force, stress in maximum force, energy in maximum force, shearing stress, shearing deformation of rocket leaf, stalk and root have determined at moisture content of 89.9%.
The strength parameters measured at root section higher than that of the stalk and leaf sections.
The lowest values were found at rocket stalk. The strength parameters of stalk section should be considered for mechanical harvesting of rocket plant to provide data for the design machines for mechanized applications.

REFERENCES

ASABE Standards., 2006. Moisture measurement e Forages. St. Joseph, MI: American Society of Agricultural and Biological Engineers (ASABE). S358.2

Asae Standards, 1985. Compression test of food materials of convex shape. St. Joseph, Mich.: American Society of Agriculture Engineering, S368.1

Bianco V.V., Boari F., 1997. Up-to-date developments on wild rocket cultivation, In: Padulosi S, Pignone D (Eds.), Rocket: A Mediterranean crop for the world. Report of a workshop 13-14 December 1996, Legnaro (Padova), Italy, 41-49.

Chen Y., Gratton J. L., Liu J., 2004. Power requirements of hemp cutting and conditioning. Biosystems Engineering, 87(4), 417-424.

Dolezalova I., Duchoslav M., Dusek, K., 2013. Biology and Yield of Rocket. Notulae Botanicae Horti Agrobotanici Cluj-Napoca, 41(2), 530.

Igathinathane C., Womac A. R., Sokhansanj S., 2010. Corn stalk orientation effect on mechanical cutting. Biosystems engineering, 107(2), 97-106.

Mohamedien S., 1995. Conservation and utilization of rocket in Mediterranean countries. Rocket cultivation in Egypt, p. 61-62. In: Padulosi S (Ed.). Rocket Genetic Resources Network. Report of the First Meeting, 13-15 November 1994, Lisbon, Portugal.

Morales M., Janick J., 2002. Arugula: A promising specialty leaf vegetable, In: Janick J, Whipkey A (Eds.). Trends in new crops and new uses. ASHS Press, Alexandria, VA, p. 418-423.

Nurzyńska-Wierdak R., 2006. The effect of nitrogen fertilization on yield and chemical composition of garden rocket (Eruca sativa Mill.) in autumn cultivation. Acta Sci PolHortoru 5:53-63.

Ozbek O., Seflek A.Y., Carman K., 2009. Some Mechanical Properties of Safflower Stalk. Appl. Eng. Agric., 25: 619-625.

Pignone D. 1997. Present status of rocket genetic resources and conservation activities, In: Padulosi S. & Pignone D. (ed.). Rocket: a Mediterranean crop for the world. Report of a workshop, 13-14 December 1996, Legnaro (Padova), Italy. International Plant Genetic Resource Instute, Rome, Italy, 2-12.

Pimpini, F., Enzo M., 1997. La coltura della rucola negli ambienti veneti. Colture protette 4:21-32.

Shahbazi F., Nazari Galedar M., 2012. Bending and shearing properties of safflower stalk. Journal of Agricultural Science and Technology, 14(4), 743-754.

Silva Dias J.C., 1997. Rocket in Portugal: botany, cultivation, uses and potential, In: Padulosi S, Pignone D (Eds.). Rocket: A Mediterranean crop for the world. Report of a workshop 13-14 December 1996, Legnaro (Padova), Italy, 81-85.

Simonton W., 1992. Physical properties of zonal geranium cuttings. Trans. ASAE, 35(6): 1899-1904.

Srivastava A.K., Goering C.E., Rohrbach R.P., Buckmaster D.R., 2006. Engineering principles of agricultural machines (2nd ed.). American Society of Agricultural and Biological Engineers, St. Joseph, USA, 185.

Taghijarah H., Ahmadi H., Ghahderijani M., Tavakoli M., 2011. Shearing characteristics of sugar cane (Saccharum officinarum L.) stalks as a function of the rate of the applied force. Australian Journal of Crop Science, 5(6), 630.

Tuzel Y., 1995. Conservation and utilization of rocket in Mediterranean countries. Rocket growing in Turkey, In: Padulosi S (Ed.). Rocket Genetic Resources Network. Report of the First Meeting, 13-15 November 1994, Lisbon, Portugal, 58-60.

21

THE EFFECTS OF DIFFERENT FERTIGATION TREATMENTS ON YIELD AND NUTRIENT UPTAKE OF WATERMELON PLANTS GROWN AS SECOND CROP IN CUKUROVA REGION

Ahmet DEMIRBAS

Cumhuriyet University, Vocational School of Sivas,
Department of Crop and Animal Production, Sivas, Turkey
Corresponding author email: ademirbas@cumhuriyet.edu.tr

Abstract

The present study was conducted to investigate the effects of different fertigation treatments (25, 50, 75 and 100 %) on yield and nutrient uptake of watermelon as the second crop and to compare with conventional practices. Experiments were conducted over the experimental fields of Çukurova University, Agricultural Faculty Soil Science and Plant Nutrition Department under field conditions. Experiment was conducted in randomized blocks split plots design with three replications. A total of 160 kg ha^{-1} nitrogen (N) as ammonium sulphate, 100 kg ha^{-1} phosphorus (P) as MKP and 200 kg ha^{-1} potassium (K) as KNO$_3$ were applied. Seedlings were planted in peat/perlite mixture (1:1 V/V) and transplanted to experimental plots following with wheat harvest in June. The watermelon plants were irrigated 9 times during the growing season in one week intervals. Current findings revealed that 75% fertigation treatment (120 kg ha^{-1} N, 75 kg ha^{-1} P, 150 kg ha^{-1} K) had the greatest yield (48.38 t ha^{-1}). Also, it increased N (%3.78), P (%0.31), Zn (45.7 mg kg^{-1}), Mn (43.1 mg kg^{-1}) and Cu (17.6 mg kg^{-1}) contents of plants.

Key words: Fertigation, watermelon, yield, nutrient uptake.

INTRODUCTION

Watermelon production had a significant place in world agriculture. Worldwide, watermelon is produced over 1.8 million hectares and annual production is around 29.7 million tons. Turkey has the second place worldwide after China in melon and watermelon production. Despite decreasing production lands in Turkey, annual production is increasing with incresing yields through proper cultural practices (Taşkaya and Keskin, 2004). The watermelon (*Citrullus lunatus* (Thunb.) is commonly grown in Turkey and fruits are consumed. Annual watermelon production of Turkey in 2008 from 123.000 ha was 3.5 million tons. In annual watermelon production Turkey, Çukurova with 678.73 thousand tons had the first place (Anonymous, 2008).

Just because of availability for production over large areas, easy marketing and high return rates per unit area; watermelon is also commonly produced in high or low tunnels (Yetişir and Sarı, 2004).

Watermelon has quite fast growing rate, short vegetation period and contains about 90-92% water. Therefore, irrigation is a must for high yield levels (Miller, 2002). Watermelon requires more frequent irrigation intervals because of excessive evaporations and low precipitation rates throughout the growing season (Doorenbos and Kassam, 1979).

Therefore, drip irrigation recommended for watermelon irrigation (Srinivas et al., 1989). Fertilizers are usually applied through drip lines together with irrigation water. This practice is called fertigation (Bar-Yosef, 1991). It is a new agricultural technique that provides fertilizer and water concurrently (Majahan and Singh, 2006; Castellanos et al., 2012). In this way, it is possible to control timing, amount and concentration of fertilizers (Hagin et al., 2002). It can provide water and fertilizer in a timely and correctly and increases nutrient uptake. Through fertigation, fertilizer contact with soil is minimized, fertilizers are directly applied to plant root zones and significant savings are achieved in water and fertilizers (Mohammad and Zuraiqi, 2003; Beyaert et al., 2007).

Fertigation is able to meet nutrient requirements of almost all plants, thus almost entire plant water requirement of watermelon is met over the low yield fields poor in nutrients and in places with irregular precipitations (Fernandes and Prado, 2004).

In this study, effects of different fertigation treatments on plant yield and nutrient uptake of watermelon grown as a second crop in Çukurova region with a semi-arid climate were investigated.

MATERIALS AND METHODS

The present research was conducted in summer season of 2012 over the experimental fields of Çukurova University Agricultural Faculty Soil Science and Plant Nutrition Department under field conditions. Soil characteristics are provided in Table 1.

Table 1. Physical and chemical characteristics of experimental soils.

Soil Property	Depth (0-30 cm)
pH (H_2O)	7.59
Lime (%)	24.1
Salt (%)	0.039
Organic matter (%)	1.30
Texture	CL
Total N (%)	0.10
Available P (kg ha^{-1})	24.2
Available K (kg ha^{-1})	1010.5
Available Fe (mg kg^{-1})	4.01
Available Mn (mg kg^{-1})	1.12
Available Zn (mg kg^{-1})	0.51
Available Cu (mg kg^{-1})	0.31

Experiments were conducted in randomized blocks split plots experimental design with 3 replications. Together with conventional method (0% fertigation), different fertigation doses as of 25% (40 kg N ha^{-1}, 25 kg P ha^{-1}, 50 kg K ha^{-1}), 50% (80 kg N ha^{-1}, 50 kg P ha^{-1}, 100 kg K ha^{-1}), 75% (120 kg N ha^{-1}, 75 kg P ha^{-1}, 150 kg K ha^{-1}) and 100% (160 kg N ha^{-1}, 100 kg P ha^{-1}, 200 kg K ha^{-1}) were experimented.

Just based on fertigation doses, remaining fertilizers were applied to soil with conventional method as follows:

- 25% fertigation - 75% conventional,
- 50% fertigation - 50% conventional,
- 75% fertigation - 25% conventional and
- 100% fertigation - 0% conventional.

In conventional method (0% fertigation) 160 kg N ha^{-1}, 100 kg P ha^{-1}, 200 kg K ha^{-1} were also applied. P and K were applied at planting and N was applied in 3 portions. Irrigations were

applied in the same fashion and same number of irrigations was performed. N was applied in ammonium sulphate form, P in MKP and K in KNO_3 form.

Crisby watermelon cultivar was used as the plant material. Seedlings were planted in peat:perlite (1/1 V/V) medium under greenhouse conditions. Then the seedlings were transplanted to field conditions as the second crop after wheat harvest on 15th of June. Each plot had 4 rows and each row had 4 plants. Therefore, each treatment had 16 plants. Row spacing was 1.80 m, on-row plant spacing was 1 m and total plot size was 16.2 m^2. There was 2 m spacing between the adjacent plots to prevent interactions among treatments. A total 9 irrigations were performed throughout the experimental period in 1 week intervals. Water was not supplied for 15 days during shoot elongation period. Leaf samples were taken from watermelon plants at the beginning of flowering. Samples were ground in a plant grinder and N content was determined with modified Kjeldahl method (Bremner, 1965). For P, K, Fe, Mn, Zn and Cu contents, 0.200 g plant samples were ashed at 550 °C for 5.5 hours in an ash oven. Then the samples were supplemented with 1/3 HCl and distilled water. Readings were performed in resultant extract at P 882 nm in a UV-spectrophotometer (Murphy and Riley, 1962). K, Fe, Mn, Zn and Cu contents were determined with an Atomic Absorption Spectrophotometer (AAS) (Güzel et al., 1992).

Watermelon plants were harvested 77 days after transplantation into field. Following the harvest, yields were determined.

Experimental results were subjected to analysis of variance (ANOVA) by using SPSS 22.0 Windows software. The difference between treatments means were tested with Tukey's test.

RESULTS AND DISCUSSIONS

The effects of different fertigation doses on yields of watermelon plants are presented in Figure 1. Current findings revealed that 75% fertigation treatment had the greatest effect on yield of watermelon plants (48.38 t ha^{-1}). It was followed by 100% fertigation treatment with a yield of 48.14 t ha^{-1}. Bhat et al (2007) carried out a study with data palm plants between the years 1996-2006 and applied 4 different fertigation doses (25, 50, 75 and 100 of recommended fertilizer dose). Researchers reported the greatest yield for 75% fertigation treatment (75:13.5:87.7 g N, P, K year^{-1}) with a yield of 37.21 t ha^{-1}. The 0% fertigation dose (conventional fertilizer application) with a yield of 44.29 t ha^{-1} had higher yield than 25 and 50% fertigation treatments and the differences

from 75% fertigation treatment were not significant. The yield of watermelon plants grown as second crop varied between 27.46–48.38 t ha^{-1}. In previous studies, based on climate conditions and water management systems, watermelon yields were reported as between 50 t ha^{-1} and 80 t ha^{-1} (Srinivas et al., 1989; Gündüz et al., 1996; Simsek et al., 2004).

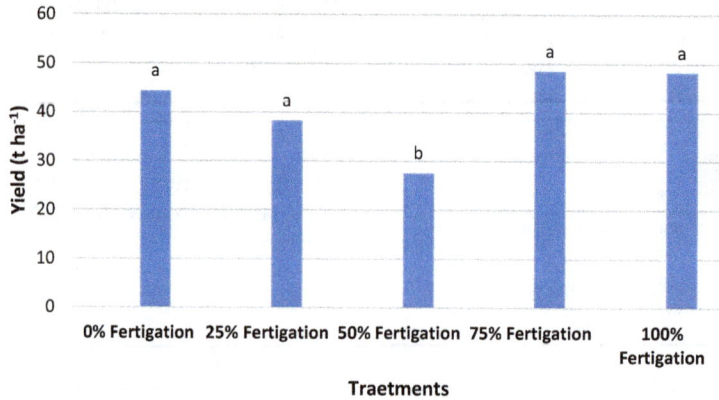

Figure 1. Effects of different fertigation doses on yields of watermelon plants (t ha^{-1})

Considering the effects of treatments on N contents of watermelon plants (Table 2), it was observed that 75% fertigation doses had the greatest N content with 3.78%. The 0% fertigation doses (conventional method) had the lowest N content with 3.43%. As compared to 0% fertigation dose, other treatments increased N contents of plants more. Similar to N contents, 75% fertigation treatments had the greatest P content with 0.31% and 0% fertigation treatment had the least P content with 0.25%. Such a case may be resulted from direct application of plant nutrients, especially hard-to-transport phosphorus to soil through fertigation. Again as it was in N and P contents, all fertigation treatments, except for 0%, increased plant K contents. Among the

fertigation treatments, 50% treatment had the greatest K content (3.62%), because N promoted only the vegetative growth and eased the uptake of other nutrients like P and K (Riley and Barber, 1971; Soon and Miller, 1977). Such a synergic impact of nitrogen on other plant nutrients also supports the P uptake within the root zone (Drew, 1975; Anghinoni and Barber, 1988). Potassium (K) and nitrogen (N) are the mostly used nutrients in watermelon (Grangeiro and Cecílio Filho, 2004; Vidigal et al., 2009; Silva et al., 2012). When these nutrients were taken through fertigation, they are distributed uniformly and thus improve fruit quality and yield levels (Hochmuth, 1992), and reduce production costs and prevent various environmental problems.

Table 2. Effect of different fertigation treatments on N, P and K contents of watermelon plants (%)

Treatments	N		P		K	
0% Fertigation	3.43	±0.02 c	0.25	±0.02 c	3.13	±0.02 d
25% Fertigation	3.75	±0.18 a	0.27	±0.00 bc	3.21	±0.04 cd
50% Fertigation	3.53	±0.06 bc	0.29	±0.01 ab	3.62	±0.15 a
75% Fertigation	3.78	±0.00 a	0.31	±0.00 a	3.52	±0.15 ab
100% Fertigation	3.70	±0.13 ab	0.29	±0.01 ab	3.38	±0.13 bc

P<0.05

Considering the effects of different fertigation treatments on microelement content of watermelon plants, it was observed that the greatest Fe content was obtained from 50% fertigation treatment with 135.5 mg kg^{-1} and it was followed by 0% fertigation treatment (conventional method) with 129.1 mg kg^{-1} (Table 3).

Table 3. Effect of different fertigation treatments on Fe, Zn, Mn and Cu contents (mg kg^{-1})

Treatments	Fe		Zn		Mn		Cu	
0% Fertigation	129.1	±1.62 ab	36.5	±0.72 c	36.4	±0.40 b	14.5	±1.71 c
25% Fertigation	118.7	±3.74 c	41.3	±0.19 b	36.2	±0.14 b	15.7	±1.06 a-c
50% Fertigation	135.5	±3.60 a	41.1	±0.36 b	37.4	±3.31 b	15.6	±0.82 bc
75% Fertigation	125.1	±2.12 bc	45.7	±2.35 a	43.1	±3.83 a	17.6	±0.34 a
100% Fertigation	108.2	±5.23 d	41.4	±1.28 b	35.9	±1.06 b	17.3	±0.33 ab

P<0.05

The lowest Fe content was obtained from 100% fertigation treatment (108.2 mg kg^{-1}). Considering the N contents, similar to N, P and K contents, the greatest value was observed in 75% fertigation treatment (45.7 mg kg^{-1}) and 0% fertigation treatment (36.5 mg kg^{-1}) did not have significant effects on Zn contents as compared to other treatments. Again 75% fertigation treatment had the greatest Mn content (43.1 mg kg^{-1}) and Cu content (17.6 mg kg^{-1}). The relations between the parameters are presented in Table 4.

Table 4. Correlation among variables tested in the experiment

	Yield	N	P	K	Fe	Zn	Mn
Yield							
N	0.185						
P	0.097	-0.392					
K	-0.360	-0.444	0.754**				
Fe	-0.584*	0.361	-0.090	0.322			
Zn	0.153	-0.181	0.869**	0.629*	-0.155		
Mn	0.231	0.150	0.609*	0.397	0.220	0.723**	
Cu	0.311	-0.460	0.800**	0.456	-0.467	0.759**	0.464

*Significant at P<0.05
**Significant at P<0.01

Positive correlation was determined between P and K, Zn, Cu concentrations (p<0.01) and Mn concentrations (p<0.05). Also between Zn and Mn, Cu concentrations (p<0.01) positive correlation was determined. Generally, the other relations were found to be insignificant (p>0.05).

CONCLUSIONS

Considering the entire results of the present study, it was concluded that nutrient supply through fertigation significantly improved yields and N, P, K, Zn, Mn and Cu contents of watermelon plants as compared to conventional method of nutrient supply (0% fertigation).
Especially 75% fertigation treatment (25% from the soil) had greater impacts on watermelon plants than the other fertigation treatments. In general, irrigation improves yields and yield components of watermelon plants grown in semi-arid climate conditions.

REFERENCES

Anghinoni I., ve Barber S.A., 1988. Corn root growth and nitrogen uptake as affected by ammonium placement. Agron. J. 80:799-802.

Anonymous, Bitkisel Üretim İstatistikleri, T.C. Başbakanlık Türkiye İstatistik Kurumu (TÜİK), Ankara, www.tuik.gov.tr (Erişim Tarihi: 30/11/2008).

Bar-Yosef B., 1991. Fertilization Under drip Irrigtion . In: Fluid Fertilizer Science and Technology (Eds. Palgrave, Derek A., Marcel Dekker) Inc. New, 285-329.

Beyaert R.R., Roy R.C., ve Coelho B.K.B., 2007. Irrigation and fertilizer management effects on processing cucumber productivity and water use efficiency. Can. J. Plant Sci. 87, 355–363.

Bhat R., Sujatha S., ve Balasimha D., 2007. Impact of drip fertigation on productivity of arecanut (Areca catechu L.). Agricultural Water Management 90, 101–111.

Bremner J.M., 1965. Method of Soil Analysis. Part 2. Chemical and Microbiological Methods. American Society of Agronomy Inc. Madison, Wise USA., 1149-1178.

Castellanos M.T., Tarquis A.M., Ribas F., Cabello M.J., Arce A., Cartagena M.C., 2012. Nitrogen fertigation: an integrated agronomic and environmental study. Agricultural Water Management,http://dx.doi.org/10.1016/j.agwat.2012.06.016.

Doorenbos J., Kassam A.H., 1979. Yield response to water. Irrigation and drainage paper no: 33. FAO-Rome, 193.

Drew M.C., 1975. Comparison of the effects of localized supply and the growth of the seminal root system in barley. II. Compensatory increases in the growth of lateral roots, and in rates of phosphate uptake, in response to a localized supply of phosphate. J. Exp. Bot. 29:435-451.

Erdem Y., Yüksel A.N., Orta A.H., 2001. The effects of deficit irrigation on watermelon yield, water use, and quality characteristics. Pak J. Biol Sci 4(7):785–789.

Fernandes F.M., Prado R.M., 2004. Fertirrigação da cultura da melancia. In: Boaretto, A.E.; Villas Boas, R.; Souza, W.F.; Parra, L.R.V. (Eds.) Fertirrigação: teoria e prática. Piracicaba, vol.1, 632-653.

Grangeiro L.C, Cecílio Filho A.B., 2004. Acúmulo e exportação de macronutrientes pelo híbrido de melancia Tide. Hortic. Bras., 22, 93-97.

Gündüz M., Kara C., Bilgel L., Değirmenci V., 1996. Determination of irrigation scheduling of watermelon in Harran plain. In: Pakyurek A (ed) The first vegetable conference proceedings. Harran, Sanlıurfa-Turkey, pp 211–216.

Hagin J., Sneh M., ve Lowengart A., 2002. Fertigation – Fertilization through irrigation. IPI Research Topics No. 23. Ed. by A.E. Johnston. International Potash Institute, Basel, Switzerland.

Hochmuth G.J., 1992. Fertilizer management for drip-irrigated vegetables in Florida. HortTechnology. 2, 27-32.

Mahajan G., Singh K.G., 2006. Response of Greenhouse tomato to irrigation and fertigation. Agricultural Water Management 84, 202–206.

Miller G., 2002. Home and Garden Information Center (Excerpted from *Home Vegetable Gardening*, EC).

County Extension Agent, Clemson University, ABD. http://hgic.clemson.edu/factsheets/hgic1325.htm.(*Eri şim Tarihi: 30/06/2008*).

Mohammad M.J., ve Zuraiqi S., 2003. Enhancement of yield and nitrogen and water use efficiencies by nitrogen drip-fertigation of garlic. Journal of Plant Nutrition. 26 (9), 1749-1766.

Murphy L., Riley J. P., 1962. A Modified Single Solution Method for the Determination of Phosphate in Natural Waters. Anal. Chem. Acta. 27:31-36.

Riley D., ve Barber A., 1971. Effect of ammonium fertilization on phosphorus uptake as related to root-induced pH changes at the root-soil interface. Soil Sci. Soc. Am. Proc. 25: 301-306.

Silva M.V.T. da, Chaves S.W.P., Medeiros J.F., de Souza M.S., de Santos A.P.F., 2012. Acúmulo e exportação de macronutrientes em melancieiras fertirrigadas sob ótimas condições de adubação nitrogenada e fosfatada. Agropecuária Científica no Semi-Árido 8, 61-70.

Simsek M, Kacira M, Tonkaz. T., 2004. The effects of different drip irrigation regimes on watermelon yield and yield components under semi-arid climatic conditions. Aust J Agric Res 55:1149–1157.

Soon, Y.K., ve Miller, H., 1977. Changes in rhizosphere due to ammonium and nitrate fertilization and phosphorus uptake by corn seedlings. Soil Sci. Soc. Am. Proc. 41:77-80.

Srinivas K., Hegde D.M., Havanagi C.V., 1989. Irrigation studies on watermelon. Irrigation Sci 10 (4):293–301.

Taskaya B., Keskin G., 2004. Kavun-karpuz. Tarımsal Ekonomi Araştırma Enstitüsü, Sayı 6, Ankara.

Vidigal S.M., Pacheco D.D., Costa E.L., da Facion C.E., 2009. Crescimento e acúmulo de macro e micronutrientes pela melancia em solo arenoso. Rev. Ceres. 56, 112-118.

Yetişir H., Sari N., 2004. Effect of Hypocotyl Morphology on Survival Rate and Growth of Watermelon Seedlings Grafted on Rootstocks with Different Emergence Performance at Various Temperatures. Turk. J. Agric., 28, 231-237.

22

ESTABLISHING THE CROP ASSORTMENT OF WATER MELON (*CITRULLUS LANATUS*) DEPENDING ON THE ELEMENTS THAT DEFINE THE PRODUCTION

Mihaela Gabriela CIUPUREANU (NOVAC)[2], Elena CIUCIUC[1], Daniela POPA[2]

[1]Research - Development Center For Field Crops on Sandy Soils Dăbuleni, Romania
[2]University of Craiova, Faculty of Horticulture, Doctoral School of Engineering Plant and Animal Resources, 13 A.I. Cuza Street, Craiova, Dolj, Romania

Corresponding author email: danapopa2013@gmail.com

Abstract

The aim of this study it was that of following the behavior in culture on field on RDCFCSS Dabuleni – Romania, on a sandy soil, of 12 cultivars of water melons of which 9 are hybrids of foreign provenance and 3 local varieties. It were made observations and determinations concerning the morphological characters and the production of the cultivars such as: the vegetative growth of plants, the flowering and the fruits binding date, the number of fruits/plants, the average weight of a fruit and the productions on cropping stages. Following the obtained results it was found that the highest production was registered on 'Baronesa' F1 cultivars and 62-269 F1 with 47.9 t/ha, followed by LF 6720 with 44.9 t/ha and 'Oltenia' with 43.6 t/ha.

Key words: water melon, temperature, fruits, production.

INTRODUCTION

Nowadays, in a competitive market conditions, the cultivators of vegetable species are becoming more interested to cultivate varieties and hybrids of foreign provenance with an acclimatization risk to the natural conditions of the specific area.

The choice of varieties and hybrids in relation to pedo-climatic conditions of the area it is possible only after a preliminary testing of those. Because the water melon fruits are consumed only in fresh conditions, the gustative qualities of those one have a particular importance.

The normal progress of the metabolic processes on vegetable plants is closely related to environmental factors. The direct factors that acts directly on plants, representing their condition of existence, are: climatic factors (light, heat, water and air) edaphic factors (the texture and the soil structure the chemistry and the trophicity of soil, the groundwater) and biotic factors (human and other living organisms) (Voican et al., 1998). From the point of view of requirements for temperature, the sandy soils area from the South of Oltenia

can be considered a very favorable area for growing water melons (Toma et al., 2007).

In the last decades the humanity is facing more and more with the effects of global change, customized on regional level. The increase of water melons productions is based on the use of biological material variety, hybrid) that are becoming more performant, that through the genetic heritage that they possessed it can adapt to the environmental conditions of the area.

Choosing the best varieties and hybrids in relation to the customer requirements and to the pedo-climatic conditions from the area, is possible only knowing in detail the specific of as many varieties and hybrids of water melons existing in our country and abroad (Maria Dinu et al., 2016).

The water melons profitable capitalize the pedo- climatic conditions specific to the sandy soils under the application of an irrigation regime well led (Marinica, 1998). For the sandy soils area, Nanu (1998), recommend the cultivars: 'Dulce de Dăbuleni', 'Crimson Sweet' and 'Red Star' F1 hybrid. The assortment of yellow and green water melons periodically renewed due to the appearance of new creations and in accordance with the market requirements. As a result, it is therefore necessary to study some cultivars (varieties and hybrids) with high adaptability to the eco-pedological specific conditions of the sandy soils in order to introduce them in culture.

The aim of this study was that to track the behavior in pedo-climatic conditions from the Southern Romania (RDCFCSS Dăbuleni) of 12 cultivars of water melon from which 3 varieties are aborigine creations and 9 foreign hybrids in order to recommend the most valuable for adaptability and productivity.

MATERIALS AND METHODS

The present study was conducted in the experimental field of RDCFCSS Dabuleni in 2016.

The experience was mono factorial and was placed after the method of randomized blocks. It has 12 variants with 4 repetitions and consists in 3 varieties and 9 hybrids.

The surface of a variant was of 20 m^2. The cultural experimental scheme was of 200 x 100 cm.

The variants specific were:
V1 – 'De Dăbuleni',
V2 – 'Dulce de Dăbuleni',
V3 – 'Oltenia',
V4 – 'Susy' F1,
V5 – 'Baronesa' F1,
V6 – 'Oneida' F1,
V7 – 'Huelva' F1,
V8 – '62-269' F1,
V9 – 'Fantasy' F1,
V10 – 'Tarzan' F1,
V11 – 'Grand Baby' F1,
V12 – 'LF 6720' F1.

The experience was established by planting the seedling on 28 April 2016, previously produced in a solar greenhouse. The maintenance works applied to the culture it was the general specific ones to a water melon crop. Regarding the crop particularities, they relate to the use of plastic for mulch and to the use of drip irrigation system. It was performed morphological determinations on plants that were focused on the haulm length and also production determinations (number of fruits/plants, the average weight of a fruit, the total production). The obtained results were calculated and interpreted statistically.

RESULTS AND DISCUSSIONS

Regarding the climatic conditions in which was placed the experience (in field, on a sandy soil) it can say that the year 2016 (table 1)it has been especially from climatic point of view, excepting May month, in all the other months have recorded average temperatures over of each month multi annual average temperature.

Table 1.The climatic conditions - April - July 2016

Month	Specification	Temperature (^{0}C)	Rainfalls (mm)
April	Minimum	0.8	60.2
	Maximum	31.4	
	Monthly average	**15.0**	
	Multiannual average	12.1	
May	Minimum	5.5	104.4
	Maximum	32.9	
	Monthly average	**16.9**	
	Multiannual average	17.4	
June	Minimum	16.9	53.2
	Maximum	37.7	
	Monthly average	**23.6**	
	Multiannual average	21.4	
July	Minimum	11.4	31.6
	Maximum	38.0	
	Monthly average	**24.8**	
	Multiannual average	23.2	

April was particularly warm registering a maximum temperature of 31.4^{0}C and a medium temperature of 15^0 C much higher from the multiannual average temperature of the month that it is of 12.1^{0}C.

The amount of precipitations from April was of 60.2 mm. The May was rich in rainfalls for the area, registering 104.4mm leading to a decrease of temperatures.

The minimum temperature of the month was of 5.5^{0}C which negatively influenced the plants of water melons hardly planted.

The maximum of the month was of 32.9^{0}C. The May monthly average was of 16.9^{0}C lower than the multiannual average of the month with 0.5^{0}C.

June and July were very warm and poor in rainfalls, the minimum temperatures exceeding the multiannual average. Thus, in June it was registered in average 23.6^0C compared to 21.4^0C which is the multiannual average with a minimum temperature of 16.9^0C and a maximum temperature of 37.7^0C, and the amount of precipitations were of 53.2 mm.

The medium temperature of July was of 24.8^0C with limits between $11.4 - 38.0^0$C on the background of some low rainfalls, respectively 31.6 mm.

During the vegetation period it were made observations regarding the dynamic of the vegetative growth of the plants on each cultivar in part (Table 2).

The length of the water melons haulm on the fruit bind time it was between 1.84-2.35 m. It have been note by vigor the cultivars: 'Huelva' F1 (2.35 m), 'Oneida' F1 (2.14 m), 'Dulce de Dăbuleni' (2.11 m), 'LF 6720' F1 (2.06 m) and 'Oltenia' (2.03 m).

The phenological stage of water melon plant flowering highlights 'Oneida' and 'Huelva' cultivars (02.06.2016) then 'Grand Baby' and 'LF 6720' (03.06.2016) and then the Romanian varieties that have a deferall against the foreign hybrids of approximately 10 days. The set of the first fruits occured in the period of 9 - 15 June on cultivars of foreign origin and in the period of 20 - 24 June on indigenous cultivars.

This difference of fruit setting between Romanian and foreign cultivars is unfavorable to the sandy area in which the water melon crops are set up mainly for the production precociously.

Table 2. Morphological and production determinations at the cultivars taken into study (minimum values)

Cultivar	Length of haulm (m)	Flowering time	Time of binding - first fruits
'De Dăbuleni	1.86	13.06	22.06
'Dulce de Dăbuleni'	2.11	10.06	24.06
'Oltenia'	2.03	10.06	20.06
'Susy' F1	1.91	2.06	9.06
'Baronesa F1	1.98	6.06	13.06
'Oneida' F1	2.14	2.06	10.06
'Huelva' F1	2.35	2.06	10.06
'62-269' F1	1.88	7.06	15.06
'Fantasy' F1	1.84	6.06	13.06
'Tarzan' F1	1.89	7.06	15.06
'Grand Baby' F1	1.86	3.06	10.06
'LF 6720' F1	2.06	3.06	10.06

The cropping was done at technological maturity, and the production was registered on each variant in part, on harvesting stages. The first harvest (table 3) was on 13 July on foreign cultivars and on 22 July on indigenous cultivars

Table 3. The dynamic of water melon production depending on cultivar (average values)

Cultivar	Production (t/ha) on:		
	13 July	22 July	1 August
'De Dăbuleni'	-	25.0	8.8
'Dulce de Dăbuleni'	-	23.1	12.2
'Oltenia'	-	21.0	22.6
'Susy' F1	13.3	17.5	1.8
'Baronesa' F1	13.9	28.5	5.5
'Oneida' F1	14.6	15.6	8.2
'Huelva' F1	18.0	18.8	5.0
'62-269' F1	19.0	25.0	3.9
'Fantasy' F1	8.8	24.1	-
'Tarzan' F1	8.2	22.0	-
'Grand Baby' F1	16.3	25.3	-
'LF 6720' F1	16.4	28.5	-

The productions obtained on 13 July were between 8.2 t/ha at 'Tarzan' F1 cultivar and 19 t/ha at '62-269' F1 cultivar. It also noted through the registered production on this date the cultivars 'Huelva 'F1 with 18 t/ha, 'Grand Baby' F1 with 16,3 t/ha and 'LF 6720' F1 with 16,4 t/ha.

At the second stage of harvest (22 July) all productions increased at all cultivars being between 15.6-28.5 t/ha. The cultivars 'Fantasy' F1, 'Tarzan' F1, 'Grand Baby' F1 and 'LF 6720' F1 completed their period of vegetation on 22 July, and the other cultivars on 1st of August.

At the last stage of harvesting, respectively at 1st of August the productions yield were between 1.8 t/ha at 'Susy' F1cultivar and 22.6 t/ha at 'Oltenia' cultivar. The average number of fruits/plants harvested was between 1.5-2 fruits/plants (table 4). The lowest number of fruits per plant was recorded at 'Fantasy' F1 cultivar, while the highest one was recorded at cultivars 'Oneida' F1, 'Huelva' F1 and 'LF 6720' F1. A healthy plant of water melon can produce 1-4 fruits per harvest.

Regarding the average weight of a fruit of water melon this ranged between 3.6-5.3 kg/fruit. The biggest fruits were recorded on cultivars 'Oltenia', 'Baronesa' F1 and '62-269' F1. These registered results are confirmed by the existent data from literature. A study conducted by Sari et al. (2007) present results according to the average weight of water melon fruits ranged between 1.885-8.033 kg. The similar results were reported also by Pakyurek şi Yanmaz (2008), in a study of a genotypes assortment of 13 water melons, where it was identified an average fruits weight between 1 – 4 kg. Ayhan et al. (2014) identified some cultivars of water melon with an average weight of fruits between 1.29-3.96 kg. Cordova et al. (1995) classified the water melons according to the weight of fruits in this way: the fruits with a weight of 4 kg are considered as being small, the fruits with a weight between 4 - 6 kg are considered as being medium and the fruits with a weight between 8 - 12 kg are considered as being giant.

Table 4. The number of fruits per plant and the average weight water melon fruits depending on cultivar

Cultivar	Number of fruits/plant	Fruit weight (kg)
'De Dăbuleni'	1.8	3.8
'Dulce de Dăbuleni'	1.8	3.8
'Oltenia'	1.6	5.3
'Susy' F1	1.7	3.6
'Baronesa' F1	1.9	5.0
'Oneida' F1	2.0	4.0
'Huelva' F1	2.0	4.0
'62-269' F1	1.8	5.2
'Fantasy' F1	1.5	4.4
'Tarzan' F1	1.6	3.9
'Grand Baby' F1	1.9	4.4
'LF 6720' F1	2.0	4.5

The production of water melon (table 5) registered at the 12 cultivars taken into study was between 32.6 – 47.9 t/ha.
The biggest productions as against the control were achieved on cultivars 'Baronesa' F1 and '62-269' F1 which realized 47,9 t/ha, followed by 'LF 6720' with 44.9 t/ha 'Oltenia' with 43.6 t/ha, 'Huelva' F1 with 41.8 t/ha and 'Grand Baby' F1 with 41.6 t/ha. It was elected the cultivar 'Oltenia' as control because it is cultivated mostly in Dabuleni area and it produce appreciable quantities of fruits. The production differences as against the control were between 4.3-1.3 t/ha, having positive differences. The smallest productions of water melon were obtained at cultivars 'Susy' F1, 'Fantasy' F1 and 'De Dăbuleni'.

Table 5. The total production at the water melon studied cultivars (average values)

Cultivar	The obtainded production (t/ha)	± Dif. As against Mt.(t/ha)
'De Dăbuleni'	33.8	-9.8
'Dulce de Dăbuleni'	35.3	-8.3
'Oltenia'	43.6	Mt.
'Susy' F1	32.6	-11.0
'Baronesa' F1	47.9	+4.3
'Oneida' F1	38.4	-5.2
'Huelva' F1	41.8	-1.8
'62-269' F1	47.9	+4.3
'Fantasy' F1	32.9	-10.7
'Tarzan' F1	30.2	-13.4
'Grand Baby' F1	41.6	-2.0
'LF 6720' F1	44.9	+1.3

DL 5% = 17.30 t/ha;
DL 1% = 23.57 t/ha;
Dl 0.1% = 31.68 t/ha.

The production of 'Oltenia' variety is remarkable for the sandy area from the Southern Romania, being the only Romanian variety with a production over 40 t/ha.
Between the foreign cultivars only five of them registered productions over 40 t/ha, three of them being precocious (62-269 F1, 'Huelva' F1 and 'LF 6720' F1).

CONCLUSIONS

From the obtained results about the behaviour in culture, in the pedoclimatic conditions from RDCFCSS Dăbuleni we notice the following cultivars:
- concerning the number of fruits/plant, 'Oneida' F1, 'Huelva' F1 and 'LF 6720' F1, and by the average weight of a water melon fruit, the cultivars 'Oltenia', 'Baronesa' F1 and '62-269' F1;
- precociousness was noted by the cultivars '62-269' F1 with 19 t/ha and 'Huelva' F1 with 18 t/ha on 13th of July.
- the biggest production was recorded on cultivars 'Baronesa' F1 and '62-269' F1 that achieved 47.9 t/ha, followed by 'LF 6720' with 44.9 t/ha; 'Oltenia' with 43.6 t/ha; 'Huelva' F1 with 41,8 t/ha and 'Grand Baby' F1 with 41.6 t/ha.

ACKNOWLEDGMENTS

The results were part of a research made in collaboration between the University of Craiova and the Research-Development Center For Field Crops On Sandy Soils Dăbuleni, for a doctoral thesis.

REFERENCES

Ayhan G., Nebahat S., Ilknur S., 2014. See Yield and Quality of Watermelon Genotypes Having Snack Food Potential. Cucurbitaceae Proceedings, Michigan, USA, 57- 62.

De Cordova F., Diez M.J., Iglesias A., Nuez F., 1995. Germplasm resources of *Citrullus lanatus* in the genebank of the Polytechnic University of Valencia, Cucurbit Genetics Cooperative 28:52–54.239

Dinu M., Soare R., 2016.The influence of cultivar on the quality of fruit the species *Cucumis melo* L. Annals of the University of Craiova, Agriculture, Montanology, Cadastre Series. Vol.XLVI, No 2, 105-111.

Marinică Gh., 1998. Cercetări privind regimul de irigare aplicat pepenilor verzi pe solurile nisipoase amenajate din sudul Olteniei. Lucrări ştiinţifice SCDCPN Dăbuleni, vol. X, 161-167.

Nanu Şt., 1998. Soiuri şi hibrizi de pepeni verzi cultivate pe solurile nisipoase din sudul Olteniei *Citrullus lanatus* (Thunb.) Matsum et Nakai. Lucrări ştiinţifice SCDCPN Dăbuleni, vol. X, 124-130.

Nanu Şt., Toma V., 2003. Soiuri şi hibrizi de pepeni galbeni cultivaţi pe solurile nisipoase din sudul Olteniei. Lucrări ştiinţifice SCDCPN Dăbuleni, Volumul XV, Ed. SITECH Craiova, 159-164.

Pakyurek A.Y., Yanmaz R., 2008. Çerezlik karpuz (*Citrullus lanatus* (Thumb.) Matsum.) yetistiriciligine uygun gen kaynaklarının toplanması ve degerlendirilmesi uzerine arastırmalar. VII th Symposium of Vegetable Crops, Yalova, 236.

Sari N., Aka Kacar Y., Yalcin Mendi Y., Solmaz I., Aktas H., 2007. Morphological and genetic characterization of watermelon genetic resources. TUBITAK Project No:104O073.

Toma V., Ciuciuc Elena, Croitoru Mihaela, Ploae Marieta, 2007. Comportarea unor cultivare de pepeni verzi în cultură altoită pe solurile nisipoase din sudul Olteniei. Lucrări ştiinţifice SCDCPN Dăbuleni, vol. XVI., 129-139.

Voican V., Lăcătuş V., 1998. Cultura protejată a legumelor în sere şi solarii. Editura CERES, Bucureşti. ISBN 973-40-0398-4.

23

BIOLOGICAL CONTROL OF TWO-SPOTTED SPIDER MITE IN PEPPER AND MELON CROPS CULTIVATED IN TUNNELS

Maria CĂLIN [1], **Tina Oana CRISTEA** [1], **Silvica AMBĂRUŞ** [1], **Creola BREZEANU** [1],
Petre Marian BREZEANU [1], **Marcel COSTACHE** [2], **Gabriela ŞOVAREL** [2], **Liliana BRATU** [2]

[1]Vegetable Research and Development Station Bacau, Calea Bârladului Street, no. 220,
Bacau County, Romania
[2]Research Development Institute for Vegetable and Flower Growing Vidra, Romania
Corresponding author email: sclbac@legumebac.ro

Abstract

The paper aims to present the attack characteristics and the results of biological control of two-spotted spider mite on pepper and melon crops cultivated in tunnels. The attack of two-spotted spider mite began in the first decade of June. The degree of attack of two-spotted spider mite on pepper plants was ascendant reaching 17.5% in the second decade of September. The degree of attack of two-spotted spider mite in melon was high, reaching 61.3% in the third decade of September. The predator, Phytoseiulus persimilis At.-H. (Arachnida: Mesostigmata: Phytoseiidae) was used to control the spider mite in peppers and melons. The variants utilized in our experiment were as it follows: V1 - one release with 50,000 ex./ha; V2 - one release with 100 thousand ex./ha; V3 – one release with 150 thousand ex./ha; V4 - Untreated control. The releases of Phytoseiulus persimilis has reduced the two-spotted spider mite degree of attack (DA%) on pepper in late August to 2.3% in V1, 2.1% in V2 and 1.7% in V3. In September the DA% was below 1% in all the three variants. In the untreated control, the pest degree of attack level increased from 9.5% in the first decade of August, up from 17.5% in the 2nd decade of September, and then began to decline. In melons, the releases of Ph. persimilis have reduced the attack degree in the third decade of August at 1.8% in V1, 1.9% in V2, 1.5% in V3. In all three variants, the degree of attack in September was less than 1%. In V4 (untreated control), DA% increased from 4.3% in the first decade of August, to 61.3% at the end of September.

Key words: biological control, Phytoseiulus persimilis, two-spotted spider mite, pepper, melon.

INTRODUCTION

The two-spotted spider mite - *Tetranychus urticae* Koch is one of the main pests of vegetable crops grown in greenhouses and tunnels (Dalpe, 2002; Zhang, 2003; Calin, 2005; Herrmann et al., 2011).

The pest is polyphagous, that is attacking 1110 host species (Herrmann, 2017), in the field, greenhouse and tunnels. In our country, this pest attacks crops of eggplants, peppers, cucumbers, beans, tomatoes, etc. Pest attack is higher in crops of eggplants and cucumbers and less in other species (Candea, 1984). The spider mites have caused little visible damage to the leaves and induced direct defense responses. According to Merjin et all., 2004 after two-spotted spider mite attack the proteinase inhibitor activity had doubled and the transcription of genes involved in jasmonate-, salicylate-, and ethylene-regulated defenses had been activated. On day four, proteinase inhibitor activity and particularly transcript levels of salicylate-regulated genes were still maintained. In addition, genes involved in phospholipid metabolism were up-regulated on day one and those in the secondary metabolism on day four. Although transcriptional up-regulation of the enzymes involved in the biosynthesis of monoterpenes and diterpenes already occurred on day one, a significant increase in the emission of volatile terpenoids was delayed until day four.

The emergence of resistance to chemical acaricides, the negative effect of chemicals on useful fauna and pesticide residues determined the usage of predator mites, as *Phytoseiulus persimilis* Athias-Henriot for the control of two-spotted spider mite (Lenteren, 2003 and 2012).

This predator arrived accidentally in Germany from Chile in 1958. Anticipating their effectiveness in pest control in 1960 research began in the UK, Netherlands, Canada, USA etc. in order to determine the influence of this predator in the control of two-spotted spider mite at various vegetable and flower crops (Rojas, 2010). The obtained results were exceptional and passed to currently reared *Ph. persimilis* (Shih 2001; Bolckmans, 2007) and its use in greenhouses and tunnels for control of two-spotted spider mite.

MATERIALS AND METHODS

During 2015 - 2016 years, experiments in tunnels were performed at Vegetable Research-Development Station Bacau - Romania, in order to study the two-spotted spider mite attack in pepper and melon crops and evaluate the effectiveness of biological control using *Phytoseiulus persimilis* predator. The relevant results from 2016 are presented in this paper.

1. Study of two-spotted spider mite attack in pepper and melon crops in tunnels

The observations were made every 10 days from May to September in Control.

For the attack estimation we used the following indicators:

-      Frequency of attack (F%),
-      Intensity of attack (I%),
-      Degree of attack (DA%).

2. Biological control of pest.

The predatory mite *Phytoseiulus persimilis* At.-H. *(Arachnida, Mesostigmata, Phytoseiidae)* was used for biological control of pest.

The effectiveness of this predator in control of two-spotted spider mite - *Tetranychus urticae* Koch. was studied in pepper collection of cultivars and melon in tunnels. When the degree of attack of mites exceeded in pepper 9.5% the predatory mites were released in the 3 variants. The trial was done in the summer period and early autumn. The average highest day temperature was between 28-32°C with peaks up to 40°C.

The trials of *Ph. persimilis* for the two-spotted spider mite control were at the following release rates:

-   V1 – 50,000 mites/ha;
-   V2 – 100,000 mites/ha;
-   V3 – 150,000 mites/ha;
-   V4 - Control.

When the degree of attack of two-spotted spider mites exceeded in melon 4.3% the predatory mites were released in the same 3 variants as mentioned above.

A variant area was 45 square meters.

The effectiveness of the predator releases was determined by decadal observations of the attack on the plant and monitoring the pest population of mite in August and September.

The effectiveness evaluation of *Ph. persimilis* on the two-spotted spider mite was performed by the Sun - Shepard method.

RESULTS AND DISCUSSIONS

1. Study of two-spotted spider mite attack in pepper and melon in tunnels

The dynamic of the degree of attack of two-spotted spider mite on pepper is showed in figure 1.

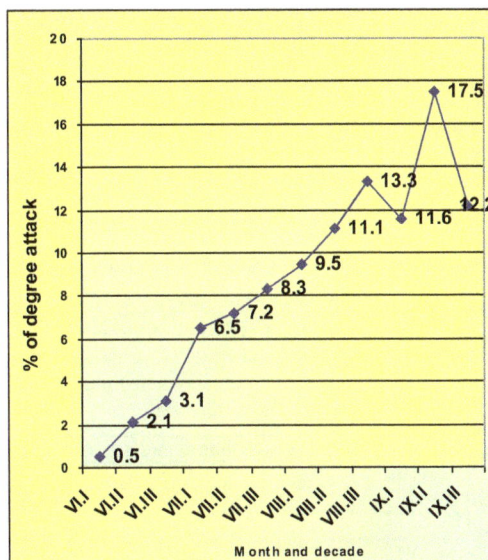

Figure 1. The attack of two-spotted spider mite in pepper

As shown in previous figure, the degree of attack in pepper has an ascendant dynamics. It started in first decade of June and increased to 17.5% in second decade of September. Then attack of pest begins to decrease reaching to 12.2%.

The dynamic of the degree of attack of the two-spotted spider mite on melon is presented in figure 2. The attack started in the first decade of June. The attack level was low until the first decade of August. Then the degree of attack had an ascendant dynamic from 4.3% in the first ten days of August, to 61.3% in the last decade of September. Under these circumstances it was necessary to apply measures for pest control, harvesting melons being carried out by the end of September.

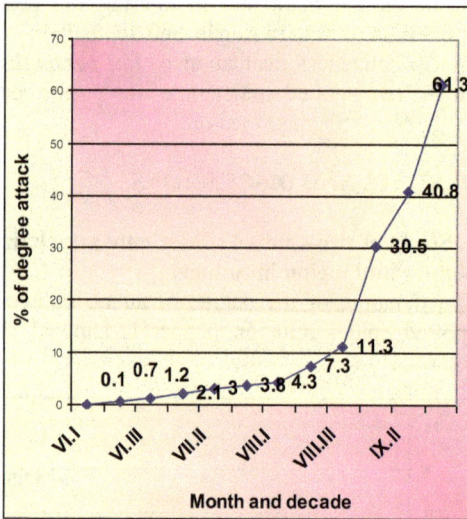

Figure 2. The attack of two-spotted
spider mite in melon

2. Biologically control of pest.

The degree of attack (DA%) of two-spotted spider mite decreased in all 3 variants of *Phytoseiulus persimilis* applications - table 1.

Table 1. The degree of attack of two-spotted spider mite
on pepper and melon plants

No. variant	DA% in month and decade					
	August			September		
	I	II	III	I	II	III
Pepper						
V1	13.5	9.5	2.3	0.7	0.4	0.1
V2	11.9	9.6	2.1	0.5	0.1	0.1
V3	9.0	6.6	1.7	0.2	0.1	0.1
V4	9.5	11.1	13.3	11.6	17.5	12.2
Melon						
V1	3.4	2.5	1.8	1.7	0.7	0.7
V2	6.1	4.5	1.9	0.8	0.7	0.7
V3	5.9	3.8	1.5	0.8	0.7	0.7
V4	4.3	7.3	11.3	30.5	40.8	61.3

The degree of two-spotted spider mite attack in V1 was reduced from 13.5% (first decade of August) to 9.5% (2nd decade of the month), 2.3% (3rd decade of August and less than 1% in September).
In V2, DA% had the same decreasing trend, which is below 1% in September. The V3,

where the release rate was 150 thousand ex./ha had the lowest values of DA% (6.6% and 1.7% in August and in less than 1% in September, table 1). In control variant (without releases), the degree of two-spotted spider mite attack increased from 9.5% in the first decade of August, up from 17.5% in the 2nd decade of September, then began to decline. The effectiveness of the release was up to 98% in variants with release rate of *Ph. persimilis* at the end of September (Figure 3).

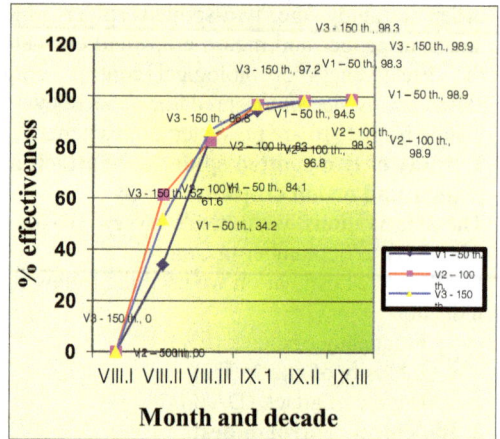

Figure 3. Effectiveness of *Phytoseiulus persimilis*
releases in control of two-spotted spider
mite on pepper crop

At melon, the pest degree attack was low (3.4% in V1, 6.1% in V2, 5.9% in V3 and 4.3% in control) in first decade of August.
After *Ph. persimilis* releases, the DA% decreased in the second decade of August as follows: V1 - 2.5%, V2 - 4.5%, V3 - 3.8%.
Descendant DA% continued for all three variants, reaching values below 1% in September. In V4 (control without predators release), the degree of attack had a sharp upward trend rising from 4.3% in the first ten days of August, to 61.3% in the last decade of September.
The effectiveness for all three releases rates of *Ph. persimilis* was up to 98% in all variants at the end of September (Figure 4).

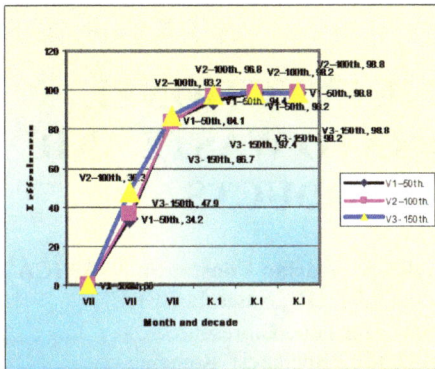

Figure 4. Effectiveness of *Phytoseiulus persimilis* releases in control of two-spotted spider mite on melon crop

CONCLUSIONS

Study of two-spotted spider mite attack on pepper and melon crops in tunnels

The degree of attack in pepper has an ascendant dynamics. It started in first decade of June and increased to 17.5% in second decade of September. Then attack of pest begins to decrease reaching to 12.2%.

In melon the attack started in the first decade of June. The attack level was low until the first decade of August. Then the degree of attack had an ascendant dynamic from 4.3% in the first ten days of August, to 61.3% in the last decade of September.

Biologically control of pest. The degree of two-spotted spider mite attack in V1 was reduced from 13.5% (first decade of August) to 9.5% (2nd decade of the month), 2.3% (3rd decade of August) and less than 1% in September. In V2 DA% had the same decreasing trend, which is below 1% in September. The V3, where the release rate was 150 thousand ex./ha had the lowest values of DA% (6.6% and 1.7% in August and in less than 1% in September). In control variant (without releases), the degree of two-spotted spider mite attack increased from 9.5% in the first decade of August, up from 17.5% in the 2nd decade of September. The effectiveness of the release was up to 98% in variants with release rate of *Ph. persimilis* at the end of September.

At melon, the pest degree attack was low (3.4% in V1, 6.1% in V2, 5.9% in V3 and 4.3% in control) in first decade of August. After *Ph. persimilis* releases, the DA% decreased in the second decade of August as follows: V1 - 2.5%, V2 - 4.5%, V3 - 3.8%. Descendant DA% continued for all three variants, reaching values below 1% in September. In V4 (control without predators release), the degree of attack had a sharp upward trend rising from 4.3% in the first ten days of August, to 61.3% in September.

The effectiveness for all three releases rates of *Ph. persimilis* was up to 98% in all variants at the end of September.

REFERENCES

Bolckmans K.J.F., 2007. Mass-rearing phytoseiid predatory mites. Proceedings of the Working Group AMRQC, C. van Lenteren, P. DeClercq, and M. W. Johnson, Eds., Bulletin IOBC Global, vol. 3, 12–15.

Calin M., 2005. Ghidul recunoaşterii şi controlului dăunătorilor plantelor legumicole cultivate în agricultură biologică, Ed. TIPOACTIV, ISBN 973-87136-3-3.

Candea E., 1984. Daunatorii legumelor si combaterea lor. Ed. Ceres, Bucuresti.

Herrmann I., Berenstein M., Amit Sade, Arnon Karnieli, David J Bonfil, Phyllis G Weintraub, 2011, Evaluation of spider mite damage to greenhouse pepper leaves by spectral assessment. Ptocedings '7th EARSeL Workshop on Imaging Spectroscopy' April 11-13, 2011, Edinburgh, Scotland, UK.

Lenteren J. C. van, 2003. Commercial availability of biological control agents," in Quality Control and Production of Biological Control Agents: Theory and Testing Procedures, J. C. van Lenteren, Ed CABI, Oxon, UK.,

Lenteren J. C. van, 2012. The state of commercial augmentative biological control: plenty of natural enemies, but a frustrating lack of uptake," BioControl, vol. 57, no. 1, 1–20.

Rojas M. G., J. A. Morales-Ramos, 2010. Tri-trophic level impact of host plant linamarin and lotaustralin on *Tetranychus urticae* and its predator *Phytoseiulus persimilis*. Journal of Chemical Ecology, vol. 36, no. 12, 1354–1362.

Shih C.I.T., 2001. Automatic mass-rearing of *Amblyseius womersleyi* (Acari: Phytoseiidae), Experimental and Applied Acarology, vol. 25, no. 5, 425–440.

Dalpe S. 2002. Pests of Greenhouse Sweet Peppers and their Biological Control. Alberta.ca Agriculture and Forestry,

Zhang Z. Q, 2003. Mites of Greenhouses Identification, Biology and Control, CABI.

24

ACHIEVEMENT OF SOME FUNCTIONAL INGREDIENTS FROM TOMATO WASTE AND WINEMAKING BY-PRODUCTS

Monica CATANĂ[1], Luminița CATANĂ[1], Enuța IORGA[1], Adrian Constantin ASĂNICĂ[2], Anda-Grațiela LAZĂR[1], Monica-Alexandra LAZĂR[1], Nastasia BELC[1]

[1]National Research & Development Institute for Food Bioresources,
IBA Bucharest, 6 Dinu Vintila, District 2, 021102 Bucharest, Romania
[2]University of Agronomic Sciences and Veterinary Medicine of Bucharest,
Faculty of Horticulture, 59 Marasti Blvd, District 1, 011464, Bucharest, Romania
Corresponding author email: mcatana1965@gmail.com

Abstract

A major problem facing the food industry is accumulation, handling and disposal of waste from the processing of raw materials. Therefore, valorisation of such waste by achievement of functional ingredients, leading to increasing of nutritional quality and antioxidant potential of foods is of real interest. Among waste and by-products from the processing industry of vegetables and fruits with valorisation potential are tomato waste, dark colour grape seed and skin (red, purple, and black). In this paper are presented results of the performed research for achievement of some functional ingredients (flours) from tomato waste and winemaking by-products (grape pomace and grape seed). Tomato waste and winemaking by-products were subjected to convective drying process at temperature of 50°C, in order to protect bioactive compounds (vitamins, phenolic compounds etc.), to a moisture which allows their milling and conversion into flours and, at the same time, their stability in terms of quality. The achieved functional ingredients were evaluated from sensory, physic-chemical and microbiological point of view. Flour obtained from tomato waste is characterized by content in carotenoids (lycopene – 225.92 mg/kg; β-carotene - 16.22 mg/kg), protein (17.62%), minerals, total fibre (59.47%), total polyphenol (18.76 mg GAE/g). Flours achieved from winemaking by-products are characterized by content of protein (10.53-14.63%), minerals, total fibre (58.06-66.06%) and total polyphenol (200.15-322.75 mg GAE/g). Antioxidant capacity of flour achieved from tomato waste was 1.62 mg Trolox Equivalents/g, and of flour achieved from winemaking by-products varied in the range (40.75–51.25 mg Trolox Equivalents/g). Microbiological analysis showed that flours obtained from tomato waste and winemaking by-products (grape pomace and grape seed) are under the provisions of the legislation in force.

Key words: tomato, waste, grape, pomace.

INTRODUCTION

Numerous studies have shown the presence of bioactive compounds in various types of agro-industrial waste, with potential application in the industry.

Their reuse would reduce environmental risks caused by disposal, besides providing a source of profitability for populations living around industrial regions (Anastasiadi et al., 2008).

Tomatoes (*Lycopersicon esculentum* L.) are cultivated worldwide for their fruits, registering an annual production of 161.8 million tonnes (FAOSTAT, 2012). Regular consumption of tomatoes and processed tomato products was correlated with a reduction in susceptibility to various cancers and cardiovascular diseases (Borguini and Da Silva Torres, 2009). These positive effects are due to antioxidant compounds present in tomatoes, such as vitamins C and E, carotenoids, polyphenols, which play a key role in the mechanism of health protection by neutralizing free radicals (Ray et al., 2011). Also, tomatoes are an important source of trace elements, namely, selenium, copper, manganese and zinc, which are cofactors of antioxidant enzymes (Martinez-Valverde et al., 2002). Due to the complex biochemical composition, consumption of fresh or processed tomatoes (juice, piuree, paste, ketchup etc.) has beneficial effects on the human body: cardioprotective (Palomo et al., 2009), anti platelet (platelet aggregation inhibition) (Fuentes et al., 2012), decreasing of triglycerides and cholesterol level in the blood (Hsu et al., 2008), reducing of oxidative stress induced by postprandial lypemia (increase of lipid level in the blood after lunch) (Burton-Freeman et al., 2012).

Studies concerning localization of antioxidant compounds in the different fractions of tomatoes (epicarp, seed and pulp) confirmed that, in all tomato cultivars under study, within the epicarp are the highest concentrations of phenolic compounds, flavonoids, lycopene and ascorbic acid. At the same time, tomato epicarp has an antioxidant activity higher than the pulp and seed fractions (George et al., 2004; Toor and Savage, 2005). Also, several studies showed that tomato seed are rich in nutrients including: carotenoids, proteins, polyphenols, minerals, fibres and oils (Liadakis et al., 1995; Persia et al., 2003; Toor and Savage, 2005; Demirbaş, 2010; Eller et al., 2010; Zuorro et al., 2013).

Million tonnes of tomatoes are annualy processed to juice, sauces, piurees, paste and tomato canned, generating high amounts of tomato peel, pulp and seed, which are industrial waste (Papaioannou and Karabelas, 2012). When tomatoes are processed and converted into ketchup, sauces or juice, waste is generated representing 3-7% of the tomato mass introduced in the manufacture process (Savatović et al., 2010).

Tomato seed contains about 24.5% crude protein and have the highest content of glutamic acid and aspartic acid (Persia et al., 2003). Del Valle et al. (2006) evaluated the chemical composition of the waste resulted from the industrial processing of tomatoes to paste, in various stages of the process flow (after pulper, after finisher, before turbopress and after turbopress). Average composition (in dry weight basis) of tomato pomace was the following: 59.03% fibres, 25.73% total sugars, 19.27% protein, 7.55% pectins, 5.85% total fat and 3.92% minerals.

Aghajanzadeh et al. (2010) have shown that the dried tomato waste contains 22.6–24.7% protein, 14.5–15.7% fat and 20.8–23.5% fibres and, at the same time, represents a source of vitamins B_1, B_2 and A. In addition, tomato waste contains essential aminoacids, and tomato seed contains high concentrations of minerals (Fe, Mn, Zn and Cu). Tomato peel contains significantly higher concentrations of lycopene and β-carotene compared to the pulp and seed (Papaioannou and Karabelas, 2012).

Majority of flavonoids is found in the peel of tomatoes.

Vitis vinifera L. production is widespread throughout the world, exceeding 68 million tonnes (FAOSTAT, 2010). As grape seeds comprise about 5% of the fruit weight (Choi and Lee, 2009), more than 3 million tonnes of grape seeds are discarded annually world-wide (Fernandes et al., 2012). Grape seeds represent a significant part in pomace, namely 38–52% of dry matter (Maier et al., 2009). Grape pomace represents a mixture of grape peel, seed and trace of pulp, resulted after wine obtaining. Grape seed and skin constituents have been shown to have health-functional activities as LDL cholesterol-lowering functional foods (Chen et al., 2011).

Composition of grape seeds is represented by 40% fibres, 16% essential oil, 11% protein, 7% complex phenolic compounds like tannins, and also sugars and minerals (Campos et al., 2008). Ca, K, Mg, Na and P are the most important minerals in grape seed (Ozcan, 2010). White grape seeds have a content of total polyphenols (on average 58.23±3.978 g/kg DM) higher compared to black grape seed (32.22±2.197 g/kg DM). Also, in grape seed, γ-tocotrienol is the most abundant (46.31±13.37 mg/kg DM), followed by α-tocotrienol (20.00±7.81 mg/kg DM) and α-tocopherol (12.45±4.85 mg/kgDM) (Lachman et al., 2013).

Grape pomace and its derivatives have use potential in diabetes management (Hogan et al., 2010). Three polyphenolic compounds in grape seed (gallic acid, catechin and epicatechin) inhibited pancreatic cholesterol esterase and may increase control on bioavailability of dietary cholesterol and cholesterol ester derivative, thus limiting the absorption of free cholesterol in blood (Ngamukote et al., 2011).

In this paper are presented results of the performed research for achievement of some functional ingredients (flours) from tomato waste and winemaking by-products (grape pomace and grape seed).

MATERIALS AND METHODS

Samples

Tomato waste resulted from tomato processing to juice within the Pilot Experiments Plant for Fruits and Vegetables Processing in IBA

Bucharest. Winemaking by-products (black grape seed and pomace) were collected after producing wine in the households in rural areas. Till processing, winemaking by-products were shipped and stored under refrigeration (3 °C). Tomato waste and winemaking by-products were subjected to dehydration process in a convection dryer at temperature 50 °C to a moisture which allows their milling and conversion into flours and, at the same time, their stability in terms of quality. Milling of dried semi-finished products was performed by using Retsch mill. The achieved functional ingredients (flours) were packed in glass containers, hermetically sealed, protected by aluminum foil against light and stored in dry and cool areas (temperature of maximum 20 °C), till to the sensory, physic-chemical and microbiological analysis). In Figure 1 are presented the flours mentioned above.

Figure 1. Flours achieved from tomato waste and winemaking by-products

Methods
Sensory analysis
Sensory analysis (appearance, taste and smell) was performed by descriptive method.

Physic-chemical analysis
Measurement of the color parameters of samples was performed at room temperature, using a HunterLab colorimeter, equipped with Universal Software V4.01 Miniscan XE Plus programme, to register CIELab parameters (the Commission Internationale de l'Eclairage - CIE), $L*$, $a*$ and $b*$: $L*$ - color luminance (0 = black, 100 = white); $a*$ - red-green coordinate (-a = green, +a = red); $b*$ - yellow-blue coordinate (-b = blue, +b = yellow).

Moisture determination was performed with Ohaus Moisture Analyzer MB45 at temperature 105 °C.

Protein content was determined by the Kjeldahl method with a conversion factor of nitrogen to protein of 6.25 (AOAC Method 979.09, 2005). Fat content was determined according to AOAC Method 963.15, and ash content according to AOAC Method 923.03 (AOAC, 2005).

In order to determine minerals samples were mineralized by calcination, with the addition of hydrochloric acid and hydrogen peroxide. The minerals sodium (Na), potassium (K), calcium (Ca), magnesium (Mg) and zinc (Zn) were determined by atomic absorption spectrophotometer (type *AAnalyst* 400, Perkin–Elmer). The minerals iron (Fe) and selenium (Se) were determined by Graphite Furnace Atomic Absorption Spectrophotometer (type *AAnalyst* 600, Perkin–Elmer).

Total dietary fibre (TDF) was determined by enzymatic method using the assay kits: K-TDFR ''Total dietary fibre'' (AOAC Method 991.43).

Lycopene and beta carotene content were performed by using spectrophotometric method developed by Nagata and Yamashita (1992).

Total polyphenol content
Total polyphenol content was conducted according to Horszwald and Andlauer (2011) with some modifications (concerning extract volumes of the used sample and reagents, using UV-VIS Jasco V 550 spectrophotometer), based on calibration curve of gallic acid achieved in the concentration range 0 to 0.20 mg/mL. The extraction of phenolic compounds was performed in methanol:water 50:50, and the absorbance of the extracts was determined at a wavelenght $\lambda = 755$ nm. Results were expressed as mg of gallic acid equivalents (GAE) per g flour (black grape seed flour, black grape pomace flour, tomato waste flour).

Antioxidant capacity
The DPPH scavenging radical assay was conducted according to Horszwald and Andlauer (2011) with some modifications (concerning extract volumes of the used sample and reagents, using UV-VIS Jasco V 550 spectrophotometer). The reaction was

performed in dark for 30 min (at ambient temperature) and after this time the absorbance was read at 517 nm. It was achieved the calibration curve Absorbance = f (Trolox concentration), in the concentration range 0-0.4375 mmol/L and the results were expressed as mg Trolox Equivalents per g flour (black grape seed flour, black grape pomace flour, tomato waste flour).

Microbiological analysis

The water activity (Aw) was determined by an instrument Aquaspector AQS-2-TC, Nagy. The measurements were performed at 25°C. Yeast and mold were determined by the method SR ISO 21527-1:2009. *Enterobacteriaceae* was determined according to the SR ISO 21528-2:2008 method and *Escherichia coli* by SR ISO 16649-2:2007 method. *Salmonella* was determined by the method SR EN ISO 6579:2003/AC:2006.

RESULTS AND DISCUSSIONS

Sensory analysis

After sensory analysis it was found that the obtained flours have specific characteristics. Tomato waste flour is in the form of orange powder with pleasant taste and smell, characteristic. Black grape seed flour and black grape pomace flour are in the form of a dark brown powder, purple tinge, respectively, with specific, pleasant taste and smell.
After the instrumental analysis of color (Figure 2) it was found that black grape seed flour (I) has the darkest color, registering the minimum value of luminance (L* = 24.38), while the tomato waste flour has the lightest color (L* = 59.13).

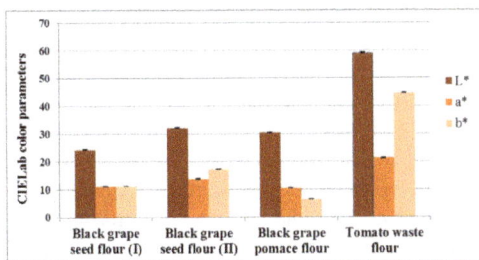

Figure 2. Color parameters of the flours achieved from tomato waste and winemaking by-products

Also, the maximum positive values of parameter a* (red coordinate) and of parameter b* (yellow coordinate) were registered for tomato waste flour.

Physic-chemical analysis

Composition of the flours achieved from tomato waste and winemaking by-products is presented in Table 1.
Water content of grape pomace flour is higher than that reported by Sousa et al., 2014 (3.33±0.04% dry basis), and of grape seed flour is lower than that reported by Aghamirzaei et al., 2015 (7.48±0.73% dry basis).

Table 1. Physic-chemical composition of flours achieved from tomato waste and winemaking by-products

Flour type	Water (%)	Ash (%)	Protein (%)	Fat (%)	Total fibre (%)
Black grape seed flour (I)	4.90±0.12	2.80±0.04	10.53±0.09	15.36±0.17	64.67±1.20
Black grape seed flour (II)	3.91±0.08	2.90±0.04	10.85±0.09	13.17±0.15	66.06±1.22
Black grape pomace flour	4.59±0.10	6.61±0.09	10.63±0.09	8.49±0.10	58.86±1.09
Tomato waste flour	7.64±0.17	4.05±0.05	17.62±0.16	10.38±0.12	59.47±1.10

Water content of tomato waste flour is higher than that reported by Majzoobi et al. (2011) (4.71% dry basis). After physic-chemical analysis it was found that the achieved flours are distinguished by content of total ash, protein and total fibre. Their ash content varied in the range 2.80-6.61% (minimum value was registered for the black grape seed flour (I), and the maximum one for the black grape pomace flour). Ash content of grape seed flour is comparable with that reported by Aghamirzaei et al. (2015) (2.45±0.18% dry basis), and ash content of grape pomace flour is higher compared to that obtained by Sousa et al. (2014) (4.65±0.05% dry basis). Ash content of tomato waste flour is comparable to that reported by Majzoobi et al. (2011) (4.53%).
Flours obtained from winemaking by-products registered close values for protein content (10.53-10.85%), lower than those obtained by Valiente et al. (1995), Llobera & Cañellas (2007) and Bravo & Saura-Calixto (1998) in grape residues (11g/100g, 12g/100g, and 14g/100g). Protein content (17.62%) of tomato waste flour obtained in this study is with 10.65% lower than that reported by Majzoobi et al., 2011 (19.72%). Grape seed flour obtained within the experiments has fat content higher than grape pomace flour. It is noted that

grape pomace flour achieved in this experimental study has a fat content significantly higher than that reported by Bampi et al. (2010) in flour grape residues (2.56 g/100g). Grape fats are concentrated, mainly, in seeds and consist in approximately 90% mono-unsaturated fatty acids, known for their beneficial properties, notably for cardiovascular system (Rockenbach et al., 2010). Tomato waste flour had a fat content 2.09 times higher than that found by Majzoobi et al. (2011) in case of powder obtained from tomato waste (4.96%).

Flours obtained from tomato waste and winemaking by-products presented a high total fibre content (58.86-66.06%), the maximum value being registered for grape seed flour (II). Values of this chemical parameter for flours obtained from winemaking by-products are higher than those reported by Aghamirzaei et al. (2015) for grape seed powder (42.74±0.6% dry basis) and Sousa et al. (2014) for grape pomace flour (46.17±0.80% dry basis). Sousa et al. (2014) mentioned that grape pomace flour is a good source of dietary fibre providing 79% insoluble fibre and 21% soluble fibre. Pérez-Jiménez et al. (2008) mentioned that dietary fibre of grapes significantly reduced lipid profile and blood pressure, and these effects were significantly higher than those produced by other dietary fibre, such as oat or psyllium fibre, probably, due to the combined effect of dietary fibre and antioxidants.

Total fibre content of tomato waste flour is lower by 18.19% than that reported by Majzoobi et al. (2011) for tomato pomace powder (72.68%).

Flours obtained from winemaking by-products and tomato waste are an important source of minerals (K, Ca, Mg, Fe, Zn și Se). Their content in minerals is presented in Figures 3 and 4. Flours achieved from winemaking by-products have a high potassium content in the range 1102.35-3406.67 mg/100g, the maxium value being recorded by black grape pomace flour. Their potassium content is higher than that reported by Gül et al. (2013) (2343.10 mg/100 g for whole flour of Öküzgözü pomace, 1587.10 mg/100g for whole flour of Narince pomace, 312.89 mg/100 g for seed flour of Öküzgözü pomace and 458.24 mg/100 g, respectively, for seed flour of Narince pomace).

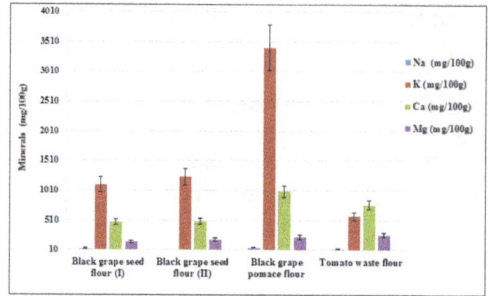

Figure 3. Mineral content (Na, K, Ca, Mg) of the flours achieved from tomato waste and winemaking by-products

Potassium content of tomato waste flour (573.09±64.76 mg/100g) is higher than that reported by Nour et al. (2015) for tomato waste (moisture content = 69.98±0.18%, K = 303.02 mg/100 g). These flours have potassium content higher than that in sodium. The results are in conformity with those obtained by Sousa et al. (2014) which states that this may lead to a balance of minerals, which favours hypertension control.

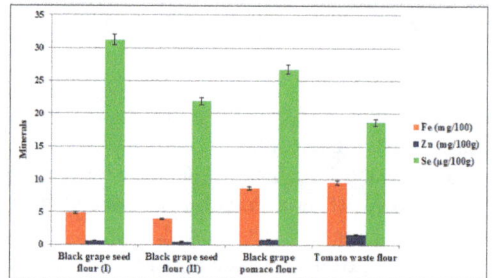

Figure 4. Mineral content (Fe, Zn, and Se) of the flours achieved from tomato waste and winemaking by-products

Calcium and magnesium content of tomato waste flour, black grape seed flour and black grape pomace flour, respectively, are higher than those reported by the other authors (Gül et al., 2013; Sousa et al., 2014).

Iron content of vegetable flours achieved in this study varied in the range 3.97-9.54 mg/100 g. Minimum value was registered for black grape seed flour (II), and the maximum one was registered for tomato waste flour. Grape pomace flour has an iron content of about 2.00 times higher than black grape seed flour. Grape pomace flour achieved in this study has an iron content of about 1.91 times higher than that reported by Lachman et al. (2013) (4.54 mg/100 g), but of about 2.6 times lower than that reported by Gül et al. (2013) (22.52 mg/100 g for whole flour of Öküzgözü pomace and 13.92 mg/100 g, respectively, for whole flour of Narince pomace). At the same time, flours achieved from blake grape seed within this study presented an iron content comparable to that reported by Gül et al. (2013) (2.86 mg/100 g for seed flour of Öküzgözü pomace and 5.13 mg/100 g, respectively, for seed flour of Narince pomace). Iron is an essential element for almost all living organisms as it participates in a wide variety of metabolic processes, including oxygen transport, deoxyribonucleic acid (DNA) synthesis, and electron transport (Abbaspour et al., 2014). Flours achieved from winemaking by-products registered a low Zn content (0.42 mg/100 g-0.75 mg/100 g), compared to those presented by other authors (Sousa et al., 2014, 0.98±0.702 mg/100 g in case of grape pomace flour and, respectively Lachman et al., 2013, 1.1 mg/100 g in case of grape seed). Tomato waste flour recorded the highest Zn content (1.56 mg/100 g). Zn is an important element of the immune system. Also, Bashandy et al. (2016) showed that the protective effect of zinc can be attributed to its antioxidant and antiinflammatory properties.

Selenium content of vegetable flours achieved within this study (18.65-31.23 µg/100 g) is comparable to that reported by Lyons et al. (2005) in case of grain (0.5-72 µg/100 g). These authors stated that the variation of selenium content is determined by the selenium content of the soil. Selenium is an essential micronutrient with an important role into human body (thyroid hormone metabolism, cardiovascular health, prevention of neurodegeneration and cancer, and optimal immune responses) (Huang et. al., 2012).

Total polyphenol content

Flours achieved from winemaking by-products and tomato waste flour are potential sources of natural antioxidants. Thus, these are characterized by the total polyphenol content and tomato waste flour contains in addition carotenoids (lycopene and β-carotene). Total polyphenol content of the achieved flours is presented in Figure 5. Total polyphenol content

of the achieved flours from winemaking by-products varied in the range 200.15-322.75 mg GAE/g, the maximum value being recorded in case of grape pomace flour. Values of polyphenol content obtained within this study are higher compared to that reported by Tseng and Zhao (2013) for flour obtained from black grape seed and peel, *Pinot Noir* cultivar (67.74 mg GAE/g, flour moisture being 5.63%; flour was obtained by lyophilisation of residues resulted after winemaking, at temperature -55 °C and vacuum of 17.33 Pa).

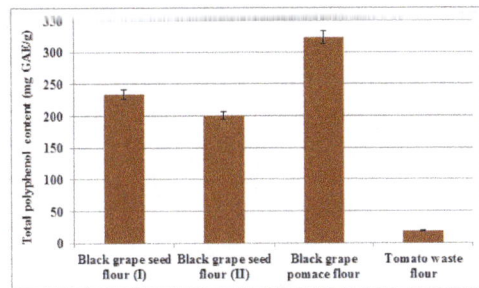

Figure 5. Total polyphenol content of flours achieved from winemaking by-products and tomato waste

At the same time, polyphenol content of the achieved grape seed flour is lower than that reported by Gül et al. (2013), for seed flour of Öküzgözü pomace (552.10 mg GAE/g) and seed flour of Narince pomace (563.27 mg GAE/g). Total polyphenol content of black grape pomace flour (322.75 mg GAE/g) achieved in this study is higher than that reported by Gül et al. (2013), for whole flour of Öküzgözü pomace (236.6 mg GAE/g) and, respectively, for whole flour of Narince pomace (65.93 mg GAE/g). Total polyphenol content of winemaking by-products is influenced by many factors: grape cultivar, climate conditions and culture area, ripening time, processing and storage conditions, as well as the used extraction methods and analytical methods (Lafka et al., 2007).

Phenolic compounds in extracts of grape pomace present antioxidant, anticancerigene and antidiabetes properties (Ruberto et al., 2007; Hogan et al., 2010; Parry et al., 2011; Zhou and Raffoul 2012; González-Centeno et al., 2013), as well as antibacterial activity against *E. coli, L. monocytogenes*, and *S. aureus* (Ozkan et al., 2004; Darra et al., 2012).

Polyphenol content of tomato waste flour (18.76±0.19 mg GAE/g; moisture = 7.64±0.17%) was significantly lower compared to that of the flours achieved from winemaking by-products. Nour et al. (2015) obtained for tomato waste (moisture content = 69.98±0.18%) a total polyphenol content of 0.866 ±0.012 mg GAE/g.

Antioxidant capacity
Antioxidant capacity of the achieved flours is presented in Figure 6.

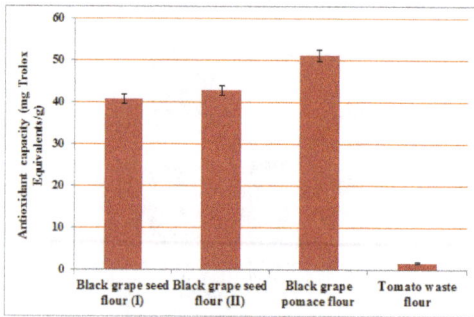

Figure 6. Antioxidant capacity of the flours achieved from winemaking by-products and tomato waste

Antioxidant capacity of flour achieved from winemaking by-products varied in the range 40.75–51.25 mg Trolox Equivalents/g, maximum value being recorded for black grape pomace flour.
Antioxidant capacity of flour achieved from tomato waste was 1.62 mg Trolox Equivalents/g.
For flours achieved from winemaking by-products and tomato waste, between the total polyphenol content and antioxidant capacity it is a linear correlation, regression coefficient R^2 being 0.9559 (Figure 7).
Results are in conformity with those of Ky and Teissedre (2015) which obtained positive correlations between the total polyphenol content and antioxidant capacity (DPPH method) for seed extract and for skin extract, respectively (R^2 = 0.87 for seed extract, R^2 = 0.79 for skin extract).
Lycopene content of the tomato waste flour was 225.92 mg/kg, and ß-carotene content, respectively 16.22 mg/kg. Lycopene content of tomato waste flour was higher than that reported by Nour et al. (2015) for tomato waste (moisture content = 69.98±0.18%): 174.12

mg/kg. ß-carotene content of tomato waste flour achieved in this study was lower than that reported by Nour et al. (2015) 32.66 mg/kg.

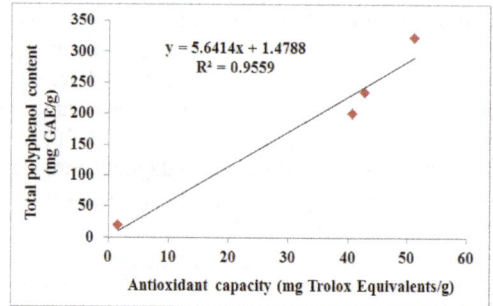

Figure 7. Correlation between the total polyphenol content and antioxidant capacity in case of flours achieved from winemaking by-products and tomato waste

Microbiological analysis
Results of the microbiological analysis of flours achived from tomato waste and winemaking by-products (grape seed, grape pomace) are presented in the Table 2.

Table 2. Microbiological analysis of flours achieved from tomato waste and winemaking by-products

Flour type	Yeast and mold (CFU/g)	Enterobacteriaceae (CFU/g)	Escherichia coli (CFU/g)	Salmonella (in 25g)	Water activity (Aw)
Black grape seed flour (I)	< 10	< 10	< 10	absent	0.338
Black grape seed flour (II)	< 10	< 10	< 10	absent	0.289
Black grape pomace flour	< 10	< 10	< 10	absent	0.274
Tomato waste flour	< 10	< 10	< 10	absent	0.344

Microbiological analysis shown that the achieved flours are in the frame of the provisions of the legislation into force. These flours show low values of water activity (0.274-0.344), which give them microbiological stability.

CONCLUSIONS

Flours achieved from tomato waste and winemaking by-products are important sources of protein, minerals (K, Ca, Mg, Fe, Zn and Se), dietary fibres and bioactive compounds. Thus, black grape pomace flour and black grape seed flour are characterized by total polyphenol content (200.15 mg GAE/g...322.75 mg GAE/g) and tomato waste flour by content of carotenoids (lycopene - 225.92 mg/kg; ß-carotene - 16.22 mg/kg). Also these flours have antioxidant potential being beneficial in a healthy diet for prevention of diseases caused by free radicals. On the other hand, flours achieved in this study are characterized by high dietary fibre content (58.86-66.06%), being important sources to increase the fibre content of foods (bakery products, pastry products, etc.). Increase of the fibre content in case of the sweet flour products is very important because it reduces their glycemic impact on the human body, thus preventing the development of diabetes mellitus and obesity. Also, dietary fibre have an important role in promoting feeling of satiety and detoxification of the human body.

Flours achieved from winemaking by-products and tomato waste can be regarded as functional ingredients and can be used to fortify food products (bakery and pastry products, especially) in order to increase the nutritional and their antioxidant potential.

ACKNOWLEDGEMENTS

This study was supported by the Ministry of Research and Innovation, by Nucleu Programme PN 16 46, contract 29 N/2016.

REFERENCES

Abbaspour N., Hurrell R., Kelishadi R., 2014. Review on iron and its importance for human health. J Res Med Sci., 19(2):164–174.

Aghajanzadeh A., Maheri N., Mirzai A., Baradaran A., 2010. Comparison of nutritive value of Tomato pomace and brewers grain for ruminants using in vitro gas production technique. Asian Journal of Animal and Veterinary Advances, 5:43-51.

Aghamirzaei M., Peighambardoust S.H., Azadmard-Damirchi S., Majzoobi M., 2015. Effects of Grape Seed Powder as a Functional Ingredient on Flour Physicochemical Characteristics and Dough Rheological Properties. J. Agr. Sci. Tech., 17:365-373.

Anastasiadi M., Chorianopoulos N.G., Nychas G.-J.E., Haroutounian S.A., 2008. Antilisterial activities of polyphenol-rich extracts of grapes and vinification byproducts. J Agric Food Chem., 57: 457-463.

Bampi M., Bicudo M.O.P., Fontoura P.S.G., Ribani R.F., 2010. Composição centesimal do fruto, extrato concentrado e da farinha da uva-do-japão. Ciência Rural, 40(11):2361-2367.

Bashandy S.A.E.M., Omara E.A.A., Ebaid H., Amin M.M., Soliman M.S., 2016. Role of zinc as an antioxidant and anti-inflammatory to relieve cadmium oxidative stress induced testicular damage in rats. Asian Pacific Journal of Tropical Biomedicine, 6(12):1056–1064.

Borguini R.G., Da Silva Torres E.A.F., 2009. Tomatoes and tomato products as dietary sources of antioxidants. Food Rev Int, 25:313–325.

Bravo L., Saura-Calixto F., 1998. Characterization of dietary fiber and the in vitro indigestible fraction of grape pomace. American Journal of Enology and Viticulture, 49(2):135-141.

Burton-Freeman B., Talbot J., Park E., Krishnankutty S., Edirisinghe I., 2012. Protective activity of processed tomato products on postprandial oxidation and inflammation: a clinical trial in healthy weight men and women. Molecular Nutrition and Food Research, 56:622–631.

Campos L.M.A.S., Leimann F.V., Pedrosa R.C., Ferreira S.R.S., 2008. Free radical scavenging of grape pomace extracts from Cabernet Sauvignon (Vitis vinifera). Bioresour. Technol., 99:8413–8420.

Chen Z.-Y., Ma K.Y., Liang Y., Peng C., Zuo Y., 2011. Role and classification of cholesterol-lowering functional foods. J. Funct. Foods, 3:61–69.

Choi Y., Lee J., 2009. Antioxidant and antiproliferative properties of –tococtrienol-rich fraction. Food Chem., 114: 1386–1390.

Darra N.E., Tannous J., Mouncef P.B., Palge J., Yaghi J., Vorobiev E., et al., 2012. A Comparative study on antiradical and antimicrobial properties of red grapes extracts obtained from different Vitis vinifera varieties. Food Nutr. Sci., 3:1420–1432.

Del Valle M., Cámara M., Torija M.E., 2006. Chemical characterization of tomato pomace. Journal of the Science of Food and Agriculture, 86:1232–1236.

Demirbas A., 2010. Oil, micronutrient and heavy metal contents of tomatoes. Food Chem, 118:504–507.

Eller F.J., Moser J.K., Kenar J.A., Taylor S.L., 2010. Extraction and analysis of tomato seed oil. J Am Oil Chem Soc, 87:755–762.

FAOSTAT. 2012. FAOSTAT agriculture production database.http://faostat. fao.org / site/ 339/default.aspx.

FAOSTAT, 2010.http://faostat.fao.org.

Fernandes L., Casal S., Cruz R., Pereira J.A., Ramalhosa E., 2012. Seed oils of ten traditional Portuguese grape varieties with interesting chemical and antioxidant properties. Food Res. Int., 50:161-166.

Fuentes E., Astudillo L., Gutiérrez M., Contreras S., 2012. Fractions of aqueous and methanolic extracts from tomato (Solanum lycopersicum L.) present platelet antiaggregant activity, Blood Coagulation and Fibrinolysis, vol. 23, pp. 109–117.

George B., Kaur C., Khurdiya D.S., Kapoor H.C., 2004. Antioxidants in tomato (Lycopersium esculentum) as a function of genotype. Food Chem, 84:45–51.

González-Centeno M.R., Jourdes M., Femenia A., Simal S., Rosselló C., Teissedre P.-L., 2013. Characterization of polyphenols and antioxidant potential of white grape pomace by-products (Vitis vinifera L.). J. Agricult. Food Chem., 61:11579–11587.

Gül H., Acun S., Şen H., Nayır N., Türk S., 2013. Antioxidant activity, total phenolics and some chemical properties of Öküzgözü and Narince grape pomace and grape seed flours. Journal of Food, Agriculture & Environment, 11(2):28 -34.

Hogan S., Zhang L., Li J., Sun S., Canning C., Zhou K., 2010. Antioxidant rich grape pomace extract suppresses postprandial hyperglycemia in diabetic mice by specifically inhibiting alpha-glucosidase. Nutrition & Metabolism, 7:71.

Horszwald A., Andlauer W., 2011. Characterisation of bioactive compounds in berry juices by traditional photometric and modern microplate methods. Journal of Berry Research, 1:189–199.

Hsu Y.M., Lai C.H., Chang C.Y., Fan C.T., Chen C.T., Wu C.H., 2008. Characterizing the lipid-lowering effects and antioxidant mechanisms of tomato paste. Bioscience, Biotechnology and Biochemistry, 72(3):677–685.

Huang Z., Rose A.H., Hoffmann P.R., 2012. The Role of Selenium in Inflammation and Immunity: From Molecular Mechanisms to Therapeutic Opportunities. Antioxidants & Redox Signaling, 16(7):705-743.

Ky I., Teissedre P.L., 2015. Characterisation of Mediterranean Grape Pomace Seed and Skin Extracts: Polyphenolic Content and Antioxidant Activity. Molecules, 20:2190-2207.

Lachman J., Hejtmánková A., Hejtmánková K., Horníčková S., Pivec V., Skala O., Dědina M., Přibyl J., 2013. Towards complex utilisation of winemaking residues: Characterisation of grape seeds by total phenols, tocols and essential elements content as a by-product of winemaking. Industrial Crops and Products, 49:445–453.

Lafka T.I., Sinanoglou V., Lazos E.S., 2007. On the extraction and antioxidant activity of phenolic compounds from winery wastes. Food Chemistry, 104:1206–1214.

Liadakis G.N., Tzia C., Oreopoulou V., Thomopoulos C.D., 1995. Protein isolation from tomato seed meal, extraction optimization. J Food Sci, 60:477–482.

Llobera A., Cañellas J., 2007. Dietary fibre content and antioxidant activity of Manto Negro red grape (Vitis vinifera): pomace and stem. Food Chemistry, 101(2):659-666.

Lyons G.H., Genc Y., Stangoulis J.C.R., Palmer L.T., Graham R.D., 2005. Selenium distribution in wheat grain, and the effect of postharvest processing on wheat selenium content. Biological Trace Element Research, 103:155-168.

Maier T., Schieber A., Kammerer D., Carle R., 2009. Residues of grape (Vitis vinifera L.) seed oil production as a valuable source of phenolic antioxidants. Food Chem., 112:551–559.

Martínez-Valverde I., Periago M.J., Provan G., Chesson A., 2002. Phenolic compounds, lycopene and antioxidant activity in commercial varieties of tomato (Lycopersicum esculentum). J Sci Food Agric, 82:323–330.

Majzoobi M., Ghavi F.S., Farahnaky A., Jamalian J., Meshahi G., 2011. Effect Of Tomato Pomace Powder On The Physicochemical Properties Of Flat Bread (Barbari Bread). Journal of Food Processing and Preservation, 35:247–256.

Nagata M., Yamashita I., 1992. Simple method for simultaneous determination of chlorophyll and carotenoids in tomato fruit. J Jpn Soc Food Sci Technol, 39:925–928.

Ngamukote S., Mäkynen K., Thilawech T., Adisakwattana S., 2011. Cholesterol-Lowering Activity of the Major Polyphenols in Grape Seed. Molecules, 16:5054-5061.

Nour V., Ionica M.E., Trandafir I., 2015. Bread enriched in lycopene and other bioactive compounds by addition of dry tomato waste. J Food Sci Technol, 52(12):8260–8267.

Ozkan G., Sagdic O., Baydar N.G., Kurumahmutoglu Z., 2004a. Antibacterial activities and total phenolic contents of grape pomace extracts. J. Sci. Food Agric., 84:1807–1811.

Ozcan M.M., 2010. Mineral contents of several grape seeds. Asian J. Chem., 22:6480–6488.

Palomo I., Gutierrez M., Astudillo L., Rivera C., 2009. "Efecto antioxidante de frutas y hortalizas de la zona central de Chile". Revista Chilena De Nutricion, 36:152–158.

Parry J.W., Li H., Liu J.-R., Zhou K., Zhang L., Ren S., 2011. Antioxidant activity, antiproliferation of colon cancer cells, and chemical composition of grape pomace. Food Nut. Sci., 2:530–540.

Papaioannou E.H., Karabelas A.J., 2012. Lycopene recovery from tomato peel under mild conditions assisted by enzymatic pre-treatment and non-ionic surfactants. Acta Biochim Pol, 59:71–74.

Pérez Jiménez J., Serrano J., Tabernero M., Arranz S., Díaz-Rubio M.E., García-Diz L., Goñi I., Saura-Calixto F., 2008. Effects of grape antioxidant dietary fiber in cardiovascular disease risk factors. Nutrition, 24(7-8):646-653.

Persia M.E., Parsons C.M., Schang M., Azcona J., 2003. Nutritional evaluation of dried tomato seeds. Poult Sci, 82:141–146.

Ray R.C., El Sheikha A.F., Panda S.H., Montet D., 2011. Anti-oxidant properties and other functional attributes of tomato: an overview. Int J Food Ferment Technol, 1:139–148.

Rockenbach I.I., Rodrigues E., Gonzaga L.V., Fett R., 2010. Composição de ácidos graxos de óleo de semente de uva (Vitis vinifera L. e Vitis labrusca L.). Brazilian Journal of Food Technology, IIISSA.

Ruberto G., Renda A., Daquino C., Amico V., Spatafora C., Tringali C., Nunziatina-De T., 2007. Polyphenol constituents and antioxidant activity of grape pomace extracts from five Sicilian red grape cultivars. Food Chem., 100:203–210.

Savatović S.M., Gordana S., Ćetković G.S., Čanadanović-Brunet J.M., Djilas S.M., 2010. Utilisation of tomato waste as a source of polyphenolic antioxidants. Acta Period Technol, 41:187–194.

Sousa E.C., Uchôa-Thomaz A.M.A., Carioca J.O.B., de Morais S.M., de Lima A., Martins C.G., Alexandrino C.D., Ferreira A.T., Rodrigues A.L.M., Rodrigues S.P., Silva J.N., Rodrigues L.L., 2014. Chemical composition and bioactive compounds of grape pomace (Vitis vinifera L.), Benitaka variety, grown in the semiarid region of Northeast Brazil. Food Sci. Technol, Campinas, 34(1):135-142.

Toor R.K., Savage G.P., 2005. Antioxidant activity in different fractions of tomatoes. Food Res Int, 38:487–494.

Tseng A., Zhao Y., 2013. Wine grape pomace as antioxidant dietary fibre for enhancing nutritional value and improving storability of yogurt and salad dressing. Food Chem., 1; 138(1): 356-65.

Valiente C., Arrigoni E., Esteban R.M., Amado R., 1995. Grape pomace as a potencial food fiber. Journal of Food Science, 60(4):818-820.

Zhou K., Raffoul J.J., 2012. Potential anticancer properties of grape antioxidants. J. Oncol. Article ID 803294.

Zuorro A., Lavecchia R., Medici F., Piga L., 2013. Enzyme-assisted production of tomato seed oil enriched with lycopene from tomato pomace. Food Bioprocess Technol, 6(12):3499–3509.

25

VITAMIN C AND TOTAL POLYPHENOL CONTENT AND ANTIOXIDANT CAPACITY OF FRESH AND PROCESSED FRUITS OF *ARONIA MELANOCARPA*

Luminița CATANĂ[1], Monica CATANĂ[1], Enuța IORGA[1], Adrian Constantin ASĂNICĂ[2], Anda-Grațiela LAZĂR[1], Monica-Alexandra LAZĂR[1], Nastasia BELC[1]

[1]National Research & Development Institute for Food Bioresources,
IBA Bucharest, 6 Dinu Vintila Street, District 2, 021102 Bucharest, Romania
[2]University of Agronomic Sciences and Veterinary Medicine of Bucharest,
Faculty of Horticulture, 59 Marasti Blvd, District 1, 011464, Bucharest, Romania
Corresponding author email: lumi_catana@yahoo.co.uk

Abstract

There are scientific evidences that a diet rich in fruits and vegetables may reduce the risk to have different chronic diseases. Berries are recommended in a healthy diet as it provides protection against degenerative diseases, cardiovascular diseases and cancer. Fruits of Aronia melanocarpa are rich sources of biologically active compounds, polyphenols (anthocyanins and procyanidins, especially) representing the most important group. Polyphenols are the main substances which give the antioxidant potential of black chokeberry fruits. In this paper are presented results of the performed research for determination of vitamin C and total polyphenol content and antioxidant capacity in case of fresh and processed fruits of Aronia melanocarpa (frozen and dried fruits, juice, jam, compote). Determination of vitamin C was performed by high performance liquid chromatography coupled with high resolution mass spectrometry, using hippuric acid as internal standard. Total polyphenol content was spectrophotometric determined, using Folin-Ciocalteu method, and assessment of the antioxidant capacity was performed using DPPH method. Vitamin C content of the samples taken into study varied in the range 7.25–98.75 mg/100g (minimum value was recorded for jam, and maximum one for fresh, unpasteurized juice). Dried fruits of Aronia registered the highest total polyphenol content (4015.25 mg GAE/100g) and antioxidant capacity (84.45mg Trolox Equivalents/g). Minimum value of antioxidant capacity was recorded for compote of Aronia, 12.25 mg Trolox Equivalents/g, respectively. Taken into consideration that fresh fruits of Aronia are available only a short time period, their processing under diverse forms is of real interest for consumers which can benefit thus of nutritional qualities and antioxidant potential of them.

Key words: Aronia melanocarpa, fruits, polyphenols, antioxidant capacity.

INTRODUCTION

Among berries, fruits of *Aronia melanocarpa*, they have gaine recently attention due to the health claims associated with their consumption (Chrubasik et al., 2010; Kokotkiewicz et al., 2010). Black chokeberry (*Aronia melanocarpa* (Michx.) Elliott belongs to the family *Rosaceae* and is native to the North America and Canada, being cultivated in Europe in the early twentieth century (Konić Ristić et al., 2013). Fruits of *Aronia melanocarpa* (Michx.) Elliott are rich sources of biologically active compounds, polyphenols (anthocyanins and procyanidins, especially) representing the most important group. Polyphenols are the main substances which give the antioxidant potential of black chokeberry fruits (Kokotkiewicz et al., 2010). Thus, black chokeberry fruits (*Aronia*

melanocarpa (Michx.) Elliott) are an important natural source of cyanidin 3-O-glycoside anthocyanins (cyanidin 3-O-galactoside, cyanidin 3-O-glucoside, cyanidin 3-O-arabinoside, and cyanidin 3-O-xyloside) (González-Molina et al., 2008), quercetin derivatives (Bermúdez-Soto and Tomás-Barberán, 2004), hydroxycinnamic acids (Zheng and Wang, 2003). Total polyphenolic content varies in the range 2-8 mg/100 d.m. and depends on the cultivar, growing conditions and harvesting time (Kähkönen et al., 1999; Hakkinen et al., 1999; Benvenuti et al., 2004; Oszmiański and Wojdyło, 2005; Hudec et al., 2006; Sueiro et al., 2006). Lidija Jakobek et al. (2012) determined polyphenols content in case of three cultivars ('Viking', 'Nero', 'Galicianka') of fruits of chokeberry (*Aronia melanocarpa*) and wild chokeberries, in

Croatia, region Slavonia during two consecutive years (2010 and 2011). Cultivars 'Viking', 'Nero' and wild chokeberries had a similar total polyphenolic content (9,012–10,804 mg kg^{-1} in the first year, 9,361–12,055 mg GAE/ kg FW in the second year).Cultivar 'Galicianka' had a lower total polyphenolic content (8,564 mg GAE/kg FW first year, 8,600 mg GAE/kg FW second year).

Besides polyphenols, fruits of *Aronia melanocarpa* are sources of sugar (10–18%), pectins (0.6–0.7%), the sugar alcohol sorbitol, and parasorboside (Weinges et al.,1998; Niedworok and Brzozowski, 2001; Wolski et al.,2007; Kulling and Rawel, 2008). Also these fruits contain small amounts of fat (0.14% fresh weight), represented especially by linoleic acid glycerides and phosphatidylinositol (Kane et al.,1991; Zlatanov, 1999). Also, Kulling and Rawel (2008) notes that fruits of *Aronia melanocarpa* contain vitamins B (B_1, B_2, B_6, niacin, pantothenic acid), vitamin C (13–270 mg/kg), β-carotene (7.7–16.7 mg/kg), minerals (4.4–5.8 g/kg as ash value), approx. 16–18% of carbohydrates (glucose, fructose, sorbitol), dietary fiber (approx. 55 g/kg) and 1–1.5% of organic acids (malic, quinic, citric).Specific almond flavor of these fruits is given by cyanogenic glycosides – amygdalin (20 mg/ 100 g fresh weight – FW) (Lehmann, 1990; Kullingand Rawel, 2008). Fruits of *Aronia melanocarpa* contain triterpenes (b-sitosterol and campesterol) and over 40 volatile compounds, the most important ones being benzaldehyde cyanohydrine, hydrocyanic acid, and benzaldehyde (Hirviand Honkanen, 1985; Zlatanov, 1999).

Fruits of *Aronia melanocarpa* (black chokeberry) demonstrate antiviral activity against influenza viruses, including an oseltamivir-resistant strain. Ellagic acid and myricetin are two components in fruits of *Aronia*, which give the anti-influenza properties (Parket al., 2013). Also, the polyphenolic-rich *Aronia melanocarpa* juice kills teratocarcinomal cancer stem-like cells, but not their differentiated counterparts (Sharif et al., 2013).

The *in vitro* experiments showed anticoagulant effect of polyphenols-rich extracts from black chokeberry and grape seeds (Bijak et al., 2011). In a pilot study, Maria Handeland et al.(2013) shown that black chokeberry juice (*Aronia melanocarpa*) reduces incidences of urinary tract infection.

Fresh fruits of *Aronia melanocarpa* can be consumed a short period time and thus to benefit by their nutritional qualities and antioxidant potential these fruits are processed under various forms: dried fruits, puree, juice, liqueur, syrup, jam, wine, compote, tea, powder (Chrubasik et al., 2010; Ochmian et al., 2012; Kapci et al., 2013; Šnebergrová et al., 2014). On the other side, fresh fruits of *Aronia melanocarpa*, have sour and astringent taste and therefore consumers prefer juice of *Aronia melanocarpa*, in combination with other fruits, such as, apples, pears and blackcurrant (Lehmann, 1990; Ara, 2002).

Anna Horszwald et al. (2013) studied the influence of drying techniques (spray drying, freeze drying and vacuum drying) on *Aronia* commercial juice, in the temperature range 40-80°C. It was found that all the obtained powders have a high content in polyphenols, in the range: 27.63±1.38 mg GAE/100 mg DM ...34.28±1.77 mg GAE/100 mg DM. Powders obtained by spray drying had the highest content of total flavonoids (5.22±0.32 mg quercetin/100 mg DM), total monomeric anthocyanins (4.80±0.13 mg Cy-3-G/100 mg DM), cyaniding-3-glucoside (21.10±0.63 mg Cy-3-G/100 mg DM) and total proanthocyanidins (59.22±3.69 mg (+)-Catechin/100 mg DM). Also, powders obtained by spray drying had the highest antioxidant capacity (251.34±18.77 μmol Trolox Equivalents/100 mg DM by ABTS; 26.49±2.34 μmol Trolox Equivalents/100 mg DM by TEAC; 248.56±11.06 μmol Trolox Equivalents/100 mg DM by FRAP).

In this paper are presented results of the performed research for determination of vitamin C and total polyphenol content and antioxidant capacity in case of fresh and processed fruits of *Aronia melanocarpa* (frozen and dried fruits, juice, jam, compote).

MATERIALS AND METHODS

Samples

Fresh and dried fruits of *Aronia melanocarpa*, compote, jam and fresh juice of *Aronia* were purchased from private producers. Fresh and dried fruits of *Aronia melanocarpa* were packed in plastic package. Ingredients of *Aronia* compote, packed in hermetically sealed glass recipients (Twist-off system), 720 mL capacity, were the following: *Aronia* fruits, water, sugar and lemon juice. *Aronia* jam was achieved by concentration of *Aronia* fruits with sugar, with adding of lemon juice and packed in hermetically sealed glass recipients (Twist-off system), 314 mL capacity. Fresh *Aronia* juice was achieved by pressing *Aronia* fresh fruits and packed in hermetically sealed glass recipients (Twist-off system), 330 mL capacity. Frozen fruits of *Aronia melanocarpa* were achieved within the Pilot Experiments Plant for Fruits and Vegetables Processing from fresh fuits purchased from private producer. Thus, fresh fuits of *Aronia melanocarpa* were sorted, washed and frozen in plastic package (net weight 250 g±3%) at – 18°C.

Methods
Vitamin C content
Determination of vitamin C was performed by high performance liquid chromatography (Accela, Thermo Scientific) coupled with high resolution mass spectrometry (LTQ Orbitrap XL Hybrid Ion Trap-Orbitrap Mass Spectrometer, Thermo Scientific) using hippuric acid as internal standard.

LC conditions:
- Column (Hypersil GOLD aQ, 150 x 2.1 mm, 3 μm);
- Column temperature: 40°C;
- Sample temperature: 4°C;
- Mobile phase A: 990 mL water: 10 mL 1M ammonium formate (aq): 1 mL formic acid;
- Mobile phase B: 990 mL methanol: 10 mL 1M ammonium formate (aq): 1 mL formic acid;
- Flow rate: 0.400 mL/min;
- Injection volume: 25 μL;

MS conditions:
Analyzer: FTMS;
Resolution: 60000;
Ionization mode: ESI-;
Specific ions were: m/z = 175.02438 (for vitamin C) and m/z = 178.05051 (for hippuric acid).
In Figure 1 is presented the calibration curve of vitamin C, achieved in the concentration range 2100-10000 μg/L.

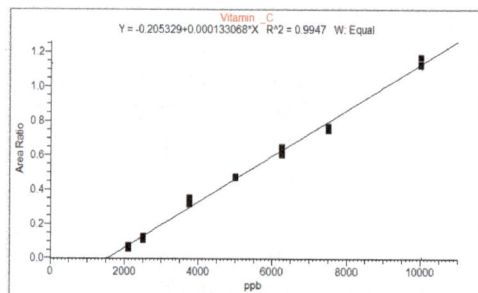

Figure 1.Calibration curve of vitamin C

Total polyphenol content
Total polyphenol content was conducted according to Horszwald and Andlauer (2011) with some modifications (concerning extract volumes of the used sample and reagents, using UV-VIS Jasco V 550 spectrophotometer), based on calibration curve of gallic acid achieved in the concentration range 0-0.20 mg/mL (Figure 2).

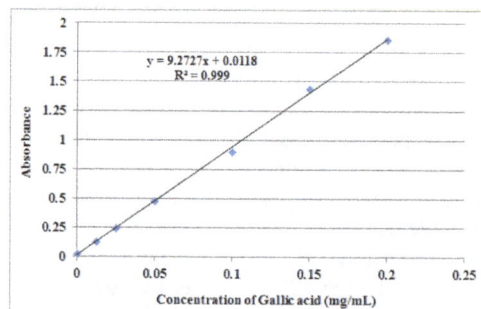

Figure 2. Calibration curve of gallic acid

The extraction of phenolic compounds was performed in methanol:water 50:50, and the absorbance of the extracts was determined at a wavelenght λ = 755 nm. Results were expressed as mg of gallic acid equivalents (GAE) per g product.

Antioxidant capacity
The DPPH scavenging radical assay was conducted according to Horszwald and Andlauer (2011) with some modifications (concerning extract volumes of the used sample and reagents, using UV-VIS Jasco V 550 spectrophotometer). The reaction was performed in dark for 30 min (at ambient temperature) and after this time the absorbance was read at 517 nm. It was achieved the calibration curve Absorbance = f (Trolox concentration), in the concentration range 0-

0.4375 mmol/L (Figure 3). Results were expressed as mg Trolox Equivalents per g product.

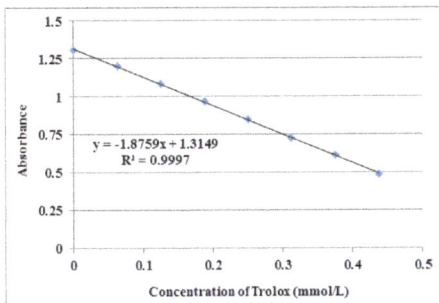

Figure 3. Calibration curve of Trolox

RESULTS AND DISCUSSIONS

Vitamin C content

Vitamin C content of fruits of *Aronia melanocarpa*, fresh and processed (frozen and dried fruits, juice, jam, compote) varied in the range: 7.25–98.75 mg/100 g (Figure 4). Minimum value was recorded for *Aronia* jam, and the maximum one for *Aronia* fresh juice (unpasteurized).

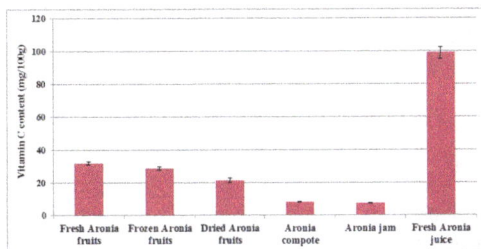

Figure 4. Vitamin C content of fresh and processed fruits of *Aronia melanocarpa*

Vitamin C content of fresh fruits of *Aronia melanocarpa* was 31.85 mg/100 g, and is higher than that mentioned by Kulling and Rawel (2008), 1.3–27 mg/100g and by Karakasova et al. (2014), 17.52 mg/100g, respectively. Also, vitamin C content of frozen fruits of *Aronia melanocarpa* (28.78 mg/100 g) is with 9.64% lower than those of the analyzed fresh fruits in this study, but higher than that mentioned by Karakasova et al. (2014), respectively, 17.15 mg/100 g. Vitamin C content of dried fruits of *Aronia melanocarpa* in this study was 1.4 times higher than that reported by Karakasova et al. (2014), respectively, 15.11 mg/100 g. *Aronia* compote

and *Aronia* jam recorded a low vitamin C content (7.96 mg/100 g, respectively, 7.25 mg/100 g), because vitamin C is very sensitive to oxygen and heat treatment. *Aronia* fresh juice analysed in this study is an important source of vitamin C (98.75 mg/100 g). Consumption of about 61 g *Aronia* juice ensure daily requirement of vitamin C for children older than 4 years and adults, respectively (60 mg vitamin C/day). The result obtained for the content of vitamin C of *Aronia* fresh juice is in line with those obtained by Djuricet al.(2015) for *Aronia* juice, obtained from fruits grown on four soil types (91.10-155.20 mg/100 mL). Frei et al. (2012) have shown that dietary supplementation with vitamin C decreased hypertension, endothelial dysfunction, chronic inflammation, and *Helicobacter pylori* infection. At the same time, vitamin C acts as a biological antioxidant that can reduce high levels of oxidative stress and may contribute to chronic disease prevention. Also, based on the performed studies, these authors concluded that 200 mg per day is the optimum dietary intake of vitamin C for the majority of the adult population, to maximize the potential health benefits of this vitamin.

Total polyphenol content

Fruits of *Aronia melanocarpa* are a valuable source of total polyphenols. Total polyphenol content of fresh and processed fruits of *Aronia melanocarpa* is shown in Figure 5. The highest total polyphenol content was recorded for dried fruits of *Aronia* (4015.25 mg GAE/100g), due to the high content of dry matter (89.8%).

The result obtained for the content of bioactive compounds is comparable with that obtained by Kapci et al. (2013) for dried chokeberry (3990±30 mg GAE/100 g, respectively, 5010±40 mg GAE/100 g).

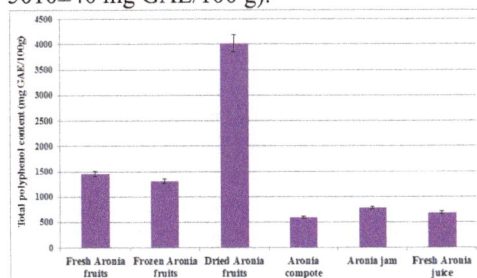

Figure 5. Total polyphenol content of fresh and processed fruits of *Aronia melanocarpa*

Also, total polyphenol content of dried fruits of *Aronia* is about 1.63 times higher than that obtained by Tolić et al. (2015) for chokeberry dried berries (2466±91 mg GAE/100 g of dry matter).

Fresh fruits of *Aronia melanocarpa* in this study had a total polyphenol content (1455.25 mg GAE/100 g) comparable to that reported by Kapci et al. (2013) for fresh chokeberry (1330±3 mg GAE/100 g), but lower than that reported by Ochmian et al. (2012) for four cultivars of chokeberry fruits ('Galicjanka'-2185 mg GAE/100 g; 'Hugin'-2340 mg GAE/100g; 'Nero' - 1950 mg GAE/100 g; 'Viking' - 1845 mg GAE/100 g).

Frozen *Aronia* fruits in this study had a polyphenol content of 1308.75±47.77 mg GAE/100g, comparable to that of fresh fruits.

Aronia compote (containing 55.25% fruits) had the lowest total polyphenol content (590.45±19.18 mg GAE/100 g), lower than that reported by Kapci et al. (2013) for this product (670±3 mg GAE/100 g). Compared to other products, *Aronia* juice had a lower total polyphenol content, this can be explained by the high water content (Shin et al., 2008). Thus, total polyphenol content of *Aronia* fresh juice studied was 688.47±22.75 mg GAE/100 g, higher than that reported by Konić Ristić et al. (2013) in case of commercial chokeberry juice (586±27 mg GAE/100 g), respectively, fresh chokeberry juice (593±33 mg GAE/100 g).

Values higher or lower for total polyphenol content of fresh or processed fruits of *Aronia melanocarpa,* reported in the literature, may result by use of various extraction methods and analytical procedures, through application of processing technologies and different conditions, respectively, the differencies between the varieties of these fruits (Denev et al., 2012).

The phenolic compounds are the most important class of bioactive compounds from the fruits of *Aronia melanocarpa,* which are also responsible for many of its medicinal properties (Kulling and Rawel, 2008). Thus, Sikora et al. (2012), in a human study shown that introduction in diet of extract of fruits of *Aronia melanocarpa* had as effects decrease of the lipid levels and significant inhibition of platelet aggregation.

Oprea et al.(2014) have shown that the addition of *Aronia* juice in the normal diet of healthy rats for 10 days, it was correlated with the reduction of values of some markers of oxidative stress and a decrease of blood glucose with 6.85%. Also, administration of *Aronia* juice in case of rats suffering alloxan induced-diabetes resulted in a significant reduction of blood glucose (42.83%).

Recent research undertaken by Daskalova et al. (2015) on animals have shown that treatment with juice of *Aronia melanocarpa* significantly reduced low-density lipoprotein fraction, with pro-aterogenic properties and a decrease of total cholesterol by 16.5%. In case of animals taken into study, dietary supplementation with *Aronia* juice has reduced atherogenic risk and also had a protective effect on the cardiovascular system. However, it was found that *Aronia* juice delay aortic changes that occur with age.

Antioxidant capacity

Antioxidant capacity of fruits of *Aronia melanocarpa*, fresh and processed (frozen and dried fruits, juice, jam, compote) varied in the range:12.25–84.45 mg Trolox Equivalents/g (Figure 6). Minimum value was recorded for *Aronia* compote, and the maximum one for dried *Aronia* fruits. Antioxidant capacity of dried *Aronia* fruits in this study is 2.32 times, respectively, 2.76 times higher than that reported by Kapci et al. (2013) for dried chokeberry (36.3±1.2 mg Trolox Equivalents/g, respectively, 30.5±1.0 mg Trolox Equivalents per g). Antioxidant capacity of frozen *Aronia* fruits (27.39±1.18 mg Trolox Equivalents/g) is with 10.05% lower than that of fresh fruits.

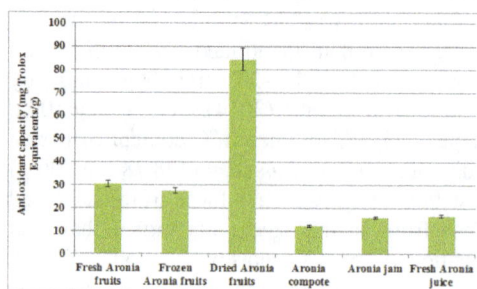

Figure 6. Antioxidant capacity of fresh and processed fruits of *Aronia melanocarpa*

On the other side, antioxidant capacity of fresh *Aronia* fruits is 2.69 times higher than that reported by Kapciet al. (2013) for fresh chokeberry fruit (11.3±0.5 mg Trolox Equivalents/g). Chokeberry fruits have one of the highest *in vitro* antioxidant activities among fruits (Denev et al., 2012).

In this study, *Aronia* jam and fresh *Aronia* juice recorded the following values for antioxidant capacity: 15.88±0.55 mg Trolox Equivalents/g, respectively, 16.55±0.58mg Trolox Equivalents per g. Antioxidant capacity of *Aronia* compote taken into study was 2.55 times higher than that reported by Kapci et al. (2013) for this product (4.8±0.1 mg Trolox Equivalents/g).

Between the total polyphenol content of fruits of *Aronia melanocarpa*, fresh and processed (frozen and dried fruits, juice, jam, compote) in this study and the antioxidant capacity was registered a linear correlation, regression coefficient, R^2, being 0.9988 (Figure 7).

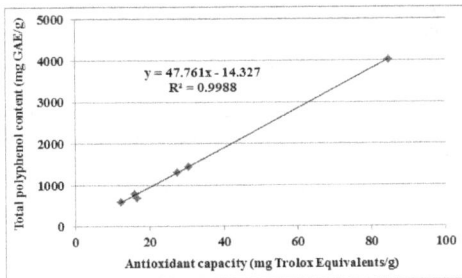

Figure 7. Correlation between total polyphenol content and antioxidant capacity in case of fresh and processed fruits of *Aronia melanocarpa*

Results are in conformity with those of Zheng and Wang (2003), which mentioned a direct correlation between total polyphenol content of fruits of *Aronia* and their antioxidant capacity.

CONCLUSIONS

In this study were evaluated vitamin C content, total polyphenol content and antioxidant capacity (DPPH method) of fruits of *Aronia melanocarpa* fresh and processed (frozen, dried, juice, jam, compote).

Fresh *Aronia* juice recorded the highest vitamin C content (98.75±3,6 mg/100 g). Also, fresh and frozen *Aronia* fruits are valuable sources of vitamin C.

Dried *Aronia* fruits shown the highest total polyphenol content (4015.25±164.63 mg GAE/100 g), followed by the fresh and frozen *Aronia* fruits. Also, fresh *Aronia* juice, *Aronia* jam and compote have an important total polyphenol content.

Dried *Aronia* fruits recorded the highest value of the antioxidant capacity (84.45±4.90 mg Trolox Equivalents/g), followed by fresh and frozen *Aronia* fruits. Fresh *Aronia* juice, *Aronia* jam and *Aronia* compote recorded relatively close values for antioxidant capacity.

Fresh and processed fruits of *Aronia* are valuable because of the content in bioactive compounds and their antioxidant capacity.

ACKNOWLEDGEMENTS

This study was supported by the Ministry of Research and Innovation, by Nucleu Programme PN 16 46, contract 29 N/2016.
This work was supported by a grant of the Romanian National Authority for Scientific Research and Innovation, CNCS-UEFISCDI, project number PN-II-RU-TE-2014-4-0749.

REFERENCES

Ara V., 2002. Schwarzfruchtige Aronia: Gesund – und bald „in aller Munde"?. Flüssiges Obst,10:653–658.

Benvenuti S., Pellati F., Melegari M., Bertelli D., 2004. Polyphenols, anthocyanins, ascorbic acid, and radicals scavenging activity of Rubus, Ribes, and Aronia.J Food Sci,69:164–169.

Bermúdez-Soto M.J., Tomás-Barberán F.A., 2004. Evaluation of commercial red fruit juice concentrates as ingredients for antioxidant functional juices. European Food Research and Technology,219(2):133–141.

Bijak M., Bobrowski M., Borowiecka M., Podsędek A., Golański J., Nowak P., 2011. Anticoagulant effect of polyphenols-rich extracts from black chokeberry and grape seeds. Fitoterapia,82:811–817.

Chrubasik C., Li G., Chrubasik S., 2010. The clinical effectiveness of chokeberry: A systematic review. Phytotherapy Research,24:1107–1114.

Daskalova E., Delchev S., Peeva Y., Vladimirova-Kitova L., Kratchanova M., Kratchanov C., Denev P., 2015. Antiatherogenic and Cardioprotective Effects of Black Chokeberry (*Aronia melanocarpa*) Juice in Aging Rats, Evidence-Based Complementary and Alternative Medicine, Volume 2015, Article ID 717439, 10 pages.

Denev P.N., Kratchanov C.G., Ciz M., Lojek A., Kratchanova M.G., 2012. Bioavailability and antioxidant activity of black chokeberry (*Aronia melanocarpa*) polyphenols: *in vitro* and *in vivo* evidences and possible mechanisms of action: a review. Compr Rev Food Sci., 11(5):471–89.

Djuric M., Brkovic D., Miloševic D., Pavlovic M., Curėic S., 2015. Chemical Characterisation of the Fruit of Black Chokeberry Grown on Different Types of Soil. REV. CHIM. (Bucharest), 66(2):178-181.

Frei B., Birlouez-Aragon I., Lykkesfeldt J., 2012. Authors' perspective: What is the optimum intake of vitamin C in humans?.Crit Rev Food SciNutr., 52(9):815-29.

González-Molina E., Moreno D.A., García-Viguera C., 2008. Aronia-enriched lemon juice. A new highly antioxidant beverage.Journal of Agricultural and Food Chemistry, 56(23):11327–11333.

Hakkinen S., Heinonen M., Karenlampi S., Mykkanen H., Ruuskanen J., Torronen R., 1999. Screening of selected flavonoids and phenolic acids in 19 berries.Food Res Int,32:345–353.

Handeland M., Grude N., Torp T., Slimestad R., 2014.Black chokeberry juice (Aronia melanocarpa) reduces incidences of urinary tract infection among nursing home residents in the long term a pilot study. Nutrition Research., 34:518-525.

Hirvi T., Honkanen E., 1985. Analysis of the volatile constituents of black chokeberry (Aronia melanocarpa). J Sci Food Agric,36:808–810.

Horszwald A., Andlauer W., 2011. Characterisation of bioactive compounds in berry juices by traditional photometric and modern microplate methods.Journal of Berry Research,1:189–199.

Horszwald A., Heritier J., Andlauer W., 2013.Characterisation of Aronia powders obtained by different drying processes. Food Chemistry,141:2858–2863.

Hudec J., Bakoš D., Mravec D., Kobida L., Burdova M., Turianica I., Hluše J., 2006. Content of phenolic compounds and free polyamines in black chokeberry (Aronia melanocarpa) after application of polyamine biosynthesis regulators.J Agric Food Chem,54:3625–3628.

Jakobek L., Drenjančević M., Jukić V., Šeruga M., 2012. Phenolic acids, flavonols, anthocyanins and antiradical activity of "Nero", "Viking", "Galicianka" and wild chokeberries.Scientia Horticulturae,147:56–63.

Kane M.E., Dehgan B., Sheehan T.J., 1991. In vitro propagation of Florida native plants: Aronia arbutifolia. Proc Fla State HortSoc,104:287–290.

Kapci B., Neradová E., Čížková H., Voldřich M., Rajchl A., Capanoglu E., 2013. Investigating the antioxidant potential of chokeberry (Aronia melanocarpa) products.Journal of Food and Nutrition Research,Published online 23 October 2013.

Kähkönen M.P., Hopia A.I., Vuorela H.J., Rauha J., Pihlaja K., Kujala T.S., Heinonen M., 1999. Antioxidant activity of plant extracts containing phenolic compounds. J Agric Food Chem,47:3954–3962.

Karakasova I., Babanovska-Milenkovska F., Stojanova M.,Karakasov B., 2014.Comparison of Qualities Properties of Fresh, Frozen and Solar Dried Chokeberry Fruits, 25th International Scientific-

Experts Congress on Agriculture and Food Industry, DOI: 10.13140/RG.2.1.4506.1209.

Kokotkiewicz A., Jaremicz Z., Luczkiewicz M., 2010. Aronia plants. A Review of traditional use, biological activities, and perspectives for modern medicine.Journal of Medicinal Food,13:255–269.

Konić Ristić A., Srdić-Rajić T., Kardum N., GlibetićM.,2013.Biological activity of Aronia melanocarpa antioxidants pre-screening in an intervention study design.Journal of the Serbian Chemical Society,78(3):429–443.

Kulling S.E., Rawel H.M., 2008. Chokeberry (Aronia melanocarpa) - a review on the characteristic components and potential health effects.Planta Medica, 74:1625–1634.

Lehmann H., 1990.Die Aroniabeere und ihre Verarbeitung. FlüssigesObst, 57:340–345.

Niedworok J., Brzozowski F., 2001. The investigation of a biological and phytotherapeutical properties of the Aronia melanocarpa E anthocyanins [in Polish].Post Fitoter,1:20–24.

Ochmian I., Grajkowski J., Smolik M., 2012.Comparison of Some Morphological Features, Quality and Chemical Content of Four Cultivars of Chokeberry Fruits (Aronia melanocarpa).Not Bot Horti Agrobo, 40(1):253-260.

Oprea E., Manolescu B.N., Fărcăşanu I.C., Mladin P., Mihele D., 2014. Studies Concerning Antioxidant And Hypoglycaemic Activity of Aronia melanocarpa Fruits.FARMACIA, 62, 2.

Oszmiański J., Wojdyło A., 2005. Aronia melanocarpa phenolics and their antioxidant activity.Eur Food Res Technol,221:809–813.

Park S., Kim J.I., Lee I., Lee S., Hwang M.W., BaeJ.Y., Heo J., Kim D., Han S.Z., Park M.S., 2013. Aronia melanocarpa and its components demonstrate antiviral activity against influenza viruses.Biochemical and Biophysical Research Communications,440:14–19.

Sharif T., Stambouli X., Burrus B., Emhemmed F., Dandache I., Auger C., Etienne-Selloum N., Schini-Kerth V.B., Fuhrmann G., 2013.The polyphenolic-rich Aronia melanocarpa juice kills teratocarcinomal cancer stem-like cells, but not their differentiated counterparts. Journal of Functional Foods,5:1244-1252.

Shin Y., Ryu J., Liu R.H., Nock J.F., Watkins C.B., 2008. Harvest maturity, storage temperature and relative humidity affect fruit quality, antioxidant contents and activity, and inhibition of cell proliferation of strawberry fruit. Postharvest Biol Technol., 49:201–9.

Sikora J., Broncel M., Markowicz M., Wojdan K., Chahalubinski M., Mikiciuk-Olasik E., 2012. Short-term supplementation with Aronia melanocarpa extract improves platelet aggregation, clotting and fibrinolysis in patients with metabolic syndrome. Eur J Nutr, 51(5):549–56.

Šnebergrová J., Čížková H., Neradová E., Kapci B., Rajchl A., Voldřich M., 2014.Variability of characteristic components of Aronia.Czech J. Food Sci.,32:25–30.

Sueiro L., Yousef G.G., Seigler D., De Mejia E.G., Grace M.H., Lila M.A., 2006. Chemopreventive potential of flavonoid extracts from plantation-bred and wild *Aronia melanocarpa* (black chokeberry) fruits. J Food Sci, 71:480–488.

Tolić M.T.,Jurčević I.L., Krbavčić I.P., Marković K., Vahčić N., 2015.Phenolic Content, Antioxidant Capacity and Quality of Chokeberry (*Aronia melanocarpa*) Products. Food Technol. Biotechnol., 53(2):171–179.

Weinges K., Schick H., Schilling G., Irngartinger H., Oeser T., 1998.Composition of an anthocyan concentrate from *Aronia melanocarpa* Elliot—X-ray

analysis of tetraacetylparasorboside. Eur J Org Chem, 1:189–192.

Wolski T., Kalisz O., Prasał M., Rolski A., 2007. Black chokeberry-*Aronia melanocarpa* (Michx.) Elliot-the rich source of antioxidants [in Polish]. Post Fitoter, 3:145–154.

Zheng W., Wang S.Y., 2003. Oxygen radical absorbing capacity of phenolics in blueberries, cranberries, chokeberries, and lingonberries. Journal of Agricultural and Food Chemistry, 51(2):502–509.

Zlatanov M.D., 1999. Lipid composition of Bulgarian chokeberry,black currants and rose hip seed oil.J Sci Food Agric,79:1620–1624.

26

PRESENT STATUS AND FUTURE PROSPECTS OF GEOTHERMAL ENERGY USE FOR GREENHOUSE HEATING IN TURKEY

Hasan Huseyin OZTURK

Cukurova University, Faculty of Agriculture Engineering of Agricultural Machineries and Technologies, 01330 Adana, Turkey

Corresponding author email: hhozturk@cu.edu.tr

Abstract

In order to obtain the highest yield of the expected product are grown in greenhouses, it is necessary to heat the greenhouse during periods of low temperatures. Conditions of our country, heating costs are one of the most important factors affecting the profitability of the greenhouse. Greenhouse heating costs vary depending on the product type, growing season and the region, accounted for 40% and 80% of the total cost. Due to the high costs of fossil fuels used for heating greenhouses heating applications cannot be done on a regular in many greenhouse of our country, heating is done only to protect plants from frost. Greenhouse heating applications, utilization of alternative energy sources instead of fossil energy sources is a priority need in order to today's energy assets and to protect the environment. In this study; current situation and problems of geothermal greenhouses in Turkey were assessed and the necessary suggestions were made to improve the utilization of geothermal resources in the greenhouse heating.

Key words: geothermal energy, greenhouse heating, strategy development.

INTRODUCTION

In the world of today's industry, the usage of energy and energy resources have crucial value. While the amount of natural resources (especially, fossil fuel resources) has been decreasing, the damage to the natural environment as many type of environmental pollutions has been increasing. Additionally, the technical improvements for the energy conversion can not be carried out as effective as needed. In order to determine the level of future energy production and consumption in developed and developing countries, there are many factors to be considered, such as population growth, economic productivity, consumer habits and technological advances. The style of energy sectors management will play an important role for the future of energy production, consumption and distribution. Careless use of energy resources and their scarcity, resulting unwanted side effects, so energy consumption must be planned and evaluated carefully and accurately.

Geothermal energy is of vital importance in terms of preventing environmental problems like greenhouse effects and acid rains arising from using and consuming fossil fuels. This is primarily because of natural superiority of geothermal energy as of environment when compared with other energy types. On the other hand, important developments have been achieved in terms of solving possible environmental problems that could occur as a result of geothermal energy use. This, in turn increased the importance of geothermal energy with regards to environment. Geothermal energy being one of our domestic resources must be evaluated in preference to other resources in order to decrease the dependency to petroleum in meeting our country's gap in energy and to prevent foreign currency loss. Geothermal energy is an inexhaustible energy resource as others like hydraulic, solar and wind energy resources. For that reason, when compared with fossil energy resources that are certainly exhaustible, geothermal energy resources are long-lasting and inexhaustible energy resources.

Bertani (2016) has analyzed the major activities carried out for geothermal electricity generation An increase of about 1.8 GW in the five year term 2010-2015 has been achieved (about 17%), following the rough standard linear trend of approximately 350 MW/year. Lund and Boyd (2015) reviewed the worldwide applications of geothermal energy for direct utilization. The report is based on country update papers received from 70 countries and regions of which 65 reported some direct utilization of geothermal energy. The thermal energy used is 592.638 TJ/year (164.635 GWh/year), about a 39.8% increase over 2010, growing at a compound rate of 6.9% annually. The distribution of thermal energy used by category is approximately 55.2% for ground-source heat pumps, 20.2% for bathing and swimming (including balneology), 15% for space heating (of which 89% is for district heating), 4.9% for greenhouses and open ground heating, 2.0% for aquaculture pond and raceway heating, 1.8% for industrial process heating, 0.4% for snow melting and cooling, 0.3% for agricultural drying, and 0.2% for other uses. Energy savings amounted to 352 million barrels (52.8 million tons) of equivalent oil annually, preventing 46.1 million tons of carbon and 149.1 million tons of CO_2 being released to the atmosphere, this includes savings for geothermal heat pumps in the cooling mode (compared to using fuel oil to generate electricity). Considerable work has been conducted on geothermal fields in Turkey and the application of geothermal energy for district heating with respect to efficient and economic use of energy for sustainable development (Mertoglu et al., 2003). Comprehensive studies have been performed on geothermal energy use in agricultural production systems. Greenhouse heating with geothermal energy has been investigated by several authors (Rafferty 1986; Popovski 1988; Bakos et al., 1999; Popovski and Vasilevska, 2003). Optimization of geothermal energy use in agriculture is reflected in two ways, i.e. an increase in productivity at the existing level of energy inputs or conserving the energy without affecting the productivity. In many studies the influence of the direct energy input on energy efficiency was analyzed. Few publications have dealt with the problems related to geothermal energy use in greenhouses. Therefore, in the present study the present status and future prospect were analyzed to develop the geothermal energy use in greenhouse cultivation.

In greenhouse heating applications, using alternative energy resources rather than fossil energy resources is a primary necessity for conserving our energy wealth and preventing the environmental pollution. Some of the alternative energy resources that are used in greenhouse heating are; solar energy, geothermal energy and low temperature heat energy from the wastes of industrial facilities. In this study, the existing situation and problems of geothermal greenhouses in Turkey were evaluated and necessary proposals were given to increase the use of geothermal energy resources in greenhouses.

GEOTHERMAL ENERGY AND GREENHOUSE SECTOR IN TURKEY

Geothermal Energy Potential of Turkey

Turkey is located in the Alpine-Himalayan organic belt is an important region in terms of geothermal resources. In terms of geothermal resources, it is among the first seven countries in the world. Due to its location on Alpine Tectonic Zone, Turkey has a significant potential for geothermal energy. Our country is on the very effective zone in the point of tectonic with graben in the west, basin regime in the middle, compressive tectonic in east and North Anatolian Fault in the north. As a result of fracture and weakness zones and magma actions that reach from these zones to the shallow depths within the shell and/or the surface of earth, magmatic and volcanic events occur. Geothermal systems are developing with the help of geological and meteorological phenomena as a geothermal fluid. Theoretical and determined geothermal energy potential in our country are summarized in Table 1.

Table 1. Potential of Turkey's Geothermal Energy (Ozturk, 2015)

Energy	Theoretical potential	Determined potential
Electricity	4 500	200
Thermal energy	31 100	2 250

The areas containing high-temperature geothermal fluids are located in the west part of Turkey due to the grabens resulting the young tectonic activities. With the effect of volcanism and faulting low and medium temperature fields are located Central Anatolia, Eastern Anatolia and north side of Turkey with the North Anatolian Fault.

Geothermal Energy Use

Although it varies depending on the fluid temperature and regions conditions, geothermal energy use can be examined under two general categories:

1) Electricity generation
2) Direct use of geothermal energy

Heat exchanger systems, as wellhead and borehole heat exchangers, can vary in design depending on the area of the feature. The efficiency, continuity or success of the heating system is based on used accordance with technology. Geothermal water that contains chemical materials at a level causing them not to be directly used and whose heat exchangers and heat energies are transferred into supply network water clean enough to utilize, should be removed from the surroundings in order to avoid environmental pollution.

The benefits provided by the direct use of geothermal energy are:

✓ Conversion efficiency is high.
✓ Can be utilized in low temperature geothermal resources.
✓ Can be utilized from wells for research purposes
✓ Project implementation period is short.
✓ Drilling costs are cheaper in shallow depths.
✓ Geothermal fluid can be transported to long distances.

Depending on the geothermal fluid's chemical properties, heating systems show significant differences. If geothermal fluid does not cause problems as chemical content property, it can be used directly circulated through radiators and proper pipes system in radiator surface. However, if the fluid contains too much minerals and is likely to cause chemical problems (scaling, corrosion and so on), heat of the fluid is transferred to water with low chemical content (water used in city supply networks) via heat exchanger. Thus, the heating provided by the heated water does not cause problems in the system.

The direct use of geothermal energy can be divided into four groups:

1) Housing and workplaces
2) Industrial applications
3) Agriculture and related fields
4) Thermal and health tourism

Greenhouse Sector in Turkey

Greenhouse sector requires high investment and industrial activity. According to TurkStat data (TUİK, 2016), in Turkey in 2015, it has a total of 66,400 hectares of greenhouse area. Total greenhouse area of 30,900 hectares of plastic greenhouses, constitutes 8097 hectares of glass greenhouses. Greenhouse areas are more concentrated in southern provinces. The biggest reason is that southern provinces are warmer than in other cities in the winter. According to the Greenhouse Registration System (OKS), in Turkey, 48% of the greenhouse enterprises have the areas between 2-3 acres in 2013. The ratio of the greenhouse enterprises that have area more than 10 acres is only 1.8%. There are 9000 acres of greenhouses with modern conditions. In modern greenhouses, the average size is 27 acres (Ozturk et al., 2015).

A significant portion of the crops grown in greenhouses are the vegetables (95%) and the least portion is constituted by the fruits (5%). Between the vegetable species that grown in greenhouses, tomato production takes first place with the amount of 50%; cucumber, watermelon, green pepper and eggplant follow it. Production share of other vegetables such as melon, bean, zucchini and lettuce is gradually increasing. The most important type of vegetable, cultivation in low plastic tunnels is watermelon. In our countries greenhouse agriculture, widely produced fruits are banana, strawberry, grapes and nectarines. Cut flower production is increasing rapidly in Turkey and it is made in the regions of Marmara, Aegean and Mediterranean. In Turkey, with the 6.5 million tons of greenhouse production, 10 million TL GDP was provided in 2014 (Ozturk and et al., 2015).

GEOTHERMAL ENERGY USE FOR GREENHOUSE HEATING IN TURKEY

The Importance of Geothermal Energy in Heating Greenhouses

To provide optimal environmental conditions in terms of quality, quantity and development time of the products grown in greenhouses, mainly used for the out-of-season produce, it is required heating in the winter's cold period and ventilation in summer's hot period. Keeping under control the ambient temperature, that has an effect mainly on the yield and quality of plant growth and development, is an important factor in greenhouse technology. To obtain maximum yield expected from the products cultivated in greenhouses, it is necessary to heat the greenhouse during periods of low temperatures. Even though it is required excess quantity of energy to heat the greenhouses in winter and summer seasons in Northern European countries, only the needed heating applications are not made enough during cold winter nights due to the suitable ecological conditions at most of the Mediterranean countries. In this case, negations are encountered, regarding the quality, quantity and harvest time of the product.

By controlling the climate conditions, it made greenhouse to spread wider in a year when the agricultural production process is the most important issue in the production of heating. In our countries conditions, heating costs are one of the most important factors affecting the profitability of the greenhouse. In greenhouse enterprises, heating costs depend on growing season, region and product type and they constitute 40% to 80% of the total cost (Ozturk, 2015). Due to the high costs of fossil fuels used for heating greenhouses, a regular heating can't be done in many greenhouses in the our country and the protecting plants from frost heating is only done. Not to be done regularly heating bring some concerns together such as low yield, limitations on the types of production, necessity of using drugs and hormones for agricultural struggle. However, a sufficient heating yield to provide the temperature at which the plant needs can increase by 50-60%. For this reason, in heated greenhouses using geothermal energy, required temperature is provided more economically for plant growth and fertilization and the relative humidity of greenhouse indoor air is controlled with the necessary ventilation, diseases can't occur and yield can increase.

According to the Ninth Development Plan (2007-2013) Mining Special Commission Energy Raw Materials Subcommittee of the Geothermal Working Group Report, including existing state and 2013 projections about Turkey's geothermal electricity generation and geothermal heating, it is stated that a qualification of geothermal energy to be emphasized in particular. In the same report, it is indicated that the use of geothermal energy for electricity generation is limited yet and there has been a steady increase in consumption for heating purposes in recent years, so 635 acres of geothermal heated greenhouse presence, has been targeted as 5000 acres in 2013. But according to the Ministry of Food, Agriculture and Livestock Protected Registration System (OCS), as of September 2013, Turkey's geothermal greenhouse uses of assets has reached to a total of 3202 hectares in 10 provinces. Although increased significantly in recent years, but reached the 64% of the target.

In Tenth Development Plan (2014-2018) Mining Special Commission Geothermal Working Group Report, greenhouse warming target is determined as 600 hectares for 2018 and 1500 hectares for 2023. The use of geothermal resources for heating in greenhouses has to be addressed as an innovation that provides economic and environmental benefits for the agricultural sector. The process of adoption of innovation is a multivariate and complicated process related on the one hand with innovation and on the other hand the systems and individuals that used the innovation. Before studies will be launched to increase the geothermal greenhouse area, first individual characteristics and resources of the target audience, its communication channels, time and social environment to be examined when considering. Besides, increases in national and international food prices increased transfers of the capital from non-agricultural sector to agriculture sector by reversing the situation. In addition, growing interest to the alternative energy resources made geothermal greenhouse investments attractive, especially using modern production technology (hydroponics etc.). Environment, making an agricultural production that does not harm human and animal health, preservation of natural resources, traceability and sustainability and manpower using knowledge in greenhouse activities for ensure reliable product supply has become the most important factor. However, investments initiated by parties who aren't making feasibility studies before starting agricultural activities couldn't be reached to desired goals and leads to the formation of excess capacity for the country. To this end, education-extension and publicity studies and planning of human resources, that ensure required capacity at all parties from production to marketing, are required.

Greenhouse Areas Heated with Geothermal Energy in Turkey

Greenhouse heating applications made with geothermal energy in Europe began in Hungary and Yugoslavia in the 1960s. In our country, while heating greenhouses have gained great importance in the utilization of geothermal energy more effectively, applying technical and economic aspects are faced with some problems. Geothermal heating is one of the places where the most widely used of geothermal resources; geothermal greenhouse heating is becoming increasingly important in Turkey. Greenhouse heating system with geothermal energy was applied for the first time in Turkey in the 0.45 hectares areas of the Denizli-Kızıldere. To more accurately and efficiently take advantage of geothermal resources, low enthalpy will need to make use of advanced technology.

According to the 2013 OKS records in Turkey, greenhouses heated by geothermal energy fields are given in Table 2. In 10 provinces in 3202 decares area, is carried out under cover production using geothermal energy. Almost half of these areas are located in Izmir (24.48%) and Manisa (23.42%). While maximum business is located in Kutahya with 46 business, it is followed by Sanlıurfa and Denizli provinces with 26 business. As shown in Table 2, a very small part of the overall greenhouse is heated with geothermal energy in Turkey. While the value of apparent heat capacity that geothermal drilling issued is 3000 MW in 2002, it is reached to 7000 MW with an increase of 230% in 2012. A 10-year period between 2002-2012 years, our greenhouse space heated by geothermal energy has increased by 406% (Hasdemir et al., 2014).

Distance between geothermal resources and businesses vary too much between businesses. The most important factor in the transport of hot water is the temperature of resources. High temperature water can be used in heating by moving to longer distances. Hot water used for heating is used by bringing from the wells average 541 m away (Hasdemir et al., 2014). Greenhouse businesses use hot water from geothermal wells in heating by exchanger system or directly circulating in greenhouses. 67.21 % of greenhouse businesses use hot water directly and 32.79% of them use heat exchanger system (Hasdemir et al., 2014). Hot water taken into greenhouse largely use as a space heating on soil. Only one company was seen heating with underground pipes (Table 3).

Table 2. Greenhouse Areas Heated Through Geothermal Energy (Hasdemir et al., 2014)

Provinces	Number of business	Greenhouse area (da)	The ratio in total geothermal greenhouse area (%)
Afyon	6	358	11.18
Aydın	17	173	5.40
Denizli	26	456	14.24
İzmir	15	784	24.48
Kırşehir	1	97	3.03
Kütahya	46	125	3.90
Manisa	7	750	23.42
Nevsehir	1	61	1.91
Şanlıurfa	26	373	11.65
Yozgat	2	25	0.78

Table 3. Heating Systems in Greenhouses Heated with Geothermal Energy

Properties	Number of business	Ratio (%)
Direct heating	18	67.21
Heat exchanger use	40	32.79
Above ground heating system	121	99.18
Below ground heating system	1	0.82
Reinjection	21	17.21
Refiner	41	33.61
Dumping to the land	60	49.18

67.21 % of available geothermal resources have additional greenhouse warming potential. Despite the availability of geothermal fluid brought to surface, the reasons why the businesses cannot do additional greenhouses are stated by themselves as follows: 32.93 % for insufficiency of land, 29.27% for not being able to acquire official permit, 19.51% insufficiency of capital, 4.88% for high costs, 2.44 % for marketing problems, 1.22% insufficiency of workforce and technical reasons, 8.54% for other reasons.

RESULTS AND DISCUSSIONS

Legislation on Geothermal Resources Prioritizes Electricity Production

Due to problems that Turkey face in energy sector, the tendency is towards using local resources and for this reason certain subsidies and incentives were provided particularly in recent years to investors in these areas. In this context Article 6, Paragraph one of *"Law on the Use of Renewable Energy Resources for Electricity Production"* states that "For those who has the production license subject to Renewable Energy Resources Support Mechanism which are or will be in business from the enforcement of this law on 18/5/2005 to 31/12/2015, the prices annexed to this law in annex 1 as a chart will be applied for 10 years". This means that purchase of the energy produced by using renewable resources is guaranteed for 10 years. It is pledged in the document that the electricity produced through renewable resources is going to be purchased at the rates of 10.5 $cent/kWh, and for 13.2 $/kWh in cases where local manufacturing is utilized (Ozturk, 2015).

Additionally, due to the fact that licenses are provided for a limited period of 3+1 years and the high fluid temperatures figured after drilling creates a tendency in investors to launch Geothermal Energy Plants (GEP). Although there is no limitation for other areas, the uncertainties in these areas causes investors to steer their investments in electricity production. The investments are directed to electricity production because, when they want to use geothermal energy in greenhouse sector there is uncertainty in the number of people and the area that will utilize greenhouse sector; when they want to use it in household heating, there is the uncertainty of potential problems with local administrations and the people who will use it. Because that the GEP's in operation before end of 2015 will receive guaranteed purchases from the states, the owners of geothermal resources want to complete their energy investments before 2015. They plan their investments in other areas after completion of energy investments. When the temperatures of geothermal resources are not sufficient for energy production, they can be utilized in greenhouses. The owners of geothermal plants may use the remaining heating potential in greenhouses after energy production, if they wish to do so.

The Licensed Geothermal Areas

The biggest obstacle for geothermal resources to be used in greenhouses is that after a process of sale of geothermal areas through procurements, the owners and the people who would like to use these areas in greenhouses are different. The owners do not encounter problems when they would like to use the resources in greenhouses, however when others would like to use it the same way, they are either reluctant to give these for greenhousing or they sell it for too high prices. Because that geothermal investments are in abundance and the number of people who will perform greenhouse sector or that the areas where greenhouses are established are separated increasing costs of deploying geothermal energy makes it further difficult to use in greenhouse sector. In cases where the greenhouses would be established collectively closer to the geothermal resources, the transfer of geothermal energy to these areas will be less costly and thus the owners approach may change. In this regard, the purchasers of geothermal plants need to know the proximity of those greenhouses of whose owners are willing to use geothermal heating in their greenhouses. This way it will be easier for owners of geothermal energy to calculate the annual operation costs and determine the price of energy for greenhouse sector. For this reason, the greenhouses should be collectively established near the geothermal resources.

When preparing the law for the use of geothermal resources in other areas, it is necessary to consider giving initiative to the local administrations close to geothermal resources to define priorities based on needs of the region and to give them the opportunity to utilize the geothermal resources. Even if there is a change in law now, this will be beneficial for the areas up for future procurement. In case there is no change in current law, introducing a provision in the contract of procurement for the use of geothermal energy in other areas would provide ground for more efficient utilization of the geothermal resources. There is currently no sanction or enforcement for the owners of the plants to use geothermal energy other than electricity production. Another difficulty is the determination of the price per hectares of geothermal energy provided by the owners. If the pricing is not reasonable, this is a problem for geothermal greenhouse sector. Measures are necessary in order to determine the price of geothermal energy and to avoid future conflicts.

Due to the fact that the geothermal areas are licensed, the initiative to decide on where to use the energy is held by owners. It is necessary to allow the social use by local administrations of blocked areas which surround the geothermal resources for their protection. It is absolutely necessary to make changes for local communities to use the geothermal resources. It does not appear to be a fair process to preserve the use of these areas which were held by the state until today only to the use of renters.

Finding Land and Geothermal Resources for Greenhouses

As a result of the use of geothermal resources after the introduction of the legislation, the number of entrepreneurs willing to invest in this area is increasing. Entrepreneurs mostly face difficulties to find areas for and to use geothermal energy for greenhouses. Sometimes they find the area for greenhouse sector but cannot find the energy, and sometimes otherwise. Consequently, it is important to keep entrepreneurs in business by promoting geothermal energy owners in investing in other areas and in greenhouse sector through incentives, subsidies or low-interest credit opportunities.

The owners of geothermal resources may face problems in finding greenhouse areas when they want to invest themselves. This is because there is not enough land owned by state or available land is private properties. When they want to buy land for investing in greenhouse sector, the prices may raise up to 10 fold. This affects geothermal greenhouse sector negatively.

The Cost of Searching and Developing Geothermal Resources

The high costs of geothermal investments make it more difficult for it to be utilized in geothermal greenhouse sector. Due to the fact that the drill operations for geothermal resource searching is costly and that the cost of re-injection wells and re-injection is high, investors tend to lean towards geothermal power plants which have lower costs and quick returns. The heavy metals, particularly the phototoxic boron in plants, have negative consequences for the environment which increases the costs due to compulsory reinjection and this effects greenhouse sector negatively.

Greenhouse Costs

Costs of modern technological hydroponic culture greenhouses change depending on the technology used and the cover material. The costs of PC sided and PE roofed steel construction and computer controlled greenhouses starts from 50-60 €/m², and of steel construction glass greenhouses starts from 75-100 €/m² (Ozturk, 2015).

Because that the current modern hydroponic culture greenhouses in Turkey are imported or patented outside of the country and that almost all of automation systems are imported as well, the costs are high. Thus it is important to develop local technologies in greenhouse automation systems. The high costs of greenhouses effects greenhouse sector and hence geothermal greenhouse sector negatively.

Problems Arising from Agriculture based Specialization Organized Industry Zones

The best means of greenhouse sector is through the establishment of Agriculture Based Specialization Organized Industry Zones in terms of more efficient use of geothermal source, healthier production, monitoring potential of production and healthier organization of marketing. There are problems both in legal and local terms in establishment of these Zones. These problems also negatively influence geothermal greenhouse sector

Published in 10.10.2009 at the Official Gazette numbered 27402 "The Application Regulation on Establishment of Agriculture Based Specialization Organized Industry Zones" was executed by Ministry of Science Industry and Technology and Ministry of Agriculture and Rural Affairs later in 12/03/2012 by Ministry of Food Agriculture and Livestock. In order for greenhouse sector to be performed in larger areas, the elimination of the rule regarding the requirement of 75% treasury land in which Zones to be established for greenhouse sector will pave the way for greenhouse sector effectively. However, the future of rural population whose lands are already expropriated is concerning.

Many producers in this situation will opt for migrating to cities in which they will encounter adaptation and economic problems. In this case, the high income groups will invest in greenhouse sector in these Zones. This is why the geothermal resources have to be planned in a way to positively influence the lives of local communities rather than negatively. In the regulation, people whose lands were expropriated should be given priority places in Zones. Because that these are small-scale family businesses, it is difficult for them to invest in greenhouses with their own assets, thus they should be provided with 7-10 years credit opportunities without repayment in the first 2-3 years or grants.

As a result of the developments in greenhouse sector in recent years, it is largely modern technology hydroponic culture greenhouse sector. When these modern greenhouses grow, the costs decrease. This is why, if there is a plan to establish such a greenhouse for economic investment the lower limit should be at least an area of 2-2.5 hectares. It is stated that for a Zone to be established there should be at least 50 hectares of land and 30 persons should be involved in the project.

For a modern greenhouse sector in Zones to be economically meaningful it is necessary either to increase the area of Zones required or to decrease the number of people required to be involved. It is necessary to determine the sizes of lands depending on the technology applied in greenhouses.

However, if there is hydroponic culture greenhouses in Zones than the greenhouses should be planned at the minimum 2-2.5 hectares. It is important to take into account the clear water and the geothermal resource to be utilized in greenhouses in the process of establishment.

Administration Shares of Facilities Utilizing Geothermal Energy

In the relevant law and regulations, it is stated that 1% equivalent of the gross production of those facilities directly or indirectly utilizing geothermal fluids or gasses stated in the Law will be collected as administrative cost. In this situation, 1% equivalent of the gross production of greenhouses utilizing geothermal energy will be paid as Special Administration cost. 1% equivalent is a considerably high payment for geothermal greenhouses. 1% equivalent payment should be for those businesses which utilize the geothermal resources directly and whose sole input is geothermal energy such as electricity production, CO_2 producer facilities. For those facilities utilizing geothermal energy as a secondary or tertiary source, 1% should either not be paid or decreased.

Turkish Development Law and the Law on Construction Inspection

Greenhouses are regulated under 'Turkish Development Law No. 3194' and subjected to license. While greenhouses without geothermal heating are subjected licensing process of Ministry of Food Agriculture and Livestock and State Hydraulic Works, the ones with greenhouse heating go through licensing process of below institutions:

1. Provincial Directorate of Culture and Tourism,
2. Provincial Directorate of Science, Industry and Technology,
3. Provincial Directorate of Food, Agriculture and Livestock,
4. State Hydraulic Works,
5. Provincial Directorate of Environment and Urbanization,
6. Special Provincial Directorate of Administration,
7. Electricity Distribution Company,
8. Provincial Gendarmerie Command
9. Directorate of Highways

Legal Status of Greenhouse Sector

New legal arrangements are needed to regulate the legal status of greenhouse sector also addressing all the aspects of it; from planning to construction and production.

This can be resolved through new regulations to be introduced by The Ministry of Food

Agriculture and Livestock. Other relevant Laws and Regulations applicable to this subject also should be revised.

Greenhouses should be in accordance with 1:100.000 scaled Provincial Environment Oder Plans and excluded from the scope of the *Development Law*. Greenhouses should be also exempted from the Law on Construction Inspection. These greenhouses should be classified as agricultural structures under the new regulations to be implemented and all pre-construction and post-construction works or inspections, including planning and authorization should be centralized under the mandate of a single authority. The best implementation approach for this matter would be centralizing all these processes under the mandate of the Ministry of Food Agriculture and Livestock.

Government Incentives Available for Geothermal Greenhouse Production

Modern greenhouse production systems are expected to face a rapid growth with increased use of technology which improves yield and profitability. In order to achieve the targets of agricultural sector in the future, appealing incentives are offered by the government for greenhouse investments in our country. Greenhouse production is the third biggest sector of our country in terms of food and agriculture investments prominent after livestock and dairy sector. In 2012, greenhouse production was listed as the most invested sector following livestock and other relevant sectors.

Despite the diverse geothermal capacity that our country possess, as a result of the insufficient levels of use, geothermal sector have been endorsed for incentives. In this regard applicable legal arrangements are as follows:

➢ The Communiqué on "Support for Agriculture Investments (No#2011/9)" published in the Official Gazette No#27871 dated 11.03.2011 within the scope of Support Programs for Rural Development Investments.

➢ "The Regulation Amending the Regulation on Pasture Land" promulgated in the Official Gazette No#27857 dated 28.02.2011.

➢ The Communiqué on "The Procedure and Principles to be Pursued Regarding Public Domain Allotments for Technology, Geo-

thermal Greenhouses and Organic Farming Investments" published in the Official Gazette No#26511 dated 03.05.2007.

➢ The regulation for "Implementation of Agro-Industries and Organized Greenhouse Sites" published in the Official Gazette No#27402 dated 10.11.2009.

➢ "The General Communiqué on National Estate" published in the Official Gazette No#27211 dated 26.04.2009 (I/N: 324)

➢ "The General Communiqué on National Estate" published in the Official Gazette No#27901 dated 10.04.2011 (I/N: 335)

➢ "The Decree on Concerning State Aids in Investments" published in the Official Gazette No#27290 dated 16.07.2009.

Within the framework of "Support Program for Rural Development Investments", renewable energy integrated greenhouse investments projects with a budget up to 3 million at max are remunerated by 50% through Financial State Aids. Furthermore, within the scope of the National Real Estate General Communiqué (number:352) published in the Official Gazette No# 27211 dated 26/04/2009, state land are allocated for entrepreneurs that shall invest on geothermal greenhouse systems. Accordingly, with respect to the changes made on the Law on Pastures, pasture lands are also available for geothermal greenhouse investment allocations.

Technology Transfer in Geothermal Greenhouses in Turkey and Sources of Agricultural Extension Information

Technology transfer, extension and adoption of innovations in geothermal greenhouses in Turkey are realized through different extension methods such as agricultural extension programs, demonstrations, field days, farmer meetings, farmer courses, farmer examination trips, incentive competitions, conferences, panels, other similar activities and mass extension media produced to be used in these activities. The information sources that are effective in decision making processes of enterprises using and not using geothermal energy resources are classified in three groups. These information sources were determined as; informal sources like the producer himself, his neighbors or relative and formal sources like Provincial/District Agricultural Directorates, agricultural consultants and technical staff of the enterprises and pesticide and fertilizer distributors that provide input to agriculture and the media organizations. While the enterprises that are using geothermal resources are using formal information sources, the enterprises not using geothermal resources are using informal information sources.

CONCLUSIONS

In accordance with the Protected Cropping Registry System of Ministry of Food Agriculture and Livestock, as of September 2013, total land allocated for protected cropping with geothermal heating in Turkey reached 3202 da in 10 different provinces. Although a significant increase can be noted in recent years, only 64% of the set targets could be achieved. As indicated in the report of Geothermal Working Group of Specialization Commission for Mining, under Tenth Development Plan (2014-2018); the target for geothermal greenhouse heating was determined as 600 ha for year 2018 and it was indicated as 1500 ha for year 2023. It is fundamental to consider geothermal greenhouse heating systems as an innovation that endeavor economic and environmental benefits for agriculture sector. The adoption process of innovation is while eminently related to the innovation itself it is also a sophisticated process encompassing the system that the innovation will be implemented and the users. Before implementing the work connected with the efforts to increase greenhouse areas with geothermal heating, it is essential to consider individual features of the target group and resources including communication channels, time and social environment.

Organized Greenhouse Sites offering cost-efficient, secure, traceable, competitive, modern and planned protected cropping with strong brand equity, clustered in regions which are rich in terms of geothermal resources (such as Afyonkarahisar, Aydın, Denizli, Diyarbakır, İzmir, Kırşehir, Konya, Kütahya, Manisa, Şanlıurfa and Yozgat) are aimed to be established. Having a geothermal capacity of 31,500 MW, Turkey possess enough resources in terms of reaching the targeted 600 ha of land for year 2018 and 1500 ha of geothermal heated greenhouse area for year 2023.

However, said resources should be efficiently managed and strategies should be in place to ensure the establishment of sustainable geothermal greenhouses.

Making use of alternate energy resources instead of fossil fuels for greenhouse heating systems is a priority concern in terms of conserving the energy resources of today's World and preventing environmental pollution. In Turkey, although effective use of geothermal energy for greenhouse heating has become more important, there are still some tackles resulting from the implementation in terms of economic and technical aspects. As geothermal heating known as a common way of adopting geothermal resources, geothermal greenhouse heating has become significantly important in Turkey. Recycled liquid injected back into Geothermal Power Plants for reinjection can also be used for greenhouse heating and legal arrangements should be made in this regard. It is fundamental to consider geothermal energy resources as an innovation that endeavor economic and environmental benefits for agriculture sector.

REFERENCES

Bakos G.C., Fidanidis D., Tsagas N.F., 1999. Greenhouse heating using geothermal energy. Geothermics, 28:759-765.

Bertani R. 2016. Geothermal power generation in the world 2010–2014 update report. Geothermics 60:31-43

Hasdemir M., Hasdemir M., Gül U., Yasan Ataseven Z., 2014. Geothermal Greenhouses in Turkey. Project Final Report, TEPGE Publication No: 227, ISBN: 978-605-4672-60-8.

Lund J.W., Boyd T., 2015. Direct utilization of geothermal energy 2015 worldwide review. Geothermisc, 60:66-93.

Mertoglu O., Bakir N., Kaya T., 2003 Geothermal applications in Turkey. Geothermics, 32:419-428.

Ozturk H.H., 2015. Geothermal Greenhouses. Umuttepe publications, Publication No: 150, ISBN: 978-605-5100-56-8, Kocaeli, Turkey.

Popovski K., 1988. Factors influencing greenhouse heating and geothermal heating systems. Geothermics, 17:173-189.

Popovski K., Vasilevska S.P., 2003. Prospects and problems for geothermal use in agriculture in Europe. Geothermics, 32:545-555.

Rafferty K., 1986. Some considerations for the heating of greenhouses with geothermal energy. Geothermics, 15:227-244.

TUIK., 2016. Turkish Statistic Foundation, Ankara, Turkey.

27

COMPARISON OF THE COSTS OF MATING DISRUPTION WITH TRADITIONAL INSECTICIDE APPLICATIONS FOR CONTROL OF CODLING MOTH IN APPLE ORCHARDS IN TURKEY

Orkun Baris KOVANCI

Uludag University, Faculty of Agriculture, Department of Plant Protection,
Gorukle campus, 16059, Bursa, Turkey
Corresponding author email: baris@uludag.edu.tr

Abstract

Mating disruption is an alternative control tactic that prevents male insects from finding females, resulting in lower pest density and less crop damage. However, the relatively high cost of mating disruption compared to the conventional chemical control may be an impediment to its adoption by growers worldwide. Therefore, this study aimed at comparing the costs of mating disruption with insecticides for control of codling moth, Cydia pomonella L., in apple orchards in Turkey in 2013 and 2014. Experimental orchards consisted of semi-dwarf 'Gala' and 'Fuji' apple cultivars. Codling moth populations, the number of insecticide applications and management costs varied between cultivars and years. When averaged over cultivars, mating disruption decreased the total number of sprays for apple pest complex by 40.70% and 56.60% in 2013 and 2014, respectively. All control costs related to the number of insecticide sprays, the application of pheromone dispensers, labour, machinery, fuel and other pheromone-based expenses such as pest monitoring were analyzed. Based on partial budgeting analysis, mating disruption treatments lowered insecticide and machinery costs but increased labour costs compared with conventional treatments. The cost of mating disruption ranged from $193.70 higher than the conventional treatment in cv. 'Gala' in 2013 to $ 96.00 less than the conventional treatment in cv. 'Fuji' in 2014. A break-even analysis showed that a price decrease of 22.22% and 70.37% for pheromone dispensers would be required to convince growers to use mating disruption in cvs. 'Gala' and 'Fuji' in 2013, respectively. However, the cost of mating disruption programme was similar or less than a conventional insecticide programme in 2014. The reduction of initial pest density, as well as the improvement of biological control, could lead to the development of more cost-effective and efficacious mating disruption programmes in subsequent years.

Key words: Cydia pomonella L., chemical control, cost analysis, economics, pheromones.

INTRODUCTION

Codling moth *(Cydia pomonella* L.) is a key deciduous fruit pest that poses a great economic threat to growers worldwide. Apart from apple, it also attacks pear, walnut, and quince (Witzgall, 2008). Damage is done by larvae, which feed on the fruit skin and bore deeply into the fruit. Larvae can cause up to 100% damage in untreated orchards (Elkins et al., 2005). Damaged apples are culled before packing, making it more challenging for growers to maintain profitability. Codling moth management mainly relies on chemical control. Two to three cover sprays are commonly used to target hatching eggs and larvae of the first generation codling moth (Kovanci, 2015). In Turkey, growers begin to spray insecticides such as diflubenzuron, methoxyfenozide, novaluron or thiacloprid with

an air-blast sprayer at 150 degree-days after the biofix in pheromone traps. Likewise, second generation codling moth larvae are also treated with at least two insecticide sprays at 800 degree-days.

However, the increasing resistance of codling moth populations to insecticides, coupled with adverse effects on beneficial insects, have led to control failures in the field (Reyes et al., 2007). In addition to high costs of chemicals, spraying equipment and gas, insecticide resistance has already increased control costs with traditional insecticide applications. Additional indirect costs of pesticide use on the human health and environment remain to be evaluated. Thus, there is a great need for alternatives to chemical control.

Mating disruption is an alternative control tactic that prevents male insects from finding females, resulting in lower pest density and less crop damage (Cardé and Minks, 1995). In this technique, growers apply large quantities of pheromone dispensers to cause no or delayed mating by disrupting chemical communication in the orchard. Mating disruption has proven to be a viable alternative method to control key pests including the codling moth, Oriental fruit moth (*Grapholita molesta* Busck.) and pink bollworm (*Pectinophora gossypiella* Saunders) (Cardé and Minks, 1995).

Unlike chemical control, mating disruption has no known toxicity or adverse effects on natural enemies so far. Besides, it has the potential to reduce or eliminate the need for insecticide treatments. However, pheromone dispensers are more costly than insecticides. Placement of pheromone emitters is labour intensive since they must be hand-deployed high in the canopy in most orchards (Elkins and Shorey, 1998).

In this context, the paper presents a detailed cost analysis of mating disruption versus conventional chemical control in 2013 and 2014. The objective of this study was to determine if the adoption of mating disruption would be financially feasible for control of codling moth in apple orchards in Turkey.

MATERIALS AND METHODS

Data on codling moth population, insecticide and pheromone use and cullage rates were collected on two apple orchards in 2013 and 2014. The orchards were located in Deydinler village of Inegol town (40.03° N, 29.53' E) near Bursa, Northwestern Turkey. One orchard contained semi-dwarf cv. 'Gala trees' on M.9 rootstock, while the other orchard contained semi-dwarf cv. 'Fuji trees' grafted on M.26 rootstock. Trees were trained with the tall spindle at a spacing of 1.7 x 3.3 m, resulting in approximately 1750 trees per ha. Fuji trees had central leader training with 2 x 4.5 m spacing (1100 trees/ha).

Each orchard was 10-ha in size and allocated into two 5-ha plots for mating disruption and chemical control. Conventional insecticide plots were separated by at least 500 m from mating disruption plots.

In the chemical control plots, Fuji trees received a total of 5 and 6 applications for control of codling moth in 2013 and 2014, respectively, whereas Gala trees had only a total of 3 and 4 sprays. A total of two cover sprays were made with diflubenzuron (Dimilin 48 SC, Hektas, Turkey) at 20 ml/ 100 l water (300 ml/ha) against eggs and larvae of first generation codling moth on May and June in both years. Depending on years, one to four thiacloprid (Calypso 240 OD, Bayer, Turkey) at 40 ml/ 100 l water (400 ml/ha) and methoxyfenozide (Prodigy 240 SC, Dow Agro Sciences, Turkey) treatments at 60 ml/100 l water (400 ml/ha) were applied to control second generation codling moth in July and August in 2013 and 2014.

In mating disruption plots, thiacloprid was applied once to control codling moth in mid-May before pheromone application. Pheromone dispensers (Isomate C Plus®, Sumitomo, Turkey) were hung on trees by hand at 1000 dispensers/ha in early-June each year. Each dispenser was baited with 190 mg of codling moth pheromone containing (E, E) - 8, 10-Dodecadien-1-ol. Four pheromone traps (Pherocon CM, Trece, USA) were used for monitoring codling moth adults in each plot.

Points of indifference between mating disruption and insecticide sprays at different prices for pheromone dispensers were determined by break-even analysis. For this purpose, cullage records were obtained from growers in both years. Data was evaluated using t-tests at 95% confidence level (Williamson et al., 1996).

In order to predict percent damage by codling moth larvae, fruit assessments were made by picking 100 fruit randomly from each of the 10 trees per treatment. Fruits with stings, entries and live larvae were counted and analyzed by ANOVA. The presence of codling moth larvae was confirmed by cutting fruits.

The technique of partial budgeting was used to examine the cost differences between two management alternatives for codling moth. The US dollar ($) amounts were provided using an exchange rate of 1TL = 0.5$ at the time of the study. The price per unit of pheromone and insecticide materials were multiplied by their amount of use to calculate the final cost per ha. In budget assumptions for machinery, predicted fuel and lubrication costs of a 60 h.p. tractor with an air-blast sprayer were at $ 2.00 per hour of operation. The same sprayer had estimated repair costs at $ 1.00 per hour. An average spray may need 1 machine hour per ha. Hence, fuel and lube cost for spraying in insecticide plots were calculated as 1 hour x $2.00/hour x number of sprays. Similarly, one insecticide spray in mating disruption plots costs $ 2.00 (1 hour x $ 2.00/hour x 1 spray).

The labour for machinery and placing pheromone dispensers in the orchard were also calculated. Machine labour cost for insecticide sprays was calculated using the following formula, where the value 1.10 represents the need for 10% more man hours to refill or clean the sprayer (Williamson et al., 1996): Machine labour cost = Machine hours x 1.10 x $ 2.00/hour x number of sprays.

The labour for hanging pheromone dispensers at 1000 dispensers/ha by hand was estimated to cost $ 4.00 per hour. The installation takes about 5 hours per ha to complete. Thus, pheromone applications would cause an estimated increase of $ 20.00 in labor costs to deploy pheromone dispensers.

Each pheromone dispenser cost $ 0.27, with a total of $ 270.00 per ha. Other pheromone related expenses included monitoring of codling moth and other pests with pheromone traps or visual inspection. A pheromone trap package for codling moth costs $ 30 per ha. Fixed or overhead costs were considered unchanged between operations using mating disruption and conventional insecticide.

RESULTS AND DISCUSSIONS

Mean codling moth catch in pheromone traps averaged across cultivars in 2013 and 2014 in Bursa, Turkey is shown in Figure 1.

Codling moth populations were higher in 2014 than in 2013. Pheromone traps in mating disruption plots caught fewer moths compared with insecticide plots in both years.

Insecticide applications for all mite and insect pests of cv. 'Gala' and 'Fuji' apples in mating disruption and insecticide plots in 2013 and 2014 are given in Table 1.

'Fuji' cultivar required 1.5 to 2 times more insecticide applications than cv. Gala as this variety was more susceptible to aphids, spider mites and codling moth (Yiem 1993; Joshi et al., 2015). When averaged over cultivars, mating disruption decreased the total number of

sprays for apple pest complex by 40.70% and 56.60% in 2013 and 2014, respectively.

Depending on the years and cultivars, a total of four to 10 sprays was eliminated in the mating disruption plots. Insecticide applications for codling moth were reduced from three to six in the conventional programme to one in the mating disruption programme.

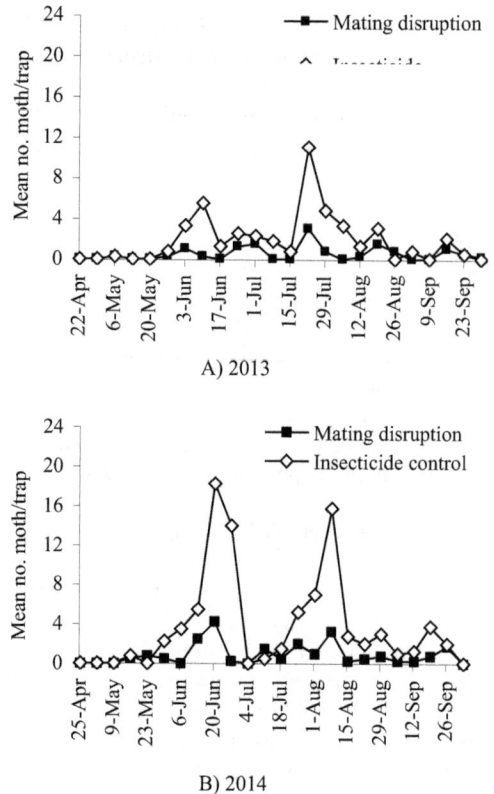

A) 2013

B) 2014

Figure 1. Mean codling moth catch in pheromone traps averaged across cultivars in mating disruption and insecticide plots in 2013 and 2014 in Bursa, Turkey

Depending on the years and cultivars, a total of four to 10 sprays was eliminated in the mating disruption plots. Insecticide applications for codling moth were reduced from three to six in the conventional programme to one in the mating disruption programme.

There were significant cost differences between conventional control and mating disruption in terms of materials, labour, and machinery (Table 2).

Table 1. Number of insecticide applications for all insect and mite pests of cvs. 'Gala' and 'Fuji' apples in mating disruption and insecticide plots in 2013 and 2014 in Bursa, Turkey

Variety	Insect pests[1]	2013			2014		
		Insecticide	Mating disruption	Difference	Insecticide	Mating disruption	Difference
'Gala'	Apple sawfly	1	1	0	2	2	0
	Aphids	2	1	1	2	1	1
	Budworm moths	0	0	0	0	0	0
	Codling moth	3	1	2	4	1	3
	San Jose scale	1	1	0	2	0	2
	Spider mites	2	1	1	2	1	1
	Total	9	5	4	12	5	7
'Fuji'	Apple sawfly	2	1	1	1	2	-1
	Aphids	4	3	1	5	3	2
	Budworm moths	1	1	0	1	0	1
	Codling moth	5	1	4	6	1	5
	San Jose scale	2	2	0	2	1	1
	Spider mites	4	3	1	3	1	2
	Total	18	11	7	18	8	10

[1] Apple sawfly = *Hoplocampa testudinea* (Klug); Aphids = *Dysaphis plantaginea* (Pass.), *D. devecta* (Walker), and *Eriosoma lanigerum* (Hausmann); Budworm moths = *Spilonota ocellana* (Den. & Schiff.) and *Hedya nubiferana* (Haworth); Codling moth = *Cydia pomonella* (L.); San Jose scale = *Quadraspidiotus perniciosus* (Comstock); Spider mites = *Panonychus ulmi* (Koch), *Tetranychus urticae* Koch, and *T. viennensis* Zacher

Table 2. Cost of materials, labour and machinery per ha compared between mating disruption and insecticide treatments consisting of cvs. 'Gala' and Fuji apples in 2013 and 2014 in Bursa, Turkey

Year	Cultivar	Treatment	Total no. applications[1]	Materials ($/ha)	Labour ($/ha)	Machinery ($/ha)	Total	Difference	Break-even analysis ($/dispenser)
2013	'Gala'	Insecticide	9	202.50	19.80	47.00	278.30		
	'Gala'	Mating disruption	6	420.00	31.00	15.00	472.00	-193.70	-0.19
	'Fuji'	Insecticide	18	382.50	39.60	150.00	590.10		
	'Fuji'	Mating disruption	12	532.50	44.20	61.00	649.70	-59.60	-0.06
2014	'Gala'	Insecticide	12	300.00	26.40	76.00	414.40		
	'Gala'	Mating disruption	6	368.00	31.00	19.00	424.00	-9.60	-0.01
	'Fuji'	Insecticide	18	397.50	39.60	170.00	625.10		
	'Fuji'	Mating disruption	9	442.50	37.60	40.00	529.10	96.00	*[2]

[1] Mating disruption treatments included one pheromone dispenser application plus insecticide applications.
[2] Since mating disruption was more cost-effective, there was no need to calculate a decrease in dispenser.

Mating disruption treatments lowered insecticide and machinery costs but increased labour costs compared with conventional treatments. Our findings are in agreement with those of Williamson et al. (1996).

Savings in insecticide expenditures varied from $ 120.00 in cv. 'Gala' to $ 255.00 in cv. 'Fuji'. Inconsistent with our results, Elkins et al. (2005) reported an average savings of $ 99 per ha in pesticide costs per year in pear orchards treated with pheromone puffers for control of codling moth. However, material costs in mating disruption plots were higher than insecticide plots due to the high cost of pheromone dispensers. The time-consuming installation procedure of pheromone dispensers caused a slight increase in labour costs in mating disruption blocks in most cases except for cv. 'Fuji' in 2014. Therefore, the cost of mating disruption ranged from $ 193.70 higher than the conventional treatment in 'Gala' in 2013 to $ 96.00 less than the conventional treatment in 'Fuji' in 2014. Similar to our cost calculations with cv. 'Gala' in 2013, Williamson et al. (1996) found in 'Red Delicious' apple orchards that codling moth mating disruption was $ 188.22 per ha more costly than conventional control on average. The cost difference could be perceived as the price of switching from conventional chemical control to mating disruption. Apparently, management costs change from crop to crop, cultivar to cultivar, and year to year. Brumfield et al. (2004) demonstrated a lower of cost mating disruption for Oriental fruit moth, *Grapholita molesta* (Busck), in peach orchards containing cv. 'Redhaven', 'John Boy' and 'Encore', but not in those with cv. 'Bounty'. However, it is important to note that the researchers did not calculate machinery costs and labour in their study. Codling moth damage and cullage data were analyzed to compare the cost differences between the two treatments. The percent total culls in 2013 were 27.40% and 30.80% in insecticide and mating disruption plots, respectively, while they increased to 35.10% and 36.00% in 2014 in the same order. In apples, codling moth is responsible for about 10% of cullage, with an average of 2-3% damage (Hansen and Schievelbein, 2002). In fact, levels of percentage damage by codling moth larvae in traditional insecticide plots were 0.90% and 1.50% in 2013 and 2014, respectively (Table 3). On the other hand, percent larval damage was higher than the economic threshold of 2% for codling moth (Kovanci et al., 2010) in mating disruption plots with 2.20% and 3.70% infestation recorded on apples in 2013 and 2014, respectively. Our results showed that mating disruption did not cause any significant increase in cullage.

Table 3. Mean percent fruit damage by codling moth larvae averaged over cultivars in mating disruption and insecticide plots in Bursa, Turkey in 2013 and 2014.

Year	Treatment	Damage[b](%)			
		Sting	Entry	Larvae	Total
2013	Mating disruption	0.5 a	0.9 a	0.8 a	2.2 a
	Insecticide	0.5 a	0.3 a	0.1 a	0.9 a
2014	Mating disruption	1.5 a	1.1 a	1.1 a	3.7 a
	Insecticide	0.8 a	0.4 a	0.3 a	1.5 a

A break-even analysis was made based on the difference in costs between mating disruption and insecticide applications. To determine a break-even situation, cost differences for each cultivar and year were divided by the number of pheromone dispensers applied (1000/ha). Our findings indicated that about 22.22% price decrease in pheromone dispensers was needed to convince growers to use mating disruption in

cv. 'Gala' apples in 2013. Even higher price reduction of up to 70.37% was necessary to achieve a break-even situation in cv. 'Fuji' apples in the same year. These results confirm the previous break-even analysis by Williamson et al. (1996), who suggested a 30-73% decrease in pheromone dispenser prices to 'Red Delicious' apple growers. To promote mating disruption, Turkish government offer subsidies of $ 125 per ha to encourage growers. However, there was no need for decrease in pheromone dispenser prices in 2014 because mating disruption programme was as cost-effective as insecticide programme.

This favourable change in costs of the two management programmes between years may have been caused by an increase in beneficial insects in apple orchards (Calkins, 1996). Evidently, the aphidophagous seven-spotted lady beetle, *Coccinella septempunctata* L., and *Typhlodromus athiasae* Porath and Swirski, the important predator of the European red mite, were more abundant in pheromone-treated orchards. In contrast, mating disruption may increase the risk of damage by some pests such as apple sawfly, *Hoplocampa testudinea* (Klug), which was previously suppressed by cover sprays. For example, two sprays were applied to control this pest in cv. 'Fuji' apples in mating disruption plots in 2014, while only one spray was applied in insecticide plots.

CONCLUSIONS

Based on partial budgeting analysis, mating disruption treatments lowered insecticide and machinery costs but increased labour costs compared with conventional treatments.

Codling moth mating disruption reduced the overall amount of insecticides used for managing apple pest complex. Despite these savings, the cost of mating disruption for codling moth averaged about $ 126.55 more than traditional insecticide applications in apple orchards in 2013.

A break-even analysis for the same year showed that a mean price decrease of 46.30% for pheromone dispensers would be required to convince growers to use mating disruption. However, on average, mating disruption programme cost $ 43.20 less than a conventional insecticide programme in 2014.

Cost differences between years could be explained by varying codling moth populations, and different number of insecticide sprays.

The decrease in initial pest density, accompanied by enhanced biological control, may help us to develop more cost-effective mating disruption programmes in subsequent years.

ACKNOWLEDGEMENTS

The author is grateful for the travel grant by the Commission of Scientific Research Projects of Uludag University, Turkey.

REFERENCES

Brumfield R.G., Martin L.S., Polk D., Hamilton G. (2004). Costs of conventional versus low input peach production in the Eastern United States. Acta Horticulturae, 638:473-478.

Calkins C.O., 1996. Areawide IPM as tool for the future. In: Lynch S., Greene C., Kramer-LeBlance C. (Eds.), Proceedings of the 3rd National IPM Symposium / Workshop: Broadening support for 21st Century IPM. USDA, CA, USA, Mis. Pub. no. 1542:154-158.

Cardé R.T., Minsk, A.K., 1995. Control of moth pests by mating disruption: successes and constraints. Annual Review of Entomology, 40:559-585

Elkins, R.B., Shorey, H.H., 1998. Mating disruption of codling moth (*Cydia pomonella*) using "puffers". Acta Horticulturae, 475:503-512.

Elkins R.B., Klonsky K.M., DeMoura R.L., 2005. Cost of production for transitioning from conventional codling moth control to aerosol-released mating disruption ("puffers") in pears. Acta Horticulturae (ISHS), 671:559-563.

Joshi N.K., Rajotte E.G., Myers C.T., Krawczyk G., Hull L.A., 2015. Development of a susceptibility index of apple cultivars for codling moth, *Cydia pomonella* (L.) (Lepidoptera: Tortricidae) oviposition. Frontiers Plant Science, 2015, 6: 992.

Hansen J.D., Schievelbein S., 2002. Apple sampling in packing house support the systems approach for export quarantine. Southwestern Entomologist, 27:277–282.

Kovanci O.B., Walgenbach J.F., Kennedy G.G., 2004. Evaluation of extended-season mating disruption of the Oriental fruit moth *Grapholita molesta* (Busck) (Lep., Tortricidae) in apples. Journal of Applied Entomology, 128:664-669. Kovanci O. B., Kumral N.A., Larsen T.E., 2010. High versus ultra-low volume spraying of a microencapsulated pheromone formulation for codling moth control in two apple cultivars. International Journal of Pest Management, 56:1-7.

Kovanci O.B., 2015. Co-application of microcapsulated pear ester and codlemone for mating disruption of *Cydia pomonella*. Journal of Pest Science, 88: 311-319.

Reyes M., Franck P., Charmillot P-J., Ioriatti C., Olivares J., Pasqualini E., Sauphanor B., 2007. Diversity of insecticide resistance mechanisms and spectrum in European populations of the codling moth, *Cydia pomonella*. Pest Management Science, 63:890-902.

Williamson E.R., Folwell R.J., Knight A.L., Howell J.F., 1996. Economics of employing pheromones for mating disruption of the codling moth, *Carpocapsa pomonella*. Crop Protection, 15:473-477.

Witzgall P., Stelinski L., Gut L., Thomson D., 2008 Codling moth management and chemical ecology. Annual Review of Entomology. 53:503-522.

Yiem M.S., 1993. Antixenosis and antibiosis of apple cultivars, Fuji, Starkrimson, Horei and Golden Delicious against *Tetranychus urticae* Koch. RDA Journal of Agricultural Science, Crop Protection 35, 409-413.

28

RESEARCHES REGARDING THE IMPLEMENTATION OF FOOD SAFETY MANAGEMENT SYSTEM ON THE FRUIT DRYING PRODUCTION PROCESS

Adrian CHIRA, Lenuța CHIRA, Elena DELIAN

University of Agronomic Sciences and Veterinary Medicine of Bucharest,
59 Mărăşti Blvd, District 1, 011464, Bucharest, Romania,
Email: achira63@yahoo.com
Corresponding author email: achira63@yahoo.com

Abstract

HACCP was originally developed as a microbiological safety system in the early days of the US manned space programme in order to guarantee the safety of astronauts' food. Up until that time most food safety systems were based on end product testing and could not fully assure safe products as 100% testing was impossible. A pro-active, process-focused system was needed and the HACCP concept was born. HACCP is a system that identifies evaluates and controls hazards which are significant for food safety. It is a structured, systematic approach for the control of food safety throughout the commodity system, from the farm to the plate. It requires a good understanding of the relationship between cause and effect in order to be more pro-active and it is a key element in Total Quality Management (TQM). This paper aims to address this subject, basing the approach as closely as possible on the Codex Code of General Principles on Food Hygiene on the fruit drying production process, which emphasises the importance of GMP/GAP/GHP as sound foundations to incorporate the HACCP approach and develop a user friendly Food Safety Management System. The main identified hazards are moulds and mycotoxin, which can keep under control by adequate monitoring of CCPs – fruit drying and the end storage product.

Key words: CCP, HACCP, food safety.

INTRODUCTION

On the producer-user line (from manipulation to processing) there are a high number of factors that can affect fruits quality (Bonsi R., 2001).

Considering these products as primary product for the fruit dried products or as finite product in the case of their fresh consume, the major preoccupations are in relation with pesticides level and others chemical contaminants (fertilisers), as well as to preserve the hygiene during harvesting, manipulation, processing and storage.

To reduce these risks, it is necessary that the small producers, as well as the high-specialised companies, to apply prevented methods as HACCP type and not those based on the end control of products (that can affect the consumer healthy) and can induce significantly economic losses (Aversano, F 2006).

MATERIALS AND METHODS

The fruit dried product has been obtained in the Technological laboratory of the Faculty of Horticulture Bucharest, using an electrical oven and fresh fruits (apples from cultivars: Jonathan and Golden Delicious).

A HACCP study was performed based on the following working stages:

1. the presentation of the specifications about product;
2. the production technological flow description;
3. the potential risk identification and evaluation;
4. the critical control points (CCP) determination;
5. establish the critical limits;
6. the monitoring of the CCP parameters;
7. corrective actions, implemented if the critical limits in CCP have been excelled;

The laborious study was finished by elaboration of the HACCP Plan, a base document, which represents a guide to follow, with a view to keep under control the relevant risks that could affect the safety of fruits dried products.

RESULTS AND DISCUSSIONS

Risk identified during the processing of fruit dried products is concerned especially to: pesticides residue provided from the fruits, as a consequence of the chemical treatments, nitrates provided by the excessive fertilization and micro-organisms (yeast, moulds) presented on the fruits or on the technological equipment, because of the inadequate hygiene (Table 1).

As a consequence of this study, there were identified two Critical Control Points:
- Primary matter reception, for the risks generated by the pesticides and nitrates;
- Fruits drying, for the risks generated by yeast and moulds;

Table 1. Hazard analysis

Processing step	Fruit dried products				Preventive /Control measures
	HAZARD				
	KIND OF HAZARD	G	P	RC	- Training of the workers
Fruit reception	B) Clostridium sp.	high	low	3	- Supplier assessment
	B) E. Coli	medium	low	2	- Analytical analysis
	B) Aspergillus flavus	medium	low	2	
	C) pesticides residue	high	low	3	
	C) heavy metal	medium	low	2	
	C) nitrit, mycotoxin	medium	low	2	
	B) Salmonella sp.	high	low	3	
	B) Clostridium	high	low	3	- Training of the workers
	B) E. Coli	medium	low	2	- Analytical analysis
Fruit drying	B) Aspergillus flavus	medium	low	2	- Process monitoring
	B) Bacillus sp	medium	low	2	
	B) Staphyloccocus	high	low	3	

Legenda:
B = biological   C = chemical   P = physical   G = gravity
P = probability   RC = risk class

Data presented in Table 2, emphasis that for these risks, there were established the critical limits and the specifically parameters (product content of pesticides, nitrite and mycotoxin, temperature or NTG) were controlled.

HACCP system, predicts also the critical limits surpass situation, therefore, there were predicted the corrective actions too, to determine the effect removing and the elimination of the causes which generated the manifested risk.

To assure the product traceability on all the production and selling process, it acts to register in specifically forms, which are useful as well to HACCP system revision.

To apply the HACCP Plan, as it was realized, determines to maintain under control the relevant risks, for the food safety of the fruit dried products and to grant an adequate product for the people consume.

Table 2. HACCP Plan

N r cr t	Pro-cess step	Relevant hazard	Control measures	Criti-cal con-trol point	Criti-cal limits	Monitoring			Correction/ Corective actions	Re-cords
						Responsa-bility	Frequency	Method		
1	Raw mate-rial recep-tion	Pesticide residues Myco-toxin Nitrate	Supplier assess-ment Labora-tory analysis	CCP 1	According Reg UE 1881/2006	Laborato-ry technician	2 weeks before purchasing	Cromato-graphic	Fruits rejection Supplier selection Personnel training	Test report
2	Fruit dry-ing	Yeast Bacteria Mould	Drying schedule	CCP 2	T= 60 –70 C degrees Product water content – max 12%	Drying operator	Continuous	Drying diagram	Product rejection Process resume Personnel training	Drying report

CONCLUSIONS

On the fruits dried products technology, there have been identified two Critical Control Points: at primary raw material reception and at the drying step.
The established monitoring system allows maintaining the relevant risks under control, for the hygienically quality of the analysed product.

REFERENCES

Aversano F., Pacileo V., 2006. Prodotti alimentari e legislazione. Edagricole Bologna.
Bonsi R., Galli C., 2001. Il Metodo HACCP. Calderini, Edagricole, Bologna.

29

THE EFFECT OF FROST AND HALE ON THE PEACH TREE CULTIVARS FROM R.S.F.G. CONSTANȚA

Cristina MOALE[1], **Leinar ȘEPTAR**[1], **Corina GAVĂȚ**[1], **Cristina PETRIȘOR**[2]

[1]Research Station for Fruit Growing (R.S.F.G) Constanta, 25 Pepinierei Street,
907300 Valu lui Traian, Romania
[2]Research and Development Institute for Plant Protection, 8 Ion Ionescu de la Brad Blvd,
District 1, Bucharest, Romania
Corresponding author email: moalecristina@yahoo.com

Abstract

One of the problems which occurred during the last years concerning all fruit-growing species is determined by climate changes. Some phenomena related to climate stress occur in a chronic manner (low fertility, weak structuring of soils, etc.) or periodically (droughts, excess of humidity in the soil, etc.) or occasionally (early or late frosts, hale, etc.); their unfavourable influence depends both on the intensity and the duration of the stress as well as on the specific phenophase of crop plants. Due to the climate changes which occurred during the last couple of years, it was observed that the resistance of peach tree cultivars differs greatly from one year to the next. The present studies were carried out over a period of three years on plantations of ripe peach trees and nectarine trees from R.S.F.G. Constanța. Branch samples belonging to 7 peach tree cultivars ('Springcrest', 'Springgold', 'Collins', 'Cardinal', 'Redhaven', 'Southland' and 'Jerseyland') and 3 nectarine tree cultivars ('Cora', 'Delta' and 'Romamer2') were harvested and analysed three days after the frost occurred. The paper presents the manner in which certain peach tree and nectarine tree cultivars reacted to the effect of the frost which occurred in 2012, 2013 and 2014 and the effect of hale (July 11th, 2014) on the peach tree production. The greatest losses caused by frost were recorded in the winter of 2012: 90% fruit buds affected at the 'Springgold' cultivar, 94% fruit buds affected at the 'Springcrest' cultivar and 62% fruit buds affected at the 'Redhaven' cultivar. The losses caused by the hale which occurred on the 11th of July 2014 reduced the production of the 'Redhaven' cultivar by 40% and that of the 'Southland' cultivar by 80%. The carried out studies and the obtained results demonstrate both the importance of choosing the assortment of cultivars according to favourable areas as well as the importance of placing anti-hale nets upon establishing fruit-growing plantations.

Key words: climate changes, late frosts, Prunus persica, 'Redhaven', 'Southland'.

INTRODUCTION

In this paper is presented the manner in which the frost and the hail influenced the fruit production of certain peach tree and nectarine tree cultivars cultivated in Dobrogea between 2014-2014.

The frosts which occur in March and April after a relatively warm period are more dangerous than those which occur during the obligatory resting period (December-January). The fruit buds in the pink button stage can resist to temperatures as low as -3.9°C for 2-3 hours; the opened flowers can tolerate a temperature of -2.8°C, while the newly tied fruits can resist to temperatures as low as -1.1°C (Chira et al., 2005). Nevertheless, the major climatic changes which have taken place during the last few years have had a significant negative influence over the triggering of the

flowering, the tying of the fruit and, evidently, over the peach tree and nectarine tree production. The climate changes problems should not be ignored in this case and might be a relevant subject for further researches.

Previous research papers have revealed that the impact of climatic changes upon fruit-growing species can already be felt. For instance, by the end of the 90's, the flowering of the trees in Germany occur several days earlier (Chmielewschi et al., 2004 and 2005). The vegetative season in Europe became longer by 10 days in the past 10 years (Chmielewschi and Rotzer, 2002).

Due to the early flowering of the trees, in certain regions of Europe there was an increase in the risk of damage caused by late frosts (Anconelli et al., 2004; Sunley et al., 2006; Legave and Clazel, 2006; Legave et al., 2008; Chitu et al., 2004 and 2008) or by the disorders in the pollination and fruit setting processes (Zavalloni et al., 2006).

According to the estimations of the weather forecasts, there have been presented in the frame of the 4th report of the International Committee for Climatic Changes in 2007, the whole Europe and implicit Romania will face in future with a process of global warming, characterized by increasing of temperatures with -0.5 - 1.5° C for the period 2020 – 2029 and with -2 – 5° C for the period 2029 – 2099. In the 2090-2099 periods, Romania will confront with pronounced drought during the summer time. Researches from many countries, in the frame of climatic research methodology have the approached aspects regarding climatic changes effects on growth and development of some fruit tree species (Chmielewski and Rotzer et al., 2002; Olensen 2002; Sunley et al.2006, Chitu et al., 2010; Sumedrea et al, 2009). Climatic changes occurred also in Romania, they have determined meteorological phenomena, which are manifesting with augmented amplitude and intense frequency (severe drought, intense flooding, tornados and hail).

MATERIALS AND METHODS

The research was carried out in the period 2012-2014 at R.S.F.G. Constanţa in Valu lui Traian. The studied material was represented by the experimental plots from R.S.F.G. Constanţa, where the peach tree and nectarine tree cultivars can be found. A number of 10 such cultivars were studied (with a different ripening period), out of which 7 were peach tree cultivars – 'Springold', 'Springcrest', 'Cardinal', 'Collins', 'Redhaven', 'Southland', 'Jerseyland' and 3 were nectarine tree cultivars – 'Cora', 'Delta' and 'Romamer 2'. The trees were planted in 1986, the utilised parent stock being *Prunus persica*; the planting density is of 625 trees/ha (planting scheme 4m x 4m) and the trees' shape of the head is that of a free palmette. As far as the soil concern on which the plantation is situated is a calcareous cernoziom with a claylike texture and only slightly alkaline pH (8.2) throughout its entire profile. In addition, the overall climatic conditions were favourable to the growth and fructification of the trees, with exception of the

years 2012 – 2014, when a very strong frost was registered in both January and February, leading to the loss of some of the fruit buds, while the hail on July 11th, 2014 affected the production of the 'Redhaven' and 'Southland' cultivars. With regard to these cultivars we observed the main fructification phenophases: the beginning of the blossoming, upon the appearance of the pink button; the beginning of the flowering, upon the appearance of the first open flowers; the ending of the flowering, when most of the flowers have lost their petals. The duration of the flowering phenophase at a certain cultivar can vary according to the action of the maximum temperatures during the day and the intensity of the wind, correlated with the degree of differentiation of the trees (i.e. the amount of flowers per tree). The intensity of the flowering was ranked on scale from 0 to 5, 0 being used when the cultivars displays no flowers at all, while 5 is used when the cultivar displays a plethora of flowers. The hardening of the core was determined by means of piercing it with a needle at regular intervals, usually 2 days. The process was carried out progressively, in the same day for all the observed cultivars. The harvesting maturity is largely influenced by a series of climatic and agro-technical factors, such as: temperature, drought, quantity of fruit per tree, shape of the head, density of the trees, etc. The observations and determinations were carried out 3-5 days after the climatic accidents recorded in 2012, 2013 and 2014, respectively and the production was assessed after the hail occurrence on July 11th, 2014. The hail, with a dimension of approximately 5-20 mm, seriously damaged the fruit production of some of the peach tree cultivars, more exactly those who had not been harvested until July 11th, 2014. The climatic data were recorded with the aid of an automatic meteorological station (the WatchDog type) and were processed as daily averages. We observed the manner in which certain peach tree and nectarine tree cultivars reacted to the change in the climatic conditions recorded during the winter of the previously mentioned years. We noticed that the resistance of peach tree cultivars differs from one year to the next because of the climatic changes that have occurred during the past few years and it depends on the gravity of climatic accidents.

The minimum and maximum temperatures during winter alternate and together with the gravity of climatic accidents lead to the weakening of the trees.

RESULTS AND DISCUSSIONS

The triggering of the main fructification phenophases in the years 2012-2014 occurred between rather wide limits, according to the characteristics of the cultivar and the climatic characteristics of the studied years.
In the period 2012-2014, the blossoming of the fruit buds of the peach trees occurred between the following limits: between 18.03 and 29.03 for the 'Springold' cultivar, between 21.03 and 27.03 at the 'Springcrest' cultivar, between 24.03 and 30.03 at the 'Collins' cultivar,

between 24.03 and 29.03 at the 'Cardinal' cultivar, between 28.03 and 03.04 at the 'Redhaven' cultivar, between 24.03 and 03.04 at the 'Jerseyland' cultivar and between 27.03 and 04.04 at the 'Southland' cultivar. The blossoming at the peach tree occurred between 18.03 and 04.04 (17 days) in the studied years 2012-2014. (Table 1).

The beginning of the flowering. For all the studied cultivars the beginning of the flowering in the period 2012-2014 was recorded; however, the cultivars entered this phenophases at different times, albeit not necessarily significant (a few days from one cultivar to the next), so that cross pollination was fully ensured. The limits for this phenophase were 26.03 and 21.04.

Table 1. The main stages of peach fructification in the 2012-2014 periods

No.	CULTIVAR	Year	The swelling of the flowering buds	The flowering			Intensity	The hardening of the stone	Harvesting maturity
				Beginning	Ending	Duration (days)			
1	SPRINGOLD	2012	18.03	26.03	16.04	20	2	04.06	26.06
		2013	25.03	06.04	21.04	15	2	10.06	27.06
		2014	29.03	03.04	12.04	9	4	07.06	01.07
		Limits	18.03-29.03	26.03-06.04	12.04-21.04	9-20	2-4	04.06-10.06	26.06-01.07
2	SPRINGCREST	2012	21.03	29.03	16.04	19	2	04.06	28.06
		2013	27.03	08.04	24.04	22	4	08.06	07.07
		2014	22.03	05.04	16.04	12	3	10.06	09.07
		Limits	21.03-27.03	29.03-08.04	16.04-24.04	12-22	2-4	04.06-10.06	28.06-09.07
3	COLLINS	2012	24.03	30.03	11.04	12	3	02.06	18.07
		2013	29.03	08.04	21.04	13	4	10.06	16.07
		2014	30.03	09.04	20.04	11	3	12.06	27.07
		Limits	24.03-30.03	30.03-09.04	12.04-30.04	11-13	3-4	02.06-12.06	16.07-27.07
4	CARDINAL	2012	26.03	04.04	17.04	13	2	06.06	13.07
		2013	29.03	09.04	23.04	14	3	10.06	18.07
		2014	24.03	20.04	28.04	8	2	08.06	25.07
		Limits	24.03-29.03	04.04-20.04	10.04-28.04	8-14	2-3	06.06-10.06	13.07-25.07
5	REDHAVEN	2012	02.04	05.04	20.04	12	4	08.06	29.07
		2013	28.03	11.04	19.04	8	5	10.06	02.08
		2014	03.04	20.04	30.04	10	4	07.06	12.07
		Limits	28.03-03.04	05.04-20.04	19.04-30.04	8-12	4-5	07.06-10.06	12.07-02.08
6	JERSEYLAND	2012	24.03	05.04	18.04	13	4	07.06	17.07
		2013	29.03	09.04	16.04	7	5	09.06	15.07
		2014	03.04	18.04	27.04	9	4	10.06	19.07
		Limits	24.03-03.04	05.04-18.04	16.04-27.04	7-13	4-5	07.06-10.06	15.07-19.07
7	SOUTHLAND	2012	27.03	06.04	13.04	7	5	09.06	04.08
		2013	29.03	08.04	17.04	9	5	11.06	30.07
		2014	04.04	21.04	27.04	6	5	07.06	06.08
		Limits	27.03-04.04	06.04-21.04	13.04-27.04	6-9	5	07.06-11.06	30.07-06.08

The ending of the flowering. In the studied period 2012-2014 the ending of the flowering occurred between 12.04 and 21.04 for the 'Springold' cultivar, between 16.04 and 24.04 for the 'Springcrest' cultivar, between 12.04 and 30.04 for the 'Collins' cultivar, between

10.04 and 28.04 for the 'Cardinal' cultivar, between 19.04 and 30.04 for the 'Redhaven' cultivar, between 16.04 and 27.04 for the 'Jerseyland' cultivar, between 13.04 and 27.04 for the 'Southland' cultivar. The dates were recorded as the days when the flowers lost their

last petals. The duration of the flowering at the peach tree (average for the three studied years) expressed in number of days varied between 6 days (the 'Southland' cultivar in 2014) and 22 days (the 'Springcrest' cultivar in 2013).
The intensity of the flowering. In 2012 the following cultivars displayed a weak intensity of the flowering: 'Springold' - 2, 'Springcrest' - 2, 'Cardinal' - 2 and 'Collins' - 3.
The hardening of the core. This phenophase occurred in the first half of the month of June

(between the 6th and the 11th) in the years 2012, 2013 and 2014.
The harvesting maturity. Each ripening period had large variation limits from one year to another, depending on how the climatic factors determine the type of vegetation in a specific year: early, late or extra late. The harvesting maturity of the fruit had as variation limits the 26th of June and the 6th of August.
At the nectarine trees, the blossoming occurred between 16.03 and 04.04 (Table 2).

Table 2. The main stages of nectarine fructification in the 2012-2014 periods

No.	CULTIVAR	Year	The swelling of the flowering buds	The flowering				Intensity	The hardening of the stone	Harvesting maturity
				Beginn-ing	Ending	Duration (days)				
1	CORA	2012	16.03	29.03	14.04	15	5	05.06	19.06
		2013	27.03	06.04	23.04	17	5	11.06	27.06
		2014	25.03	10.04	28.04	18	5	07.06	28.06
		Limits	**16.03-27.03**	**29.03-10.04**	**14.04-28.04**	**15-18**	**5**	**05.06-11.06**	**19.06-28.06**
2	DELTA	2012	20.03	29.03	16.04	19	5	04.06	23.06
		2013	29.03	08.04	30.04	22	5	08.06	20.06
		2014	21.03	05.04	16.04	12	5	10.06	06.07
		Limits	**20.03-29.03**	**29.03-08.04**	**16.04-30.04**	**12-22**	**5**	**04.06-10.06**	**20.06-06.07**
3	ROMAMER 2	2012	24.03	28.03	11.04	13	5	04.06	08.07
		2013	04.04	06.04	30.04	24	5	10.06	11.07
		2014	30.03	02.04	24.04	22	5	14.06	13.07
		Limits	**24.03-04.04**	**28.03-06.04**	**11.04-30.04**	**13-24**	**5**	**04.06-14.06**	**08.07-13.07**

The limits for the beginning of the flowering in the studied years 2012-2014 were 29.03 and 08.04. The duration of the flowering (average for the three analysed years) expressed in number of days varied between 12 days (the 'Delta' cultivar in 2014) and 24 days (the 'Romamer 2' cultivar in 2013). All the studied cultivars displayed a large abundance of flowers and obtained the grade 5 in all the three studied years. The hardening of the kernel occurred in the first half of the month of June (between the 4th and the 14th). The harvesting maturity of the fruit had as variation limits the 19th of June and the 13th of July, period in which there are no other nectarine types on the market. This constitutes a great advantage for retailers through the income that can be realised. The cultivars become ripe at a difference of 3-5 days one from the other.
As we can notice in Figure 1a, January of 2012 was the coldest month, during which 9 days recorded daily average temperatures ranging from -10.2 °C and -17.6 °C. These values,

together with those that were extremely varied in February (7 days with daily average temperatures -10.4 -16.4 °C la °C) and 8 consecutive days of hoarfrost, the ice on the branches caused the loss of 19% - 94% of the fruit buds at the studied cultivars.
Figure 1b. reveals the fact that the coldest month in the period September 2012 - April 2013 was January 2013, when the recorded values were as low as -13.7°C (January 10th, 2013). These values did not significantly influence the loss of fruit buds at the peach tree cultivars (local observations).
In the period October 2013 - March 2014 (Figure 1c.) the lowest temperature was recorded in January: -17.6 °C (January 30th, 2014); another day when the recorded temperature was low (-9.4° C) was February 5th, 2014. The low temperatures recorded during this period affected the 'Cardinal' cultivar (57%) and the 'Jerseyland' cultivar (70%).

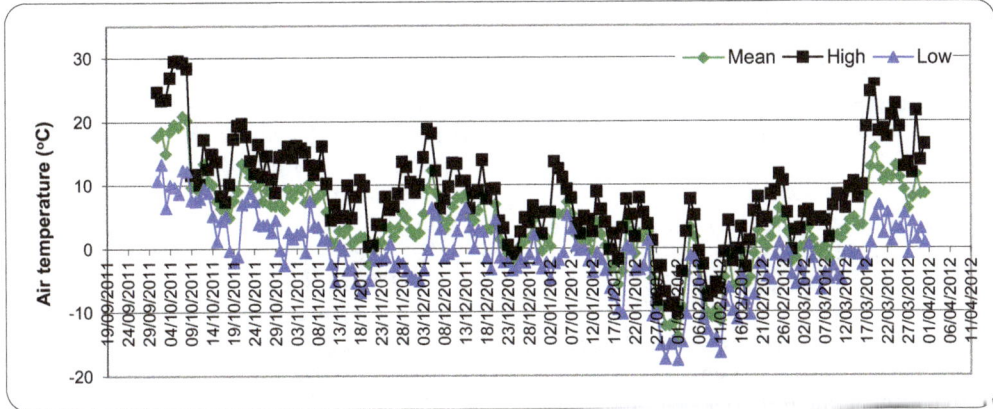

1 a

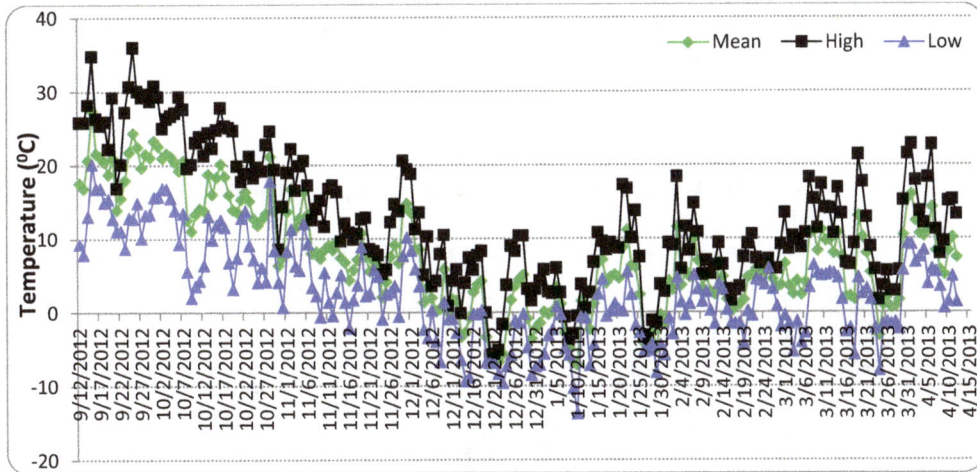

1b

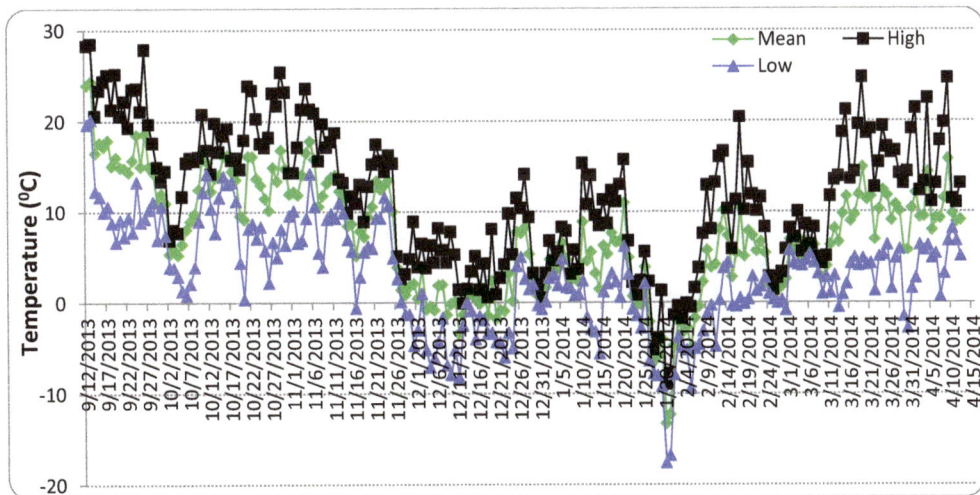

1c

Figure 1. a,b,c,. Air temperature (°C) in the cold period October 2011 – March 2012 (a),
October 2012 – March 2013 (b), October 2013 – March 2014 (c) at Valu lui Traian, Constanța

The observations were carried out with the aim of assessing the losses of fruit buds because of temperature variations during winter and the low temperatures during the day.

Thus, for the 'Springold' cultivar the losses recorded for 2012 were of approximately 90%, 14% for 2013 and 49% for 2014, there being difference from one cultivar to another. The winter frost caused losses for the 'Springcrest' cultivar of 94% in 2012, 21% in 2013 and 48% in 2014.

For the 'Cardinal' cultivar, the losses were of 66% in 2012, 19% in 2013 and 57% in 2014. We must bear in mind the fact that the losses caused by the winter frost of 2012, together with those caused by hoarfrosts and late frosts were very severe, taking also into account the surface of the Station's orchards cultivated with this cultivar.

These losses were also caused by the warm period before the frost – in the first three weeks of January 2012 the average temperature of the air was positive, of approximately 5 °C.

For the 'Collins' cultivar the losses were of 54% in 2012, 29% in 2013 and 53% in 2014. The 'Redhaven' cultivar recorded losses of 62% in 2012, 15% in 2013 and 56% in 2014. The 'Jerseyland' cultivar recorded losses of 63% in 2012, 27% in 2013 and 70% in 2014. For the 'Southland' cultivar the recorded losses were of 48% in 2012, 21% in 2013 and 49% in 2014.

The losses caused by frost recorded by the nectarine tree cultivars were rather small: for the 'Cora' cultivar they were of 23% in 2012, 19% in 2013 and 28% in 2014, for the 'Delta' cultivar, 21% in 2012, 17% in 2013 and 14% in 2014, while for the 'Romamer 2' cultivar, the losses were of 19% in 2012, 9% in 2013 and 29% in 2014 (Figure 2).

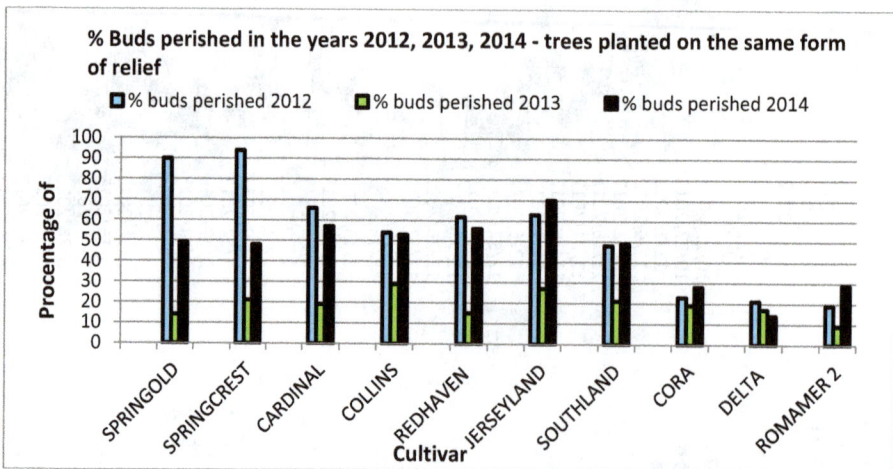

Figure 2. Procentage of peach and nectarine tree flowering buds perished due to frosts during the winter of 2012,2013 and 2014 at Valu lui Traian, Constanța

A good resistance to frost during the winter of the three studied years was remarked at the nectarine cultivars, with the following percentages: 'Cora' - 23%, 'Delta' - 17% and 'Romamer 2' - 19% (Figure 3).

In these conditions, the 'Springold' and 'Springcrest' cultivars were more than 50% damaged, while other cultivars such as 'Redhaven' and 'Southland' were less affected, the percentages being 44% and 39%, respectively. The climatic accidents recorded in January and February 2012 (sudden temperatures of -16.4°C, minimum temperature

during the day) and 8 days of hoarfrost caused the damaging of the production for the early cultivars 'Springold', 'Springcrest' and 'Cardinal', while the 'Redhaven', 'Collins' and 'Southland' were only partially affected.

At R.S.F.G. Constanța, in the second week of June 2014, more exactly on July 11th, the amount of precipitations was accompanied for 10 minutes by hail, which affected 80% of the fruit production for the 'Southland' cultivar (the fruit were just beginning to ripe) and 40% for the 'Redhaven' cultivar (Figures 4 and 5).

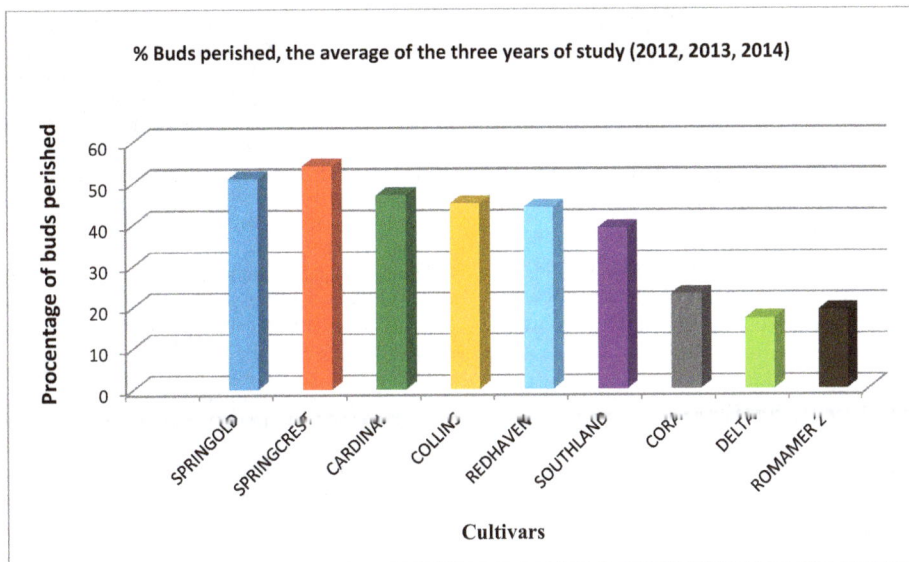

Figure 3. Procentage of peach and nectarine tree flowering buds affected by frosts
(average over the three years), Valu lui Traian

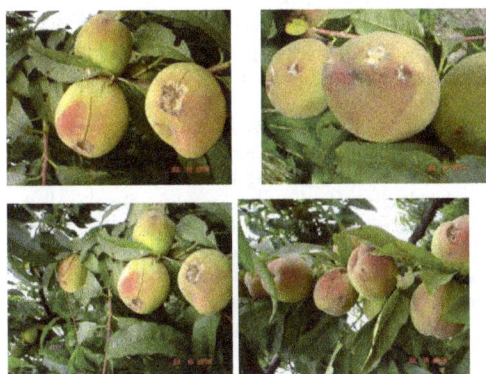

Figure 4. Fruit of the 'Southland' cultivar affected
by the hail on July 11th, 2014

Figure 5. The 'Redhaven' cultivar affected by the hail
on July 11th, 2014 (full maturity)

The hail bruised the fruit, shoots and stems, thus creating an environment for future infections and diseases. The bruises on the fruits, despite some of them becoming scars, diminished the commercial aspect and the quality of the production.

Although the south-eastern part of Romania is generally considered favourable to the culture of the peach tree, the specie has suffered greatly over the past decade because of climatic variations which manifested themselves mainly through the aggressiveness of low temperature in alternation with maximum positive temperatures. The peach tree encountered considerable losses because of temperature variations which occurred during the dormand period, in the climatic conditions of 2012, 2013 and 2014; the losses were also caused by late hoar frosts in spring, especially in the second half of March and in April, as well as by hail occurrences.

CONCLUSIONS

The novelty brought forward by the results is represented by the fact that the winter frosts from 2012, 2013 and 2014 affected the peach tree and the nectarine tree to various extents, according to the cultivar (approximately 9-94%).

The greatest production losses were recorded in 2012 – 94% for the 'Springold' cultivar and 90% for the 'Springcrest' cultivar.

The smallest losses during the three studied years were recorded by the nectarine tree cultivars 'Cora', 'Delta' and 'Romamer 2'.

The hail from July 11th, 2014, which lasted for only 10 minutes, affected the 'Redhaven' cultivar (40%) and the 'Southland' cultivar (80%).

In order to protect the trees from hail occurrences we recommend that the orchards be equipped with anti-hail nets.

Moreover, when choosing the assortment of cultivars to be cultivated in a specific area one must make sure that particular area is favourable to the setting up of fruit-growing plantations.

ACKNOWLEDGEMENTS

This research work was carried out with the support of Ministry of Agriculture and Rural Development, Project ADER 3.3.2/2015.

REFERENCES

Anconelli S. Antolini G. Facini O. Giorgiadis T.Merletto V. Nardino M. Palara U. Pasquali A.Pratizzoli W. Reggitori G.Rossi F. Sellini A. Linoni F, 2004. Previsione e difesa dalle gelate tardive – Risultati finali del progetto DISGELO. CRPV Diegaro di Cesena (FO). Natiziario tecnico N.70. ISSN 1125-7342. 64.

Cociu V., Oprea Şt., 1989. Metode de cercetare în ameliorarea plantelor pomicole. Ed. Dacia, pg. 29, 120 – 126.

Chira l, Chereji V, Roman M., 2005. Caisul şi piersicul. Editura MAST, 2005, ISBN: 973-8011-64-7: 210-211.

Chitu E., Butac M., Ancu S., Chitu V., 2004. Effects of low temperatures in 2004 on the buds viability of some fruit species grown in Maracineni area. Annals of the University of Craiova. Vol. IX (XLV), ISSN 1435-1275: 115-122.

Chiţu E., Sumedrea D., Pătineanu C., 2008. Phenological and climatic simulation of late frost damage in plum orchard under the conditions of climate changes foreseen for România. Acta Horticulturae (ISHS) 803:139-146.

Chiţu, E., Mateescu E., Petcu A., Surdu I., Sumedrea D., Tănăsescu N., Păltineanu C., Chiţu V., Mladin P., Coman M., Butac M., Gubandru V., 2010. Metode de estimare a favorabilităţii climatice pentru cultura pomilor în România. Editura INVEL Multimedia, ISBN 978-973-1886-52-7.

Chmielewski F.M., Rotzer T., 2002. Annual and spatial variability of the begenning of growing season in Europe in relation to air temperature changes. Clim. Res. 19(1), 257-264.

Chmielewski F.M., Muller A., Bruns E., 2004. Climate changes and trends in phenology of fruit trees and field crop in Germany, 1961-2000, Agricultural and Forest Meteorology 121 (1-2), 69-78.

Chmielewski F.M., Muller A., Kuchler W., 2005. Possible impacts of climate change on natural vegetation in Saxony (Germany). Int. J. Biometeorol, 50:96-104.

Legave, J.M., Clazel G., 2006. Long-term evolution of flowering time in apricot cultivars grown in southern France: wich future imtacts of global warming? Acta Horticulturae, 714: 47-50.

Legave J.M., Farrera I., Almeras T., Calleja, M, 2008. Selecting models of apple flowering time and understading how global warming has had an impact on this trait. Journal of Horticultural Science & Biotechnology, 83:76-84.

Olensen J.O., Bindi M., 2002. Consequences of climate change for European agricultural productivity, land use and policy. European Journal of Agronomy, 16, 239–262.

Sumedrea D., Tănăsescu N., Chiţu E., Moiceanu D., Marin Fl., 2009. Present and perspectives in Romanian fruit growing technologies under actual global climatic changes. Scientific Papers of the Research Institute for Fruit Growing Pitesti, Vol. XXV, ISSN 1584-2231, Editura INVEL Multimedia, Bucureşti: 51-86.

Sunley R.J., Atkinson C.J., Jones, H.G., 2006. Chill unit models and recent changes in the occurrence of winter chill and soring frost in the United Kingdom. Jurnal of Horticulturae. Science & Biotechnology, 81: 949-958.

Zavalloni C., Andersen J.A., Flore J.A., Black J.R., Beedy T.L., 2006. The pileus project: climate impacts on sour cherry production in the great lakes region in past and projected future time frames. Acta Horticulturae, 707: 101-108.

30

THE INFLUENCE OF BIO-FERTILIZATION UPON PRODUCTION LEVEL OF SOME HARVESTED APPLE CULTURES (*Vf*) IN AN INTENSIVE SYSTEM IN THE SOUTH-EAST OF ROMANIA

George DUDU[1], Sorin Mihai CÎMPEANU[1], Ovidiu Ionuț JERCA[1]

[1]University of Agronomic Sciences and Veterinary Medicine of Bucharest, 59 Mărăşti Blvd, District 1, 011464, Bucharest, Romania

Corresponding author email: george.dudu@citygarden.ro

Abstract

Knowing the relationship between orchards and bio-fertilization practices is needed to develop new production systems which bring known benefits and those of economic nature. Bio-fertilization systems applied to varieties of apple (Malus domestica Borkh.) Vf type are recommended in order to maximize profitability in the apple culture. The purpose of this study (2012-2013) was to calculate bio-fertilization influence on yield (t/ha) of apple varieties with genetic resistance to scab (Vf): 'Topaz', 'Redix', 'Rubinola', 'Goldrush' and economic indicators in the climate conditions of the Ilfov county. Experimental module was arranged by trifactorial type using the subdivided parcels method in three repetitions. At the base of the economical study the technological files of apple culture/ha was used, determining the following indicators: production costs (lei/ha), net profit (lei/ha) and net profit ratio(%). Significant increases ineconomic indicators between 10 and 20% resulted in the experimental variant Naturamin 7,5kg/ha.

Key words: biofertilization, apple varieties, economical indicators, yield.

INTRODUCTION

In Romania, the apple (*Malus domestica* Borkh.) has among other tree cultures, a leading position in terms of production and harvested areas. Because of the economical and nutritional importance the apple culture will maintain an important continuous place in the tree sector of our country. The results of research in tree domain was performed with a duration of a couple of years, this has shown the decisive influence of pedoclimatical conditions and culture technologies upon the apple production and implicitly economical efficiency. Regarding obtaining a economically profitable culture, rigours zoning must be impided on new apple varieties of *Vf* type according to the type of soil, climatic resources and their genetic needs.

In the areas with precipitation the evident efficiency of released production is shown and once with the increase in the mechanized grade and fertilization, economical spendings rise as well.

Numerous researches have shown a positive relation between genetic quality of the varieties used and technological works applied, all of these increasing the chance of growth in productivity and implicitly the economical efficiency (Țiu J.V. and colab., 2014).

Establishing the plane for technical economical activity supposed the early establishing of a specific document among which the technological file holds an important role (Merce and colab., 2000).

MATERIALS AND METHODS

The experiences had the purpose of determining the productivity of varieties taken in study with conditions of biofertilization and establishing economical efficiency for the apple culture of *Vf* type. To establish optimal technologically growth of profit, variant realized spendings were calculated for entire culture and released profit, on different hydric regime and numerous varieties *Vf* type. Observations and determinations were performed in 2012-2013 at Didactical Farm Station Belciugatele, didactical farm Moara Domneasca, with the following experimental factors: A - irrigation: a1 - unirrigated, a2 -

irrigated 2 l/h, a3 - irrigated 4 l/h; B - biofertilization: b1 – unfertilized, b2 – fertilized Naturamin 3,75 kg/ha, b3 – fertilized Naturamin 7,5 kg/ha; C - variety: c1 – Topaz, c2 – Rubinola, c3 – Goldrush.

The period of irrigation application was established by following a hydric graphic in the period of maximal requirement for plants and according to the active humidity index (AHI). The fertilizers were applied 4 times: immediately after blossom and every 3 weeks after. The biofertilizant Naturamin is a latest generation fertilizer with 80% free aminoacids with the role for biostimulation of growth and plant development in all phases, compatible with most fertilizers and pesticides and contributing to the growth and quality of production. The experimental module was of trifactorial type, arranged after the subdivide parcels method in three repetitions. Specific data of economical efficiency were calculated: cost production, income and profit rate.

RESULTS AND DISCUSSIONS

Analizing the index of economical efficiency in 2012 for the apple culture, every factor was took into the study, income was higher than the expenses related in the cultures maintenance.

As such the data from table 1 shows, that the variants less profitable from an economical point of view were irrigated variant 4l/h + unfertilized (a3b1) for Topaz Vf and Rubinola Vf and unirrigated + unfertilized variant (a1b1) for Goldrush Vf (Table 1).

With the highest procent of profit rate was obtained in the fertilizer with Naturamin 7,5 kg/ha + irrigated 2 l/h (a2b3) for all variants took into study, recording percentage of 59,3% of Topaz Vf, 56,81% of Goldrush Vf and 57,23% of Rubinola Vf varieties. Of irrigated 4 l/h+ biofertilized with Naturamin 7,5 kg/ha variant (a3b3), satisfactory percentages were released between 44,94% (Rubinola Vf), 49,57% (Topaz Vf) and 56,64% (Goldrush Vf) (Table 1).

Table 1. Economic efficiency of apples productions in 2012

Hydric regime	Fertilizer level	Production (kg/ha)	Production value (lei/ha)	Costs (lei/ha)	Profit (lei/ha)	Profit rate (%)
Topaz *Vf*						
a1(unirrigated)	b1	10.986,00	13.183,20	5.874,00	7.309,20	55,44
	b2	12.144,00	14.572,80	6.286,00	8.286,80	56,86
	b3	13.564,00	16.276,80	6.866,00	9.410,80	57,82
a2 (irrigated 2l/h)	b1	11.867,00	14.240,40	5.862,00	8.378,40	58,84
	b2	13.144,00	15.772,80	7.341,00	8.431,80	53,46
	b3	16.235,00	19.482,00	7.930,00	11.552,00	59,30
a3(irrigated 4l/h)	b1	12.130,00	14.556,00	8.188,00	6.368,00	43,75
	b2	13.950,00	16.740,00	8.477,00	8.263,00	49,36
	b3	14.864,00	17.836,80	8.995,00	8.841,80	49,57
Goldrush *Vf*						
a1(unirrigated)	b1	10.762,00	12.914,40	5.860,00	7.054,40	54,62
	b2	13.350,00	16.020,00	6.840,00	9.180,00	57,30
	b3	13.782,00	16.538,40	7.290,00	9.248,40	55,92
a2 (irrigated 2l/h)	b1	11.998,00	14.397,60	6.220,00	8.177,60	56,80
	b2	14.016,00	16.819,20	7.532,00	9.287,20	55,22
	b3	15.840,00	19.008,00	8.210,00	10.798,00	56,81
a3(irrigated 4l/h)	b1	15.120,00	18.144,00	7.998,00	10.146,00	55,92
	b2	14.872,00	17.846,40	7.990,00	9.856,40	55,23
	b3	15.980,00	19.176,00	8.315,00	10.861,00	56,64
Rubinola *Vf*						
a1(unirrigated)	b1	10.097,00	12.116,40	5.887,00	6.229,40	51,41
	b2	11.363,00	13.635,60	6.670,00	6.965,60	51,08
	b3	12.674,00	15.208,80	6.950,00	8.258,80	54,30
a2 (irrigated 2l/h)	b1	11.876,00	14.251,20	7.120,00	7.131,20	50,04
	b2	12.849,00	15.418,80	7.440,00	7.978,80	51,75
	b3	15.433,00	18.519,60	7.920,00	10.599,60	57,23
a3(irrigated 4l/h)	b1	11.870,00	14.244,00	7.995,00	6.249,00	43,87
	b2	12.863,00	15.435,60	8.380,00	7.055,60	45,71
	b3	13.562,00	16.274,40	8.960,00	7.314,40	44,94

In 2013 at the apple culture in all the variants studied, income was higher than expenses related in culture maintenance.

The table 2 shows that the least profitable variants were the once from irrigation 4 litri/h + unfertilized variant (a3b1) at Topaz *Vf* (43,46%) and Rubinola *Vf* (43,87%) and Goldrush *Vf* from irrigated 4 l/h + fertilized with Naturamin 3,75 kg/ha variant (55,37%) (a3b2).

The highest procentage of profitability rate (%) were obtained in the irrigated 2 l/h + fertilized with Naturamin 7,5 kg/ha variant (a2b3) for all varieties studied and performed values of 59,23% of Topaz *Vf*, 55,52% of Rubinola *Vf*, respectively 57,76 % of Goldrush *Vf*. At the same experimental variants was achieved satisfactory income (lei/ha) between 9760 and 12088 lei/ha (Table 2).

Table 2. Economic efficiency of apples productions in 2012

Hydric regime	Fertilizer level	Production (kg/ha)	Production value (lei/ha)	Costs (lei/ha)	Profit (lei/ha)	Profit rate (%)
Topaz *Vf*						
a1 (unirrigated)	b1	10.468,00	12.561,60	5.645,00	6.916,60	55,06
	b2	11.744,00	14.092,80	6.229,00	7.863,80	55,80
	b3	12.564,00	15.076,80	6.997,00	8.079,80	53,59
a2 (irrigated 2l/h)	b1	11.167,00	13.400,40	5.839,00	7.561,40	56,43
	b2	13.144,00	15.772,80	7.210,00	8.562,80	54,29
	b3	16.120,00	19.344,00	7.886,00	11.458,00	59,23
a3 (irrigated 4l/h)	b1	12.130,00	14.556,00	8.230,00	6.326,00	43,46
	b2	13.950,00	16.740,00	8.554,00	8.186,00	48,90
	b3	14.864,00	17.836,80	8.875,00	8.961,80	50,24
Goldrush *Vf*						
a1 (unirrigated)	b1	11.890,00	14.268,00	6.210,00	8.058,00	56,48
	b2	13.350,00	16.020,00	6.834,00	9.186,00	57,34
	b3	13.934,00	16.720,80	7.218,00	9.502,80	56,83
a2 (irrigated 2l/h)	b1	11.998,00	14.397,60	6.322,00	8.075,60	56,09
	b2	14.016,00	16.819,20	7.464,00	9.355,20	55,62
	b3	17.440,00	20.928,00	8.840,00	12.088,00	57,76
a3 (irrigated 4l/h)	b1	15.120,00	18.144,00	7.886,00	10.258,00	56,54
	b2	14.872,00	17.846,40	7.964,00	9.882,40	55,37
	b3	15.980,00	19.176,00	8.170,00	11.006,00	57,39
Rubinola *Vf*						
a1 (unirrigated)	b1	11.737,00	14.084,40	6.272,00	7.812,40	55,47
	b2	11.363,00	13.635,60	6.430,00	7.205,60	52,84
	b3	11.774,00	14.128,80	6.860,00	7.268,80	51,45
a2 (irrigated 2l/h)	b1	11.876,00	14.251,20	6.990,00	7.261,20	50,95
	b2	12.849,00	15.418,80	7.410,00	8.008,80	51,94
	b3	14.650,00	17.580,00	7.820,00	9.760,00	55,52
a3(irrigated 4l/h)	b1	11.870,00	14.244,00	7.995,00	6.249,00	43,87
	b2	12.863,00	15.435,60	8.268,00	7.167,60	46,44
	b3	13.562,00	16.274,40	8.797,00	7.477,40	45,95

Analizing the profit rate for the applied experimental factors it has concluted that in both studied years, the Topaz *Vf* and for the Rubiola *Vf* varieties, the lowest profit was irrigated 4 l/h + unfertilized (a3b1) and máximum profit was recorded in all three varieties for irrigated 2l/h + fertilized 7,5 kg/ha variant (a2b3) as well as the experimental variant.

The minimal profit for Goldrush variety was unirrigated + unfertilized (a1b1) in 2012a nd irrigated 4 l/h + fertilized 3,5 kg/ha in 2013 year. It can be concluded that values between the minimal and maximal profit between the years, the smallest differences were situated between 54,62% (2012 – a1b1) and 57,76% (2013 - a2b3) (Figure1).

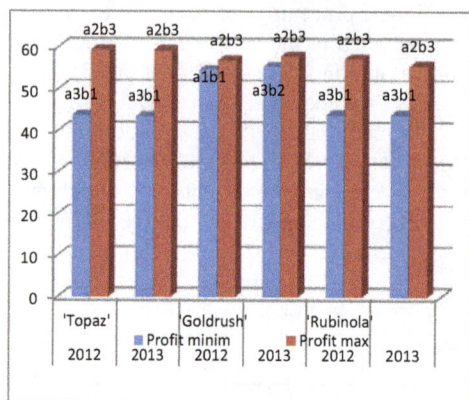

Figure 1. Experimental variants with minimal and maximal profit

CONCLUSIONS

❖ In the unfavourable climatic conditions specific for the 2012 year, technology for the apple culture regarding studied variants depended on the application of punctual irrigation at the trees root and application of four treatments with Naturamin biofertlizant products in doses of 3.75 kg/ha, respectively 7.5 kg/ha, respecting the phenophases of wetting application shown in the wetting graphic.

❖ In normal climatic conditions similar to 2013, technology for the apple culture regarding studied variants depended on the wetting application in critical phenophases and application of biofertilizing products.

❖ The high economical results were recorded due to a high level of production as a following of the influence of favourable climatical condition of 2013 in the technological links used.

❖ Maximum profit was recorded in all three studied varieties in the irrigated 2 l/h + fertilized with Naturamin 7.5 kg/ha variants (a2b3) and minimum profit for the irrigation 4l/h + unfertilized variants (a3b1).

ACKNOWLEDGEMENTS

This paper was published under the frame of European Social Fund, Human Resources Development Operational Programme 2007-2013, project no. POSDRU/159/1.5/S/132765.

REFERENCES

Axinte Stela, Axinte M., Agafiţei Alina, Axinte Lorica, Neştian I., 2005. Lucrări ştiinţifice vol. 48, seria Agronomie, USAMV Iaşi
Ceapoiu N., 2001. Pomicultura aplicată, Editura Stiinţelor agricole, Bucureşti
Elfing D.C., 1982. Crop response to trickle irrigation, Hort. rev., vol. 4, p.1-48.
Ghena N., 1979. Pomicultura generală şi specială, Editura Ceres, Bucureşti.
Goldberg D., 1975.Metode şi tehnici de irigare în Israel (traducere Soil Science Society of America).
Grumeza N, Ionescu Pr., 1970. Irigarea plantaţiilor pomicole, Editura Ceres Bucureşti
Grumeza N. şi colab., 1989. Prognoza şi programarea aplicării udărilor în sistemele de irigare, Editura Ceres, Bucureşti.
Iancu M., 1975. Cercetări privind consumul de apă din sol în livezile de măr. Lucrări stiinţifice I.C.D.P.P. Mărăcineni, vol. IV, p. 159-173.
Iancu M, 1993. Irigarea prin picurare, Hortiinform nr.17
Iancu M , 1983.Aspecte moderne de irigare in pomicultura, Sinteza in Sesiunea I.C.C.P Maracineni.
Tiu Jeni Veronica, Cîmpeanu S.M., Teodorescu R., Tudor V., Asănică A., 2014. Techno-economic efficiency of some apricotsand apples cultivars in the Moara Domneasca farm conditions, Analele Universităţii din Craiova, Vol. XIX (LV)–2014, ISSN 1453-1275, p. 379-388.

31

RESEARCH CONCERNING THE INFLUENCE OF CLIMATE ON EVOLUTION OF PHENOLOGICAL STAGES IN SWEET CHERRY TREE

Sorina SÎRBU, Gelu CORNEANU, Elena IUREA, Margareta CORNEANU

Research Station for Fruit Growing, 175 Voineşti, 707305, Iaşi, Romania

Corresponding author e-mail: sorinas66@yahoo.com

Abstract

Carrying out the phenological stages of vegetation and fructification in sweet cherry cultivars is determined by the cumulative action of rainfall registered and daily average temperatures exceeding 5°C, value considered as biological limit of this species. Research was conducted during 2011-2014 at six Romanian cultivars 'Cetăţuia', 'Cătălina', 'Maria', 'Andreias', 'Margonia' and 'George' and three introduced sweet cherry cultivars 'Van' (Canada), 'Bigarreau Burlat' (France) and 'Bigarreau Dönissen' (Germany). Were recorded number of days, active thermal balance (°C) and rainfall quantity (mm) on periods between phenological stages: from swelling buds to blooming, flowering period, from end of flowering to ripening of fruits and from ripening of fruits to leaves fall. The number of days recorded in periods between phenological stages from swelling buds to start of blooming was between 20-26 days, the active thermal balance ranged between 192 to 269 °C and rainfall quantity ranged between 41 mm to 63 mm. In the blossom period number of days ranged between 8-12 days, with the active thermal balance ranged between 97°C to 151°C and rainfall quantity ranged between 6.5mm to 30 mm. The number of days from swelling buds to start of blooming, from the end of blossom to fruit maturity and from ripening of fruits to leaves fall is positive correlated with the active thermal balance. The number of days from swelling buds to start of blooming and from ripening fruit to leaves fall is positive correlated with the rainfall quantity registered.

Key words: Prunus avium L., temperature, rainfall, cultivars, blooming.

INTRODUCTION

Sweet cherry is an important specie in Romania and occupies an area of 7,760 ha (Coman and Chitu, 2014) with the great extension in the following years by new plantations established. The period between phenological stage of swelling buds and fruit maturation is very short for sweet cherry cultivars compared with other tree species, excepting strawberry (Budan and Grădinariu, 2000). Global climate change are subject to the recent research on plants having great influence in the development of phenological stages (Ansari and Davarynejad, 2008; Balaci et al., 2008; Chmielewski et al., 2004; Inouye et al, 2003). The previous research showed that phenological stages carrying the sweet cherry are determined by the cumulative action of daily average temperatures that exceed the value of 5°C (Sîrbu et al., 2013; Istrate, 2007). Average daily temperature has a direct influence in flowering plants (Radicevic et al., 2011; Tooke and Battey, 2010; Roversi & Ughini, 1996). Temperature is very important during ripening

fruit, but from the end of flowering to strengthen of the kernel, the influence of these climatic parameter is low (Budan and Grădinariu, 2000). Sparks et al. (2000), shows that climate change affects the starting time of flowering but Darbyshire et al. (2012) show that every Celsius degree increased of temperature advancing phenophases with 4 to 7 days. Also, rainfall in the period of fruit maturity induced fruit cracking in sweet cherries and can cause heavy losses in yields and returns (Meland et al., 2014).

This paper aims to determine the number of days of active heat balance and rainfall necessary to conduct phenological stages at different sweet cherry cultivars in terms of climate change and establish correlations between the studied parameters.

MATERIALS AND METHODS

In this study during 2011-2014, six Romanian sweet cherry cultivars: 'Cetăţuia', 'Cătălina', 'Maria', 'Andreiaş', 'Margonia' and 'George' and three introduced sweet cherry cultivars

'Van' (Canada), 'Bigarreau Burlat' (France) and 'Bigarreau Dönissen' (Germany) were evaluated. All cultivars were cultivated on *P. mahaleb* L. seedlings rootstock.

Three trees presented each cultivar and were planted at spacing of 4 x 5 m, with free palmette crown shape with support system. The orchard was located on a medium sandy clay loam with medium (6%) humus content.

Herbicide spraying were mantained along trees rows and grass was cut three times during summer in alleyways. No irrigation, rainfall, frost or birds protection system provided.

Phenological data were determined through the Fleckinger system (Fleckinger, 1960): B_1 - the bud swelling: the bud rounds delicate and gains a green light at the top; F_1 – start of blooming: the flowers are open for 5%; G - the end of the flowering: the petal of flowers have fallen for 90%. The data of the fruit ripening was established in the time of marketing quality traits (colour, the content of dry matter) specific to each cultivar.

The climatic data were recorded with the AgroExpert system by the station located on the perimeter of the experimental plot of the Research Station for Fruit Growing, Iaşi - Romania. The active thermal balance ($\Sigma t°a$) is provided by the sum of average daily temperature degree, which exceeds the biological limit characteristic to the sweet cherry tree, considered to be 5°C (Istrate, 2007).

$$\Sigma t°a = \Sigma T \ atd - BL, \ \text{in which:}$$

ΣT atd = sum of average temperature of days between two subsequent phenological stages;
BL = the biological limit of fruit tree species.

The statistical analysis was performed with the XLSTAT programme (ProAcademic, 2011, Addinsoft). The differences between cultivars were determined by the Duncan's test ($p \leq 0.05$). The Pearson correlation coefficient has been calculated between the variables measured ($p \leq 0.05$).

RESULTS AND DISCUSSIONS

For studied sweet cherry cultivars, the number days during the swelling buds to start of blooming ranged between 20 ('Cetăţuia') and 26 ('Margonia') (Table 1).

Table 1. Number of ffdays, active themal balance and rainfall quantity registered during the swelling buds to start of blooming at sweet cherry cultivars (2011-2014).

Cultivar	Number days[1]			Active themal balance (°C)			Rainfall quantity (mm)		
	Av	Min	Max	Av	Min	Max	Av	Min	Max
Cetăţuia	20[b2]	11	26	191.6[b]	114.6	242.8	43.9[ab]	7.6	85.4
Cătălina	22[b]	12	28	197.7[b]	123.2	245.7	44.9[ab]	7.6	85.4
Bigarreau Burlat	22[b]	14	28	208.5[b]	148.7	253	46.0[ab]	7.6	85.4
Maria	22[b]	14	29	217.4[b]	151.4	261.2	47.4[ab]	7.6	85.4
Van	21[b]	15	23	204.7[b]	166.2	225.8	46.3[ab]	4.6	85.4
Andreias	21[b]	15	26	213.9[b]	160	261.6	41.3[b]	9.2	85.4
Bigarreau Dönissen	22[b]	17	27	222.4[ab]	183.3	255.05	48.9[a]	7.6	85.4
Margonia	26[a]	21	33	269.1[a]	233.2	309.1	62.9[a]	33.4	85.4
George	21[b]	15	26	214.2[b]	171.1	254.8	41.9[b]	9.2	85.4
LSD $_{0.05}$	3.1			25.6			20.8		

[1] Av - average; Min - minimum; Max – maximum;
[2] - Different letters after the number corresponds with statistically significant differences for P 5% - Duncan test.

Also, 'Bigarreau Burlat' is a control cultivar for other cultivars as start of blossom, according with other studies (Kazantzis et al., 2011), but 'Cetăţuia' was earlier.

Minimum days for this phenological stage was 11 ('Cetăţuia') and maximum was 33 ('Margonia'). The active termal balance as average on period 2011-2014 ranged between 191.6°C ('Cetăţuia') and 269.1°C ('Margonia'). Except 'Margonia' all others sweet cherry cultivars have not differed statistically significant among them as active termal balance during the swelling buds to start of blooming.

Rainfall quantity ranged between 41.3 mm ('Andreias') and 62.9 mm ('Margonia') but significant statistical differences registered just 'Andreias' and 'George' (table 1). For period 2011-2014, the minimum values was 4.6 mm at 'Van' but the maximum value was the same for all studied cultivars as 85.4 mm.

Number of days for entire blossom time as average during 2011-2014 ranged between 8 ('Van' and 'George') to 12 ('Cătălina') (Table 2). The minimum value was 4 at 'Cetăţuia' and 'Andreiaş' but the maximum value was at 'Cătălina' with 16 days. Active thermal balance ranged between 97.3°C ('Cetăţuia') and 145.7 ('Margonia') but significant statistical differences registered only 'Cetăţuia' and 'Van'. Minimum value was at 'Cetăţuia' with 46.2°C but the maximum value was 174.7°C at 'Bigarreau Dönissen' (Table 2).

Table 2. Number of days, active themal balance and rainfall quantity registered during the blossom time at sweet cherry cultivars (2011-2014).

Cultivar	Number days[1]			Active themal balance (°C)			Rainfall quantity (mm)		
	Av	Min	Max	Av	Min	Max	Av	Min	Max
Cetăţuia	9[ab2]	4	15	97.3[b]	46.2	131.3	28.2[ab]	0	66.6
Cătălina	12[a]	8	16	151.0[a]	147.8	153.4	29.8[a]	0	67.8
Bigarreau Burlat	9[ab]	5	15	105.9[ab]	68.3	155.1	29.3[ab]	0	72.6
Maria	9[ab]	8	11	119.1[ab]	94.6	147.8	26.4[ab]	0	67.8
Van	8[b]	5	11	99.9[b]	82.2	125	25.4[ab]	0	67.8
Andreias	10[ab]	4	12	123.9[ab]	80.2	161.3	28.7[ab]	0	71.2
Bigarreau Dönissen	11[ab]	6	15	143.2[ab]	114.5	174.7	27.6[ab]	2.6	72.6
Margonia	9[ab]	6	13	145.7[ab]	114.5	174.1	6.5[c]	0	14.8
George	8[b]	5	9	107.2[ab]	65.1	166.4	23.5[abc]	0	65.0
LSD 0.05	3.3			48.1			15.9		

[1] Av - average; Min - minimum; Max – maximum;
[2] - Different letters after the number corresponds with statistically significant differences for P 5% - Duncan test.

For studied sweet cherry cultivars, the rainfall quantity registered values between 6.5 mm to 29.8 mm as an average for 2011-2014 period with a minimum as 0 mm to 72.6 mm. Number of days recorded from the end of blossom to fruit maturity at the studied sweet cherry cultivars ranged between 32 ('Cătălina') to 67 ('George') with a minimum value at 'Cătălina' with 26 days and a maximum value at 'George' with 81 days. These results are according with other studies for sweet cherry cultivars (Sîrbu et al., 2011) which show that from the blossom to fruit ripening are needed 70 - 98 days.

Also, for this stage the studied cultivars required an active thermal balance ranged between 539.2°C ('Cătălina') to 1,226.1°C ('George') with a minimum value to 'Cetăţuia' with 485°C and a maximum value at 'George' with 1,381.6°C (Table 3).

Number of days registered from the ripening fruit to leaves fall at the studied sweet cherry cultivars ranged between 112 ('George') to 152 ('Cetăţuia') with a minimum value at 'Maria' with 99 days and a maximum value at 'Cetăţuia' with 157 days (Table 4).

Table 3. Number of days, active themal balance and rainfall quantity registered during the end of blossom
to fruit maturity at sweet cherry cultivars (2011-2014).

Cultivar	Number days[1]			Active themal balance (°C)			Rainfall quantity (mm)		
	Av	Min	Max	Av	Min	Max	Av	Min	Max
Cetățuia	33^c	28	38	541.3^c	485	586.6	67.7^b	41	104
Cătălina	32^c	26	41	539.2^c	489	600.3	75.9^b	54.6	101.4
Bigarreau Burlat	36^c	31	41	602.3^c	554.9	667.4	89.8^b	54.6	113.8
Maria	52^b	42	60	911.5^b	763.9	1027.6	114.6^b	72	148.2
Van	54^b	45	58	955.1^b	770.5	1092.5	120.8^b	62.2	182.8
Andreias	50^b	44	57	897.7^b	804.7	1006.2	113.1^b	63.4	150.8
Bigarreau Dönissen	56^b	47	65	1006.6^b	845.4	1074.7	123.6^{ab}	56	192.8
Margonia	52^b	47	59	930.5^b	845.4	1002.5	313.5^a	56	892.8
George	67^a	49	81	1226.1^a	858.7	1381.6	167.4^{ab}	62.2	345.6
LSD $_{0.05}$	8.2			155.1			190.9		

[1] Av - average; Min - minimum; Max – maximum;
[2] - Different letters after the number corresponds with statistically significant differences for P 5% - Duncan test.

Table 4. Number days, active themal balance and rainfall quantity registered during the ripening fruit
to leaves fall at sweet cherry cultivars (2011-2014).

Cultivar	Number days[1]			Active themal balance (°C)			Rainfall quantity (mm)		
	Av	Min	Max	Av	Min	Max	Av	Min	Max
Cetățuia	152^{a2}	146	157	$2,767.2^a$	2,537.3	3,031.8	271.7^a	177.0	415.4
Cătălina	149^a	146	152	$2,701.9^a$	2,537.3	3,031.8	261.8^{ab}	177.0	415.4
Bigarreau Burlat	146^a	142	150	$2,654.9^a$	2,466.1	2,983.9	240.6^{ab}	139.8	406.0
Maria	120^{bcd}	99	132	$2,338.2^b$	2,248.9	2,514.0	223.2^{abc}	131.8	368.6
Van	126^{bc}	115	138	$2,300.8^b$	2,058.6	2,432.0	215.6^{bc}	123.2	336.6
Andreias	128^b	119	132	$2,337.5^b$	2,262.4	2,497.8	223.3^{abc}	132.0	368.6
Bigarreau Dönissen	121^{bcd}	116	133	$2,209.0^b$	1,946	2,433.3	210.8^{bc}	131.8	324.0
Margonia	114^{cd}	102	121	$2,229.9^b$	1,934.6	2,528.4	210.9^{bc}	132.0	324.0
George	112^d	102	134	$2,021.9^c$	1,670.1	2,343.8	170.8^c	129.2	240.2
LSD$_{0.05}$	11.3			153.2			49.6		

[1] Av - average; Min - minimum; Max – maximum;
[2] - Different letters after the number corresponds with statistically significant differences for P 5% - Duncan test.

Also, for this stage the studied cultivars required an active thermal balance ranged between 2,021.9°C ('George') to 2,767.2°C ('Cetățuia') with a minimum value to 'George' with 1,670.1°C and maximum value at 'Cetățuia' and 'Cătălina' with 3,031.8°C. Correlating the number of days with the active thermal balance (Table 5) we observed distict significant positive correlation in the period between swelling buds to start of blooming (r=0.9139), in the period between end of blossom to fruit maturity (r=0.9995) and between ripening fruit to leaves fall (r= 0.9730).

Table 5. Correlation coefficient (r) between number days
and active thermal balance and number days and rainfall quantity

Period	Number of days - active thermal balance[1]	Number of days – rainfall quantity
I - Swelling buds -start of blooming	0.9139**	0.9353**
II - Blossom time	0.7702*	0.3137^{ns}
III - End of blossom - fruit maturity	0.9995**	0.4961^{ns}
IV - Ripening fruit - leaves fall	0.9730**	0.9158**

[1]-*-significant correlation; **- distinct significant correlation; [ns]-non-significant correlation.

CONCLUSIONS

The climate change from recent years have influenced the duration of the phenological phases of different sweet cherry cultivars.

The number of days with the rainfall quantity was distict significant positive correlated in the period between swelling buds to start of blooming (r=0.9353) and between ripening of fruits to leaves fall (r=0.9158). The action of daily average temperatures determines different blooming periods in different year conditions.

The number of days from swelling buds to start of blooming, from end of blossom to fruit maturity and from ripening fruit to leaves fall are positive correlated with the active thermal balance.

The number of days from swelling buds to the start of blooming and from ripening of fruits to leaves fall are positive correlated with the the rainfall quantity registered.

ACKNOWLEDGEMENTS

This study has partially been financed by the Ministry of Agriculture and Rural Development - Romania, Grant No. 3.1.2./2015, with title 'Management of fruit tree genetic resources *in situ* and *ex situ*'.

REFERENCES

Ansari M., Davarynejad G., 2008. The Flower Phenology of Sour Cherry Cultivars. American-Eurasian J. Agric. & Environ. Sci., 4 (1): 117-124.

Balaci R. A., Zagrai I., Platon I., Zagrai L., Festila A., 2008. The Evaluation of Produductive and Qualitative Potential of Some Sweet Cherry Varieties in the Pedoclimatic Conditions of Bistrita Area. Bulletin UASVM, Horticulture 65 (1), 502-507.

Budan S., Grădinariu G., 2000. Cireşul, Edit. 'Ion Ionescu de la Brad', Iaşi.

Chmielewski F.-M., Müller A., Bruns E., 2004. Climate changes and trends in phenology of fruit trees and field crops in Germany, 1961–2000. Agricultural and Forest Meteorology 121, 69–78.

Coman M.,Chiţu E., 2014. Zonarea speciilor pomicole în funcţie de condiţiile pedoclimatice si socio-economice ale României. Editura Invel Mutimedia, Piteşti.

Darbyshire R., Webb L., Goodwin I., Barlow E. W. R., 2012. Evaluation of recent trends in Australian pome fruit spring phenology, Int. J. Biometeorol., 57 (3), 409-421.

Inouye D. W., Saavedra F., Lee-Yang W., 2003. Environmental Influences on the Phenology and Abundance of Flowering by Androsace Septentrionalis (Primulaceae), American Journal of Botany, 90 (6): 905–910.

Istrate M., 2007. Pomicultură generală, Edit. Ion Ionescu de la Brad, Iaşi.

Kazantzis, K., Chatzicharissis, I., Papachatzis, A., Sotiropoulos, T., Kalorizou, H., Koutinas N., 2011. Evaluation of Sweet Cherry Cultivars Introduced in Greece. Lucr. st. Univ. Craiova, XVI (LII), 293-296.

Meland M., Kaiser C., Christensen Mark J., 2014. Physical and Chemical Methods to Avoid Fruit Cracking in Cherry, AgroLife Scientific Journal ,3 (1), 177-183.

Radicevic S., Cerovic R., Maric S., Dordevic M., 2011. Flowering time and incompatibility groups – cultivar combination in commercial sweet cherry (Prunus avium L.) orchard, Genetika, 43 (2), 397-406.

Roversi A., Ughini V., 1996. Influence of weather conditions of the flowering period on sweet cherry fruit set, Proc. Intl. Cherry Symp., Eds. Hampson C.R., Anderson R.L., Perry R.L., Webster A.D., Acta Hort. 410, 427 - 441.

Sîrbu S., Beceanu D., Niculaua M., Anghel R. M., Iurea E., 2011. Fruit's physico-chemical characteristics of two bitter cherry cultivars. Lucr. st. USAMV Iaşi, Seria Horticultură, 54 (1), 531-536.

Sîrbu S., Iurea E., Corneanu M., 2013. Research concerning the influence of current climate changes over the phenological stages at sweet cherry tree (Prunus avium L.), Lucr. st. USAMV Iaşi, Seria Horticultură, 56 (2), 201-207.

Sparks T.H., Jeffree E.P., Jeffree C.E., 2000. An examination of the relationship between flowering times and temperature at the national scale using long-term phenological records from the UK, International Journal of Biometeorology, 44, 82–87.

Tooke F., Battey N.H., 2010. Temperate flowering phenology. Journal of Experimental Botany, 61 (11), 2853–2862.

32

SOME PHYSICO-CHEMICAL CHARACTERISTICS OF BLACK MULBERRY (*MORUS NIGRA* L.) IN BITLIS

Volkan OKATAN[1], Mehmet POLAT[2], Mehmet Atilla AŞKIN[2]

[1]Uşak University, Sivaslı Vocational High School, 64800 Sivaslı, Uşak/Turkey
[2] Süleyman Demirel University, Faculty of Agriculture, Department of Horticulture, 32100 Isparta/Turkey

Corresponding author email: okatan.volkan@gmail.com

Abstract

In this study, physico-chemical properties (total soluble solid contents, pH, titratable acidity, vitamin C, antioxidant activity, total phenolic and total anthocyanins) of black mulberry (Morus nigra L.) fruits grown in the Bitlis province of Turkey were investigated. The total soluble solids content of black mulberry varies between 15.65 % (13-BIT-2) and 22.1 % (13-BIT-6), titratable acidity between 1.45 % (13-BIT-1) and 1.85 % (13-BIT-4), pH between 3.65 (13-BIT-2) and 4.12 (13-BIT-5), respectively. Ascorbic acid (vitamin C) was in the range from 18.40 (13-BIT-3) to 23.67 (13-BIT-6) mg/100 g fresh weight (FW). The highest total phenolic contents were found 1920 (13-BIT-2) to 2575 (13-BIT-7) mg of gallic acid equivalent (GAE) 100 g^{-1} fresh weight. Antioxidant capacity (DPPH) was in the range from 18.24 (13-BIT-1) to 23.18 (13-BIT-6) % and total anthocyanin content between 643 (13-BIT-4) and 826 (13-BIT-8) mg/100 g.

Key words: antioxidant, phenolic, Morus nigra, mulberry, selection.

INTRODUCTİON

Morus nigra L., called black mulberry is a species of flowering plant in the family *Moraceae,* native to in a wide area of tropical, subtropical, and temperate zones in Asia, Europe, North America, South America, and Africa.

The trees historically have been used for sericulture especially in east, central, and south Asia. There are at least 24 species with more than 100 known cultivars. Farmers cultivate mulberry for silkworms in the China and India but european farmers cultivate for fruit. (Pawlowska et al., 2008).

Anatolia is one of the important diversity centers of mulberries with a long cultivation history dating back to 400- 500 years ago. The most popular mulberry species with edible fruits grown in Turkey are black, white and red mulberry (Ercisli and Orhan, 2007).

Traditionally, the fruits have been processing into several products like mulberry juice, molasses, jam, vinegar and some very special products such as 'mulberry pestil', 'mulberry kome' in Turkey. All of these have significant marketing value due to its nutritive and distinct characteristics (Erturk and Gecer, 2012).

Black mulberry fruits have also been effectively used in folk medicine in Turkey for a long time to treat fever, protect liver, strengthen the joints, facilitate discharge of urine and lower blood pressure (Baytop, 1984). Dark-coloured fruits, particularly berries (currant, honeyberry, aronia, blackberry, blueberry, mulberry, etc.) are recognized as contributors to the human health. In addition, there is increasing interest in pigment components of this group of fruits that may improve human health or lower the risk of disease (Lin and Tang, 2007). Black mulberry contains bioflavonoids that are important natural antioxidants and also contains non anthocyanin phenolics that are known to have many bioactive functions, including neuroprotective effects, which may be responsible for their medicinal properties (Ercisli and Orhan, 2007).

The aim of this research was to determine the some physico-chemical characteristics of local *Morus nigra* L. cultivars grown in Bitlis region in Eastern part of Turkey. The obtained results can be used in the registration process of these local cultivars and may be taken into consideration in the selection of parents in future breeding programs.

MATERIALS AND METHODS

Fruit material
Eight local black mulberry cultivars were harvested in different region of Bitlis, East Anatolia, Turkey, in August 2014. The harvested fruits were then transported to the laboratory for analysis.

Methods
Total soluble solid content (TSS) was measured with a digital refractometer (Model HI-96801 Hanna, German) at 20 °C. pH measurements were done by using Hanna-HI 98103. pH meter calibrated with pH 4.0 and 7.0 buffers. Titratable acidity was determined potentiometrically by titrating the sample with 0.1 NaOH until the pH reached 8.01 and expressed as citric acid.

Determination of total phenolics
Total phenolic content were determined with Folin-Ciocalteu assay (Singleton and Rossi, 1965). For this, flesh and peels (10 g) were centrifuged at 6000 rpm after homogenized in 40 ml ethanol solution. After, diluted (1/10) 1000 µl Folin-Ciocalteu and 800 µl Na_2CO_3 solution was added upon supernatant. After 2 hours of incubation, the samples were read spectrophotometrically at the wavelength 750 nm. Water-ethanol mixture was used as blank. Gallic acid was used as standard in the calculation.

Determination of total anthocyanins
For total anthocyanin analysis, 10 g flesh and peels were homogenized in methanol solution that included HCl 1 %. After a night of standing, the samples were filtered using a filter paper. The samples were read against the blank at the wavelengths of 530 and 700 nm (Giusti and Wrolstad, 2001).

Determination of vitamin C
After pureeing and filtering, the fruit juices samples were obtained. The juices were used for vitamin C analysis. The samples were homogenized by centrifuge and 400 µL oxalic acid (0.4 %) and 4.5 ml 2,6 - diclorofenolindofenol solution were added upon supernatant. The data were read spectrophotometrically at the wavelength of 520 nm against the blank.

Determination of radical – scavenging activity
In the 1,1-diphenyl-2-picrylhydrazyl (DPPH•) assay, antioxidants were capable to reduce the stable radical DPPH• to the yellow coloured diphenylpicrylhydrazine (DPPH-H). The test is based on the reduction of an alcoholic solution of DPPH• in the presence of a hydrogen donating antioxidant due to the formation of the non-radical form DPPH-H (Gulcin, 2007). The DPPH• radical-scavenging activity was estimated after Blois (1958). Briefly, 0.1 mL of each sample extract was mixed with 0.9 mL of 0.04 mg/mL methanolic solution of DPPH•. The mixtures were left for 20 min at room temperature and its absorbance then measured at 517 nm against a blank. All measurements were carried out in triplicate. The percentage of DPPH• scavenging activity was calculated using the following equation:

$$\% \ DPPH = [(A_c - A_s)/A_c] \ x \ 100$$

where A_c was the absorbance of the negative control (contained extraction solvent instead of the sample), and A_s was the absorbance of the samples.

Statistical analysis
Five replicates including 20 fruits per replicate were used. Descriptive statistics of total soluble solid contents, pH, titratable acidity, vitamin C, antioxidant activitiy, total phenolic and total anthocyanins extracted from eight *Morus nigra* cultivars were expressed as mean ± standard error (SE). The phytochemical characteristics were statistically analyzed with One-way ANOVA with five replicates. Duncan Test determined significant differences between the evaluated cultivars. All statistical evaluations were performed using SPSS 20 program.

RESULTS AND DISCUSSIONS

Table 1 shows the results of pH, TSS (%) and titratable acidity composition. In Table 2 are presented the results regarding the analysis of vitamin C, free radical scavenging activity, total phenolic and antocyanin content of eight mulberry cultivars collected from Hizan district in Bitlis province. The analysis of variance indicated that the cultivar had a major influence on all parameters under evaluation ($p < 0.05$).
pH, total soluble solids and titratable acidity in black mulberry cultivars
pH value in black mulberry cultivars was between 3.65 (13-BIT-2) and 4.12 (13-BIT-5). The average pH of cultivars was 3.85. The pH

contents in different mulberries ranged from 3.3 to 3.8 (Uzun ve Bayir, 2009) and our pH results are generally within limits of these studies.
Total soluble solid content (TSS) in black mulberry cultivars varied from 15.65 (13-BIT-2) to 22.10 % (13-BIT-6) with an average of 18.94 %. Previous studies had shown that soluble solid content of mulberry fruits grown in different agroclimatic regions of Turkey is between 15.27–30.80 % (Lale and Ozcagiran,

1996; Aslan, 1988), and our SSC results are generally within limits of these studies.
The titratable acidity of black mulberry cultivars was between 1.45 % (13-BIT-1) and 1.85 % (13-BIT-4).
The average titratable acidity of black mulberry cultivars was 1.63 %, which is a little higher than those reported for red and white mulberries (Ercisli et al., 2010) and black mulberry (Iqbal et al., 2010).

Table 1. pH, TSS (%) and titratable acidity composition of black mulberry cultivars

Cultivars	pH	TSS (%)	TA
13-BIT-1	3.82 b	17.20 bc	1.45 d
13-BIT-2	3.65 c	22.10 a	1.75 b
13-BIT-3	3.90 ab	16.70 c	1.56 c
13-BIT-4	3.70 bc	20.50 a	1.85 a
13-BIT-5	4.12 a	18.40 b	1.62 bc
13-BIT-6	4.06 a	15.65 d	1.80 a
13-BIT-7	3.92ab	21.40 a	1.55 c
13-BIT-8	3.66 c	19.55 a	1.48 c
Mean value	3.85	18.94	1.63

Vitamin C, antioxidant activity, total phenolic and total anthocyanins content in black mulberry fruits
We found vitamin C content between 18.40 (13-BIT-3) and 23.67 (13-BIT-6) mg/100 g for black mulberry cultivars under the investigation. In the earlier work conducted on the northeast Anatolia region of Turkey, Ercisli and Orhan (2008) reported that vitamin C

contents of black mulberry cultivars varied from 14.9 to 18.8 mg/100 mL. Ercisli et al. (2010) reported the average vitamin C content in black and purple mulberries as 20.79 and 18.87 mg per 100 mL extract, respectively. Lale and Ozcagiran (1996) reported that vitamin C content in black and purple mulberries was 16.6 and 11.9 mg/100 ml extract.

Table 2. Vitamin C, DPPH, total phenolic and anthocyanins content of black mulberry cultivars

Cultivars	Vitamin C	DPPH %	Total phenolic content	Total anthocyanins content
13-BIT-1	22.65 a	18.24	2125 c	815 a
13-BIT-2	19.50 b	21.56	1920 c	793 ab
13-BIT-3	18.40 bc	20.44	2330 ab	710 c
13-BIT-4	20.28 ab	19.86	2255 ab	826 a
13-BIT-5	21.15 b	22.38	1970 d	674 d
13-BIT-6	23.67 a	23.18	2345 ab	808 a
13-BIT-7	18.87 c	20.42	2575 a	756 b
13-BIT-8	22.36 b	18.66	2215 b	643 d
Mean value	20.86	20.59	2217	753.13

The antiradical activity of black mulberry cultivars were 18.24 (13-BIT-1) to 23.18 % (13-BIT-6) (DPPH assay).
Ozkaya (2015) reported that antioxidant activity in black mulberry were 15.037-24.443 μM TE/g. The results for total phenolics content between 1920 (13-BIT-2) and 2575 (13-BIT-7) GAE mg/g. Earlier reports had

shown that the total phenolic content in mulberry fruits was between 1515–2570 GAE mg/g (Lin and Tang, 2007; Bae and Suh, 2007). The difference between mulberry cultivars and between species in terms of phenolics is supposed to be a genetic characteristic because all plants were grown under the same agroclimatic conditions.

The effect of cultivar within the same fruit species on total phenolic content is well documented by several researchers on apples and strawberries (Scalzo et al., 2005; Voča et al., 2008), sea buckthorns (Ercisli et al., 2007) and cornelian cherries (Yilmaz et al., 2009). The total anthocyanin content per fresh weight of black mulberry (*Morus nigra*) cultivars ranged from 643 (13-BIT-4) and 826 (13-BIT-8) Cy 3-glu mg/g. According to earlier reports, total anthocyanin content in purple and black mulberries was 99 and 571 Cy 3-glu mg/g (Ozgen et al., 2009).

CONCLUSIONS

The results clearly indicate the difference between the cultivars used grown in the same conditions. Antioxidant activity also varies among the different cultivars of black mulberry, and this is a reflection of the phytochemical differences between cultivars.

These local black mulberry cultivars have high vitamin C, total phenolic, anthocyanin and antioxidant capacity in fruit. It is known, positive effect on human health of these substances.

This cultivars can be used for future breeding activities to obtain more healthier black mulberry.

REFERENCES

Aslan M. M., 1988. Selection of promising mulberry genotypes from Malatya, Elazig, Erzincan and Tunceli region of Turkey, MSc Thesis, Cukurova University, Adana, Turkey.

Bae S. H., Suh H. J., 2007. Antioxidant activities of five different mulberry cultivars in Korea, LWT-Food Sci. Technol. 40: 955–962.

Baytop T., 1984. Therapy with Medicinal Plants in Turkey, Istanbul University Publication No. 3255, Turkey.

Blois M. S., 1958. Antioxidant determinations by the use of a stable free radical, Nature, 181, 1199 - 1200.

Giusti M. M., Wrolstad R. E., 2001. Anthocyanins characterization and measurement with UV visible spectroscopy. In R. E. Wrolstad (Ed.), current protocols in food analytical chemistry. Willey, New York.

Gulcin I., 2007. Comparison of in vitro antioxidant and antiradical activities of L-tyrosine and L-Dopa, Amino acids, 32: 431-438.

Ercisli S., Orhan E., 2007. Chemical composition of white (*Morus alba*), red (*Morus rubra*) and black (*Morus nigra*) mulberry fruits. Food Chem., 103: 1380–1384

Ercisli S., Orhan E., Ozdemir O., Sengul M., 2007. The genotypic effects on the chemical composition and antioxidant activity of sea buckthorn (*Hippophae rhamnoides* L.) berries grown in Turkey, Sci. Hortic. 115: 27–33.

Ercisli S., Orhan E., 2007. Some physico-chemical characteristics of black mulberry (*Morus nigra* L.) genotypes from Northeast Anatolia region of Turkey. Sci. Hortic. 116, 41-46.

Ercisli S., Tosun M., Duralija B., Voća S., Sengul M., Turan M., 2010. Phytochemical Content of Some Black (*Morus nigra* L.) and Purple (*Morus rubra* L.) Mulberry Genotypes. Food Technology & Biotechnology, 48(1).

Erturk Y. E., Gecer M. K., 2012. Economy of berries. Proceedings of 4th National Berry Symposium, 3-5 October 2012, Antalya-Turkey, 368- 385.

Lale H., Ozcagiran A., 1996. Study on pomological, phenologic and fruit quality characteristics of mulberry (Morus sp.) species, Derim, 13: 177–182 (in Turkish).

Lin J. Y, Tang C. Y., 2007. Determination of total phenolic and flavonoid contents in selected fruits and vegetables, as well as their stimulatory effects on mouse splenocyte proliferation, Food Chem. 101: 140–147.

Ozkaya Z., 2015. Uşak ili ulubey ilçesinde yetişen karadutların (*Morus nigra* L.) morfolojik, fenolojik ve pomolojik özelliklerinin belirlenmesi. Adnan Menderes Universitesi, Fen Bilimleri Enstitüsü, Yüksek Lisans Tezi s. 45.

Özgen M., Serçe S., Kaya C., 2009. Phytochemical and antioxidant properties of anthocyanin-rich Morus nigra and Morus rubra fruits, Sci. Hortic., 119: 275–279.

Pawlowska A. M., Oleszek W., Braca A., 2008. Quali-quantitative analyses of flavonoids of *Morus nigra* L. and *Morus alba* L. (Moraceae) fruits. J. Agric. Food Chem., 56: 3377–3380.

Scalzo J., Politi A., Pellegrini N., Mezzetti B., Battino M., (2005). Plant genotype affects total antioxidant capacity and phenolic contents in fruit, Nutrition, 21:207–213.

Singleton V. L., Rossi J.L., 1965. Colorimetry of total phenolics with phosphomolybdic phosphotungstic acid reagents. Am. J. Enol. Viticult., 16:144-158.

Uzun H., Bayır A., 2009. Farklı dut genoiplerinin bazı kimyasal özellikleri ve antiradikal aktiviteleri. III. Ulusal Üzümsü Meyveler Sempozyumu, 10-12 Haziran 2009, Kahramanmaraş.

Voča S., Dobričević N., Dragović-Uzelac V., Duralija B., Družić J., Cmelik Z., Babojelič, M. S., 2008. Fruit quality of new early ripening strawberry cultivars in Croatia, Food Technol. Biotechnol., 46:292–298.

Yilmaz K. U., Ercisli S., Zengin Y., Sengul E., Kafkas Y., 2009. Preliminary characterisation of cornelian cherry (*Cornus mas* L.) genotypes for their physico-chemical properties, Food Chem., 114:408–412.

PERMISSIONS

All chapters in this book were first published in SPSBH, by University of Agronomic Sciences and Veterinary Medicine of Bucharest; hereby published with permission under the Creative Commons Attribution License or equivalent. Every chapter published in this book has been scrutinized by our experts. Their significance has been extensively debated. The topics covered herein carry significant findings which will fuel the growth of the discipline. They may even be implemented as practical applications or may be referred to as a beginning point for another development.

The contributors of this book come from diverse backgrounds, making this book a truly international effort. This book will bring forth new frontiers with its revolutionizing research information and detailed analysis of the nascent developments around the world.

We would like to thank all the contributing authors for lending their expertise to make the book truly unique. They have played a crucial role in the development of this book. Without their invaluable contributions this book wouldn't have been possible. They have made vital efforts to compile up to date information on the varied aspects of this subject to make this book a valuable addition to the collection of many professionals and students.

This book was conceptualized with the vision of imparting up-to-date information and advanced data in this field. To ensure the same, a matchless editorial board was set up. Every individual on the board went through rigorous rounds of assessment to prove their worth. After which they invested a large part of their time researching and compiling the most relevant data for our readers.

The editorial board has been involved in producing this book since its inception. They have spent rigorous hours researching and exploring the diverse topics which have resulted in the successful publishing of this book. They have passed on their knowledge of decades through this book. To expedite this challenging task, the publisher supported the team at every step. A small team of assistant editors was also appointed to further simplify the editing procedure and attain best results for the readers.

Apart from the editorial board, the designing team has also invested a significant amount of their time in understanding the subject and creating the most relevant covers. They scrutinized every image to scout for the most suitable representation of the subject and create an appropriate cover for the book.

The publishing team has been an ardent support to the editorial, designing and production team. Their endless efforts to recruit the best for this project, has resulted in the accomplishment of this book. They are a veteran in the field of academics and their pool of knowledge is as vast as their experience in printing. Their expertise and guidance has proved useful at every step. Their uncompromising quality standards have made this book an exceptional effort. Their encouragement from time to time has been an inspiration for everyone.

The publisher and the editorial board hope that this book will prove to be a valuable piece of knowledge for researchers, students, practitioners and scholars across the globe.

LIST OF CONTRIBUTORS

Volkan Okatan
Uşak University, Sivaslı Vocational High School, 64800 Sivaslı, Uşak/Turkey

Hristo Dzhugalov, Valentin Lichev, Anton Yordanov and Pantaley Kaymakanov
Department of Fruit Growing, Agricultural University, Plovdiv, 12 Mendeleev Str., 4000, Bulgaria

Velmira Dimitrova and Georgi Kutoranov
Bio Tree Ltd., 7 Bansko road, 1331 Sofia, Bulgaria

Corina Gavăǧ, Liana Melania Dumitru, Cristina Moale and Alexandru Opriǧă
Research Station for Fruit Growing Constanta, No.1 Pepinierei Street, 907300, Valu lui Traian, ConstanĠa, Romania

Dorel Hoza, Adrian Asănică and Ligia Ion
University of Agronomic Sciences and Veterinary Medicine of Bucharest, 59 Mărăşti Blvd, District 1, 011464, Bucharest, Romania

Hakan Cetinkaya
Kilis 7 Aralık University, Faculty of Agriculture, Department of Horticulture, Kilis, Turkey

Muhittin Kulak
Kilis 7 Aralık University, Faculty of Arts and Sciences, Department of Biology, Kilis, Turkey

Alina Viorica Ilie and Dorel Hoza
University of Agronomic Sciences and Veterinary Medicine of Bucharest, 59 Mărăşti Blvd., District 1, Bucharest, Romania

Viorel Cătălin Oltenacu
Research Station for Fruit Trees Growing Băneasa, Bd. Ion Ionescu de la Brad, No. 4, District 1 Bucharest, Romania

Yasin Ozdemir
Ataturk Central Horticultural Research Institute, Department of Food Technology, Yalova,Turkey

Sultan Filiz Guclu
Süleyman Demirel University Atabey Vocational School, Isparta, Turkey

Ayşe BetüL Avci
Ege University Ödemiş Vocational School, İzmir Turkey

Mehmet Polat
Süleyman Demirel University, Faculty of Agriculture, Isparta/turkey

Volkan Okatan
Usak University, Vocational High School, Sivaslı/Uşak/Turkey

Burak Durna
Provincial Directorate of Agriculture, Çorum/ Turkey

Mehmet Koç
Kilis 7 Aralık University, Faculty of Agriculture, Department of Horticulture, Kilis

Kenan Yildiz
Gaziosmanpaşa University, Faculty of Agriculture, Department of Horticulture, Tokat

Saadettin Yildirim
Adnan Menderes University, Faculty of Agriculture, Department of Biosystems Engineering, Aydın

Gultekin Ozdemir and Akile Beren Sogut
Dicle University, Faculty of Agriculture, Department of Horticulture, Diyarbakir, Turkey.

Mihdiye Pirinccioglu, Göksel Kizil and Murat Kizil
Dicle University, Faculty of Science, Department of Chemistry, Diyarbakir, Turkey

Mehmet Ali Gundogdu, Alper Dardeniz and Ahmet Faruk Pekmezci
Canakkale Onsekiz Mart University, Faculty of Agriculture, Department of Horticulture, Terzioglu Campus, 17100, Canakkale, Turkey

Baboo Ali
Canakkale Onsekiz Mart University, Faculty of Agriculture, Department of Agricultural Biotechnology, Terzioglu Campus, 17100, Canakkale, Turkey

Marinela Vicuţa Stroe
Department of Bioengineering of Horticulture - Viticulture Systems University of Agronomical Sciences and Veterinary Medicine, 59 Marasti Blvd, District 1, 011464, Bucharest, Romania

Elena Tsolova
Institute of Agriculture-Kyustendil, Sofjisko shose str., 2500 Kyustendil, Bulgaria

Lilyana Koleva
University of Forestry, Sofia, 10 Kliment Ohridski Blvd, 1756, Sofia, Bulgaria

Mihaela Şerbulea and Arina Oana Antoce
University of Agronomic Sciences and Veterinary Medicine of Bucharest, Faculty of Horticulture, Department of Bioengineering of Horti-Viticultural Systems, 59, Mărăşti Blvd., District 1, 011464 Bucharest, Romania

Gheorghe Nicolaescu, Antonina Derendovskaia, Silvia Secrieru, Valeria Procopenco and Mariana Godoroja
State Agrarian University of Moldova, 44 Mircesti str., MD-2049, Chişinău, Moldova

Dumitru Mihov
Tera Vitis Ltd., Burlacu vil., Cahul dis., Moldova

Cornelia Lungu
Dionysos Mereni Joint-stock Company, 44 Mircesti str., MD-2049, Chişinău, Moldova

Fatih Mehmet Tok and Sibel Derviş
Mustafa Kemal University, Faculty of Agriculture, Department of Plant Protection, 31040 Antakya, Hatay, Turkey

Marinela Vicuţa Stroe, Toniţa Valentina Dunuţă and Daniel Nicolae Cojanu
University of Agronomic Sciences and Veterinary Medicine of Bucharest, Department of Bioengineering of Horticulture-Viticulture Systems, 59 Marasti Blvd., District 1, 011464, Bucharest, Romania

Aysun Ozturk and Yasin Özdemir
Atatürk Horticultural Central Research Institute, Department of Food Technologies, Yalova, Turkey

Barış Albayrak
Atatürk Horticultural Central Research Institute, Land and Water Resources, Yalova, Turkey

Mehmet Simşek and Kutay Coşkun Yildirim
Atatürk Horticultural Central Research Institute, Vegetables Production Department, Yalova, Turkey

Deniz Yilmaz and Mehmet Emin Gokduman
Süleyman Demirel University, Agriculture Faculty, Agricultural Machinery and Technologies Engineering Department, Doğu Campus, Isparta, Turkey

Ahmet Demirbas
Cumhuriyet University, Vocational School of Sivas, Department of Crop and Animal Production, Sivas, Turkey

Mihaela Gabriela Ciupureanu (Novac) and Daniela Popa
University of Craiova, Faculty of Horticulture, Doctoral School of Engineering Plant and Animal Resources, 13 A.I. Cuza Street, Craiova, Dolj, Romania

Elena Ciuciuc
Research - Development Center For Field
Crops on Sandy Soils Dăbuleni, Romania

**Maria Călin, Tina Oana Cristea, Silvica
Ambăruș, Creola Brezeanu and Petre Marian
Brezeanu**
Vegetable Research and Development Station
Bacau, Calea Bârladului Street, no. 220, Bacau
County, Romania

**Marcel Costache, Gabriela Șovarel and
Liliana Bratu**
Research Development Institute for Vegetable
and Flower Growing Vidra, Romania

**Monica Catană, Luminița Catană, Enuța
Iorga, Anda-Grațiela Lazăr, Monica-
Alexandra Lazăr and Nastasia Belc**
National Research & Development Institute
for Food Bioresources, IBA Bucharest, 6 Dinu
Vintila, District 2, 021102 Bucharest, Romania

Adrian Constantin Asănică
University of Agronomic Sciences and
Veterinary Medicine of Bucharest, Faculty
of Horticulture, 59 Marasti Blvd, District 1,
011464, Bucharest, Romania

Hasan Huseyin Ozturk
Cukurova University, Faculty of Agriculture
Engineering of Agricultural Machineries and
Technologies, 01330 Adana, Turkey

Orkun Baris Kovanci
Uludag University, Faculty of Agriculture,
Department of Plant Protection, Gorukle
campus, 16059, Bursa, Turkey

**Adrian Chira, Lenuța Chira and Elena
Delian**
University of Agronomic Sciences and
Veterinary Medicine of Bucharest, 59 Mărăști
Blvd, District 1, 011464, Bucharest, Romania

**Cristina Moale, Leinar Șeptar and Corina
Gavăt**
Research Station for Fruit Growing (R.S.F.G)
Constanta, 25 Pepinierei Street, 907300 Valu
lui Traian, Romania

Cristina Petrișor
Research and Development Institute for Plant
Protection, 8 Ion Ionescu de la Brad Blvd,
District 1, Bucharest, Romania

**George Dudu, Sorin Mihai Cîmpeanu and
Ovidiu Ionuț Jerca**
University of Agronomic Sciences and
Veterinary Medicine of Bucharest, 59 Mărăști
Blvd, District 1, 011464, Bucharest, Romania

**Sorina Sîrbu, Gelu Corneanu, Elena Iurea
and Margareta Corneanu**
Research Station for Fruit Growing, 175
Voinești, 707305, Iași, Romania

Mehmet Polat and Mehmet Atilla Așkin
Süleyman Demirel University, Faculty of
Agriculture, Department of Horticulture,
32100 Isparta/Turkey

Index